Microstructural Parcellation of the Human Cerebral Cortex

Stefan Geyer • Robert Turner
Editors

Microstructural Parcellation of the Human Cerebral Cortex

From Brodmann's Post-Mortem Map to in Vivo Mapping with High-Field Magnetic Resonance Imaging

Springer

Editors
Stefan Geyer
Robert Turner
Department of Neurophysics
Max Planck Institute for Human Cognitive and Brain Sciences
Leipzig, Germany

ISBN 978-3-642-37823-2 ISBN 978-3-642-37824-9 (eBook)
DOI 10.1007/978-3-642-37824-9
Springer Heidelberg New York Dordrecht London

Library of Congress Control Number: 2013941360

Printed on acid-free paper

Springer is part of Springer Science+Business Media (www.springer.com)

Preface

A fundamental goal of brain research is to elucidate the functional properties of the structural elements of the brain, at an appropriate organizational scale. One major scientific milestone in this regard was the publication of Korbinian Brodmann's famous map of the cerebral cortex in 1909. This map defines around 40 structural areas in the human cortex based on differences in cytoarchitecture (i.e., size, shape, and topographic arrangement of nerve cells). Subsequent investigators found out that these areas, defined purely anatomically by Brodmann, also correspond to functional entities of the cerebral cortex, so that, for example, Brodmann's area (BA) 4 corresponds to primary motor cortex (M1), and BA 17 to primary visual cortex (V1). Since its publication, Brodmann's map has become a "classic" in the field of neurobiology, and, despite many advances in neuroscience, his nomenclature of cortical areas is still widely used to designate functional regions. Two key problems intrinsic to this mapping strategy, however, are that cytoarchitectonic parcellation requires microscopic analysis of postmortem brain sections and cytoarchitectonic areas vary between individuals in their topography relative to the gyral anatomy of the brain. This means that correlations between microstructure (based on cytoarchitectonic analysis in postmortem brains) and function (based on, e.g., functional magnetic resonance imaging (fMRI) in living brains) have almost always been made probabilistically, with the aid of a computerized brain atlas.

It would be a revolutionary scientific breakthrough if it were possible to map the microstructural correlates of functional activations in the human cortex in a noninvasive and individual-specific way directly in vivo. However, until now, microstructural details of the cerebral cortex have been beyond the resolution of conventional structural MRI, except within the primary visual cortex, where the very prominent Stria of Gennari can relatively easily be detected at the MRI field strength of 3 T. Recently, however, high-field MRI, at a field strength of 7 T and spatial resolution below 0.5 mm, has radically changed this situation by detecting further systematic structural differences within the cerebral cortex. For instance, use of 7 T MRI can resolve the functionally important microanatomical border between primary motor (area 4) and somatosensory (area 3a) cortex in vivo. This opens up the door toward an individual-specific microanatomical brain map with the enormous potential to make direct correlations between microstructure and function in living human brains.

This brief outline spans an entire century from the publication of Brodmann's postmortem map at the beginning of the twentieth to "in vivo Brodmann mapping" with high-field MRI at the beginning of the twenty-first century. In our book, however, we would like to shed some light also on a few milestones of structural brain mapping that lie between these two "cornerstones". For this reason, the book is divided into three parts.

Part I starts with the world of "classical" cytoarchitectonic brain maps, published in the first half of the twentieth century: the famous parcellation of Korbinian Brodmann (chapter by Guy Elston and Laurence Garey) and the much lesser known map of Constantin von Economo and Georg Koskinas (chapter by Lazaros Triarhou). In contrast to Brodmann, von Economo and Koskinas provide a much more detailed verbal and pictorial description of each area's cytoarchitectonic features. We also bring back to life the almost forgotten myeloarchitectonic map (based on differences in the arrangement of myelinated fibers in preparations stained for myelin sheaths) by Cécile and Oskar Vogt (chapter by Rudolf Nieuwenhuys). Mapping the cortex with high-field MRI shows a renewed interest in myeloarchitecture, since many types of MR image contrast depend on the presence of myelin within the image voxel.

Part II covers more recent approaches that aim at mapping cortical areas noninvasively in living human brains. Bruce Fischl and colleagues use cortical folding patterns to estimate the topography of Brodmann areas in individual brains. Simon Eickhoff and Danilo Bzdok identify functional modules in the cortex in a data-driven fashion by clustering together voxels with similar co-activation patterns and separating them from voxels with different co-activation profiles.

In Part III, we arrive at the second "cornerstone," namely, "in vivo Brodmann mapping" with high-field MRI. The two chapters by Robert Turner argue for the necessity of more realistic functional and structural analysis methods that more effectively exploit the great potential inherent in high-field MRI and, together, should lead to a new understanding of the relationships between structure, function, and connectivity in the living brain. The second chapter also focuses on a discussion about the microstructural origin of the high-field MRI contrast in the cortex. Does it originate from regional variations in the arrangement of cells (cytoarchitecture) or myelin sheaths (myeloarchitecture)? Evidence is provided that the latter (i.e., myelin) is the case. This leads us to the chapter by Nicholas Bock and Afonso Silva on visualizing myeloarchitecture with MRI in the cortex of living marmoset monkeys (*Callithrix jacchus*). We conclude with the first breakthrough in high-field MR mapping in the living human brain (chapter by Stefan Geyer): the detection of the functionally important border between primary motor (area 4) and somatosensory (area 3a) cortex.

January 2013 Stefan Geyer
Leipzig, Germany Robert Turner

Contents

Part I
“Classical” Cyto- and Myeloarchitectonic Human Brain Maps

Chapter 1
The Cytoarchitectonic Map of Korbinian Brodmann: Arealisation and Circuit Specialisation

Guy N. Elston and Laurence J. Garey

Abstract Korbinian Brodmann is best known for his 1909 monograph on comparative localisation of cerebral cortex in a variety of mammals, including man. His "areas" are still widely used to delineate cortical functional regions. He identified "homologous" parts of the cortex in different animals by their structure and produced an "organic" theory of cortex based on anatomical features. He formalised the description of the cortical pattern as being composed of six basic layers, with variations between animals and between areas. He integrated phylogenesis and ontogenesis with observations of adult cortical structure, function and pathology. Later, Brodmann turned to a systematic study of human brains of different races, culminating to a paper on "anthropological" aspects of cortical anatomy in 1913. His work over his short lifetime is a rich source of quantitative information and is of importance for the interpretation of modern imaging studies, particularly involving visual or prefrontal cortex, and the search for a neuroanatomical basis for human consciousness and intelligence. With the advent of new methodologies it has been possible to probe neuron structure at the microscopic level in Brodmann's cortical areas, teasing out and quantifying elements of circuit structure and specialisation. The study of pyramidal cells, the most abundant neuronal type in cortex, has revealed significant and systematic differences in structure and integrative ability among cortical areas, which reflect the physiological characteristics of the neurons and functional complexity. Moreover, comparison of pyramidal cell structure in homologous cortical areas among species reveals different trends among different cortical areas. Pyramidal cell structure in Brodmann's area 17, for example, varies relatively little among primate species whereas pyramidal cells in granular prefrontal cortex are larger, more branched and

G.N. Elston (✉)
Centre for Cognitive Neuroscience, 4562 Sunshine Coast, QLD, Australia
e-mail: guyelston@yahoo.com

L.J. Garey
CH-1166 Perroy, Switzerland
e-mail: l.garey@sunrise.ch

S. Geyer and R. Turner (eds.), *Microstructural Parcellation of the Human Cerebral Cortex*, DOI 10.1007/978-3-642-37824-9_1, © Springer-Verlag Berlin Heidelberg 2013

more spinous in progressively larger prefrontal cortices. Pyramidal cells in prefrontal cortex in man, that region associated with higher cognitive functions, are more complex and integrate more inputs than lower order primates, bridging Brodmann's theories in prefrontal cortex and Cajal's theories on the psychic cell in present day thinking on intelligence.

1.1 Korbinian Brodmann: Life and Works

In 1909 Korbinian Brodmann published his "Vergleichende Lokalisationslehre der Grosshirnrinde in ihren Prinzipien dargestellt auf Grund des Zellenbaues", destined to become a major classic of the neurological world. It still forms the basis for localisation of function in the cerebral cortex. His "areas" are widely used to delineate cortical functional regions by neurologists and experimentalists in various animals (see Garey 1994 for a translation of Brodmann's original monograph).

Brodmann was born in 1868 in Liggersdorf, Hohenzollern and studied medicine, qualifying in 1895. After working with Oskar Vogt in 1896 in the Neurological Clinic in Alexanderbad he turned to neurology and psychiatry, and later studied pathology in Leipzig where, in 1898, he wrote a doctoral thesis on chronic ependymal sclerosis. He then went to the University Psychiatric Clinic in Jena, directed by Otto Binswanger, before transferring to the Municipal Mental Asylum in Frankfurt-am-Main from 1900 to 1901, where he met Alois Alzheimer who inspired an interest in neuroanatomy. In 1901 Brodmann rejoined Vogt and worked with him in Berlin where he began his famous studies on cytoarchitectonics of mammalian cortex, (Brodmann 1903a,b, 1905a,b, 1906, 1908a,b) in the "Journal für Psychologie und Neurologie", and which served as a basis for his 1909 monograph on comparative cortical localisation. In 1910 he moved to Tübingen and was appointed Professor in the Faculty of Medicine. While in Berlin Brodmann had lectured in courses in Munich organised by Emil Kraepelin who forecast an important contribution to neuroanatomical research from architectonics. In 1918 Brodmann received a prestigious appointment to the Psychiatric Research Institute in Munich, where Nissl was working. So he and Nissl began a very promising collaboration, cut short by Brodmann's early death less than a year later.

Before Brodmann, more than a little confusion reigned concerning the structure of the cerebral cortex. In 1858, Theodor Meynert's pupil, Berlin, described six layers in the human isocortex on the basis of variations in cell size and type, including pyramidal and granule cells. Meynert himself, from 1867, subdivided the human cortex into various functional regions. Another important early cortical localisational study was by Vladimir Betz in 1874, in which he described "giant pyramids", in the human motor cortex. In 1878 David Ferrier devoted his Croonian Lecture to cerebral localisation, and before the end of the nineteenth century numerous publications dealt with the laminar pattern of the cerebral cortex, notably by Lewis (1878, 1881), Lewis and Clarke (1878) and Hammarberg (1895). In his monograph of 1909 Brodmann considered these in detail, and pointed out their inconsistencies.

The year 1900 saw the first in the series Ramon y Cajal's studies on human cortex (Cajal 1900–1906), as well as Bolton's treatise on human visual cortex. Brodmann had little respect for Cajal's "erroneous" views on cortical lamination. Grafton Elliot Smith published a detailed atlas of human cortical localisation in 1907, referring to the work of Flechsig, Campbell and Brodmann. In 1905 Alfred Campbell's work on "Histological studies on the localisation of cerebral function" appeared based on human cerebral hemispheres, and those of chimpanzee, orang-utan, cat, dog and pig. In 1953 Constantin von Bonin commented that Campbell's localisation was not as "fine as those of the German school", referring especially to Brodmann's work.

The basis of Brodmann's localisation is the subdivision of the cortex into "areas" with similar cellular and laminar structure. He compared the human cortex with that of several other mammals, including primates, rodents and marsupials. Brodmann's observations integrated concepts of phylogenesis and ontogenesis with his observations of adult cortical structure, function and even pathology. Important support for Brodmann's concepts of functional localisation came from Otfried Foerster's electrical stimulation of human cortex in 1926, work based on Brodmann's structural studies.

Later, Brodmann continued his comparative studies, but his attention turned more toward a systematic study of human brains of different races, culminating in his paper on "anthropological" aspects of cortical anatomy (1913, translated by Elston and Garey 2004). He was not biased in his studies by the prevailing attitude that some human races were "superior" to others. As he states in his text, his motivation was scientific, without ulterior motives. Indeed, Brodmann not only presented data on the cortex of different human races, but also new data from the brains of patients suffering from pathology such as microcephaly, epilepsy, blindness and idiocy. He also presented a wealth of data on granular prefrontal cortex, agranular precentral motor cortex and the primary visual area from a diverse range of primates and non-primates. This paper is a rich source of quantitative information and emphasises the variation in cortical topography in human brains, and is of importance for the interpretation of modern imaging studies, particularly involving visual or prefrontal cortex, and the search for a neuroanatomical basis for human consciousness and intelligence (Sengpiel and Kind 2002; Schoenemann et al. 2005; Elston and Garey 2009).

1.1.1 Brodmann's Aims and Results

The subject of Brodmann's 1909 treatise was histological localisation in the cerebral cortex, related to contemporary physiological or clinical data. He set himself the task of parcellating the cortex according to common anatomical features, such as structurally similar neuronal features. His aim was to identify "homologous" parts of the cortex in different animals by their structure and produce an "organic" theory of cortex based on anatomical features. He excluded consideration of fibre architecture and myelogenesis, although admitting that they were major factors in cortical localisation. Oskar Vogt (1906) had demonstrated cortical parcellation in man using myeloarchitectonics that was compatible, but more detailed than, Brodmann's cellular localisation. The latter accepted that it could be used to subdivide cytoarchitectonic zones into smaller fields. This was a matter of degree of spatial localisation and not a major divergence. Indeed his colleague, Mauss (1908), confirmed in monkeys an overall agreement with Brodmann's cytoarchitectonic subdivisions.

For Brodmann, cortical cytoarchitectonic localisation was of three types, **elemental** (according to histological elements), **laminar** (according to cell layers) and **topographical** (according to tangentially organised "areas").

Elemental localisation depended on neuronal groups of similar structure having similar functions. Brodmann admitted that so far results were not exactly encouraging: "The difficulties in achieving such a subdivision by elements are considerably greater than may appear at first sight. First and foremost we still lack clear criteria for the recognition of anatomically equivalent cellular elements." (Brodmann 1909, Introduction). He stated that perhaps the only good example at that time came from Betz (1874) that the "motor" cortex anterior to the central sulcus was typified by "giant pyramidal" neurons, unlike the "sensory" cortex posterior to the sulcus. However, different cell types (Brodmann distinguished pyramidal cells, spindle cells, granule cells, and stellate cells) were not organised similarly over the whole cortex. They varied widely between areas. He forecast that new techniques would be needed to differentiate particular neurons functionally: "It is possible that later it will be feasible to further differentiate histologically many grossly morphologically similar cell types according to their fine structure. For this, the main necessity is new histological, and particularly staining, techniques that have a specific affinity for functionally related cells or, what amounts to the same, histochemically related cells." This sounds like a plea for not only the sort of histochemistry we know today, but also electron microscopy!

Brodmann equally emphasised the limits of **laminar localisation**. He admitted that he knew little about the significance of individual layers. He returned to his previous example of the layer of Betz giant pyramids saying that their significance remained largely obscure although it must be related to motor function from pathological observations in, for example, amyotrophic lateral sclerosis. "Above all, we have absolutely no proof that this layer represents the only motor component of the cortex, comparable with a specifically sensory one in the granular layers … Many new observations (electrical stimulation) support the idea that cortical motor activity can be produced without the intervention of this giant pyramidal layer … Above all, it is clear that the excitomotor zone stretches anteriorly well beyond the extent of this layer". He also gave the example of the stria of Gennari in the visual cortex around the calcarine sulcus, of which the spatial extent was recognisable to the naked eye. Even though this "striate area" was associated with visual activity, or even with specific parts of the retina, Brodmann remained, as always, cautious as to what constituted the "visuosensory element" within the area.

So he concluded that neither elemental nor laminar localisation constituted the whole story. He opted for **topographical localisation** of tangential cortical "areas" of homogeneous intrinsic structure, which involved a knowledge of both their structural elements and their lamination, that is to say their cytoarchitectonics.

1.1.2 The Basic Laminar Pattern of the Cerebral Cortex

Brodmann expressed his surprise at the variation in the number of cortical layers described by various authors. In man it varied from five to nine and in other animals between three and ten. "Thus, sometimes completely different layers carry identical names, while on other occasions layers that are anatomically similar, and homologous, are given different names by different authors, although it is a basic prerequisite of scientific logic that similar structures should carry similar names and that homologous patterns should have homonymous designations." (Brodmann 1909, Chap. 1). He supported the idea that the primitive pattern of cortical lamination in all mammals was six-layered, except in certain "rudimentary" cortex, such as in the rhinencephalon and cingulate gyrus. He distinguished two basic cortical patterns. Most of the cortex was **homogenetic**, derived from the embryonic six-layered type, whereas in **heterogenetic** cortex, the six-layered embryonic stage had not been demonstrated. There were two categories of architectonic transformation of homogenetic cortex: **homotypical** (maintaining the same basic six-layered pattern throughout life), and **heterotypical** (no longer having six layers in the mature brain). He cited the description by His (1904) of the original unlayered human cortical Anlage followed by differential growth in thickness, migration of neuroblasts from the "inner plate", and ingrowth of nerve fibres. After the fifth month neuroblasts became organised into layers, with deep layers V and VI first. Finally the cortex entered a six-layered phase over the whole surface, except the small heterogenetic regions.

Brodmann adopted a nomenclature based on previous observations, but which attempted to resolve the many contradictions:

I. *Lamina zonalis* – molecular layer
II. *Lamina granularis externa* – outer granular layer
III. *Lamina pyramidalis* – pyramidal layer
IV. *Lamina granularis interna* – inner granular layer
V. *Lamina ganglionaris* – ganglion cell layer
VI. *Lamina multiformis* – spindle cell layer

Local transformations in this basic six-layered pattern started around the beginning of the seventh month. There could be either a loss of layers, or the formation of sublayers. An example of loss of layers was in agranular cortex, where the inner granular layer (IV) was not present in the mature brain, as in the giant pyramidal "motor cortex". The calcarine "visual cortex" was the best example of subdivision of layers. There was division of the inner granular layer into two cell-dense laminae, a superficial inner granular sublayer (IVa) and a deep inner granular sublayer (IVc), with a cell-poor lamina, the intermediate inner granular sublayer (IVb – the stria of Gennari), between them. Cortical structure could be modified through changes of cell packing density either in the whole depth of the cortex or in a single layer. It could also be modified through changes of cell size or type in one or more layers. The relative thickness of layers, or the whole cortical thickness, could change.

1.1.3 The Comparative Anatomical Basis for the Six-Layered Cortex

Brodmann believed that all mammals had a common primitive six-layered cortex. He used brains from 62 species from all orders of mammals except cetaceans, of which a non-exhaustive list is:

Primates: man, orang-utan, chimpanzee, and various Old and New World monkeys
Prosimians: lemur
Chiropterans: flying fox, pipistrelle
Insectivores: hedgehog, mole
Carnivores: various canines and felines, kinkajou
Pinnipeds: common seal
Rodents: squirrel, mouse, rat, rabbit (although the rabbit would not be considered a rodent now)
Ungulates: hyrax, elephant, pig, goat
Edentates: three-toed sloth
Marsupials: phalanger, kangaroo, wallaby, opossum
Monotremes: echidna

For nine of these he published detailed cortical maps. We shall, however, concentrate on the best known, the human brain map, summarised from his own descriptions. It is noticeable that he was very careful to state when his various areas were easy to differentiate, and when difficult. Regional variations in cytoarchitecture sometimes resulted in sharp borders, sometimes in subtle transitions. He also drew attention to individual differences. Contrary to a widely-held view that his observations on the human brain were from a single case, he often referred to individual differences, and even thanked "Professor Benda for kindly providing human brains".

1.1.4 Brodmann's Description of the Human Brain Map

Brodmann "roughly" subdivided the hemispheres of man and gyrencephalic animals into four main lobes, but as he preferred not to speculate on homologies between lobes, especially in non-primates, he described 11 "regions" composed of several individual "areas" (Figs. 1.1 and 1.2). These were: postcentral, precentral, frontal, insular, parietal, temporal, occipital, cingulate, retrosplenial, hippocampal and olfactory (Fig. 1.1).

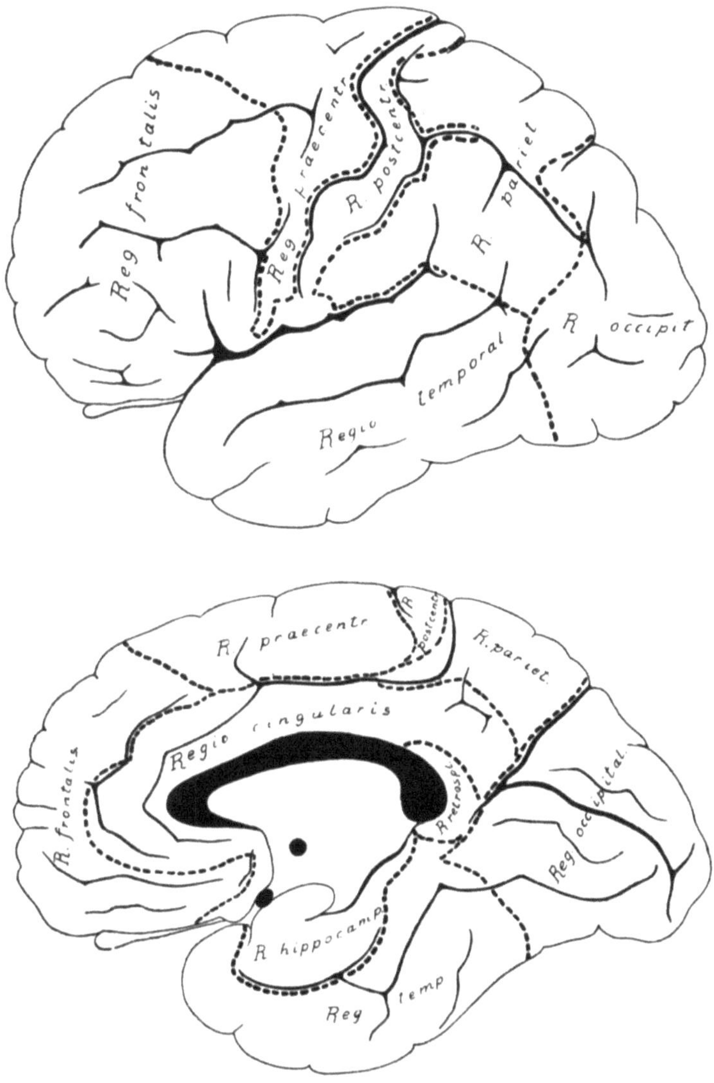

Fig. 1.1 The human cortical cytoarchtectonic regions (From Brodmann 1909). The olfactory region is not indicated

The postcentral region lies directly posterior to the central sulcus and consists mainly of the postcentral gyrus. It is subdivided into four areas: 1, 2, 3 and 43.

Area 1 – the *intermediate postcentral area* – a strip in the middle of the postcentral region between areas 2 and 3, along the apex of the postcentral gyrus and onto the medial surface, encroaching on the cortex of the central and postcentral sulci.

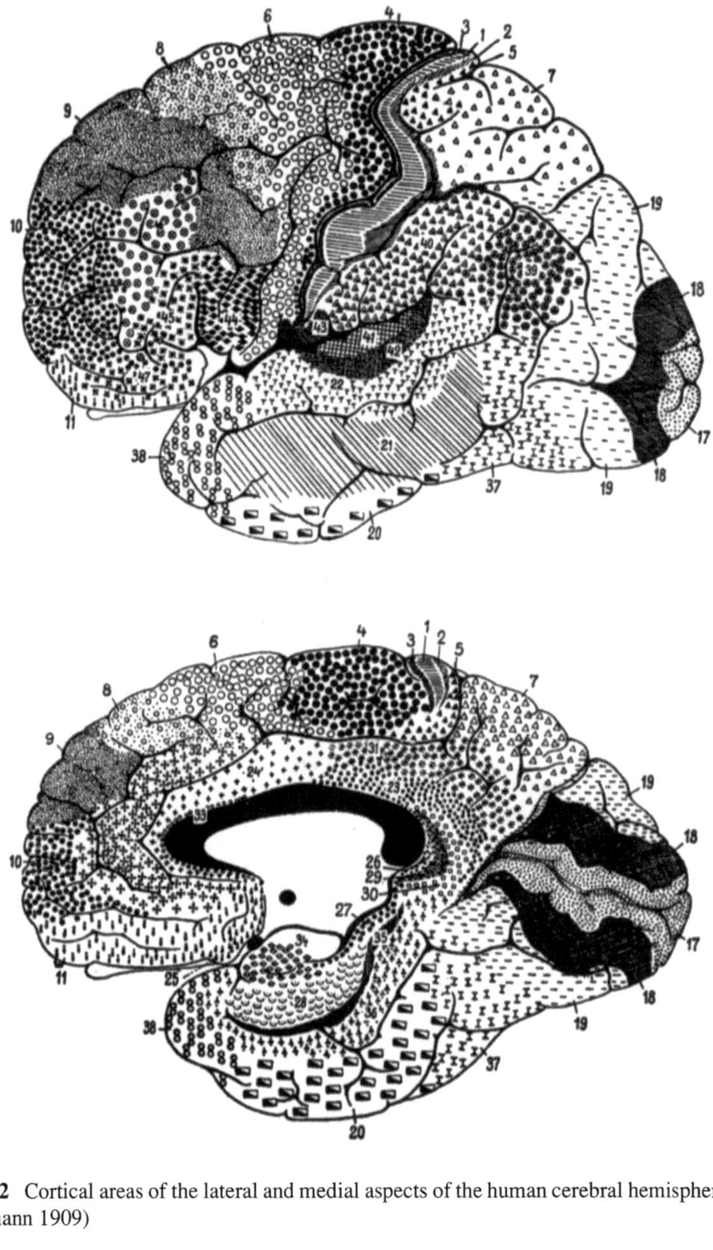

Fig. 1.2 Cortical areas of the lateral and medial aspects of the human cerebral hemispheres (From Brodmann 1909)

Area 2 – the *caudal postcentral area* – a narrow strip, mainly on the posterior aspect of the postcentral gyrus (the anterior bank of the postcentral sulcus). Its borders are not always sharp or constant. There are individual differences, as in the sulcal pattern.

Area 3 – the *rostral postcentral area* – covers the anterior of the postcentral gyrus (the posterior bank of the central sulcus). It has a variable width along the central sulcus. Its border with giant pyramidal area 4 anteriorly is sharp but in some brains its other borders can be less clear. At its medial and lateral ends, area 3 encroaches on the precentral gyrus, pushing area 4 anteriorly.

Area 43 – the *subcentral area* – is at the junction of the pre- and postcentral gyri at the inferior end of the central sulcus. Its anterior border is sharp and coincides with the anterior subcentral sulcus. It extends into the depths of the Sylvian fissure where it has a distinct boundary with the insular cortex.

The precentral region lies directly anterior to the central sulcus and is characterised by the lack of an inner granular layer. Dorsally, its anterior border crosses the precentral gyrus and encroaches on the superior and middle frontal gyri. Its anterior borders are clear but vary between individuals. Its posterior border is well demarcated from the postcentral region, and particularly from area 3, in the central sulcus. Areas 4 and 6 are characterised by the lack of an inner granular layer, and area 4 by the Betz giant cells.

"**Area 4** – the *giant pyramidal area* – is one of the most strikingly differentiated and cytoarchitectonically delimitable structural regions of the whole human cerebral cortex." It is wedge-shaped, along the central sulcus, narrowing from superior to inferior on the precentral gyrus and the adjacent part of the paracentral lobule. Superiorly it includes the whole width of the precentral gyrus but ventrally is restricted to the posterior half of this gyrus. Its borders are variable, especially in the paracentral lobule. There are local and individual differences in the number, size and distribution of giant pyramids: their size and number decrease from superior to inferior. "I must definitely classify as erroneous the idea, proposed by Elliot Smith, that the anterior subcentral sulcus is "a limiting furrow" for area 4".

Area 6 – the *agranular frontal area* – is broad superiorly, narrowing inferiorly and laterally, and covers the whole vertical extent of the frontal lobe. Medially it occupies the anterior part of the paracentral lobule and the superior frontal gyrus. Laterally, it is on the superior and middle frontal gyri, and further inferiorly the whole precentral gyrus except where it is occupied by area 4.

The frontal region is the most extensive of the human cerebral cortex; it includes the frontal lobe anterior to the precentral region, around 20 % of the total cortical area. All its areas contain an inner granular layer. Posteriorly it has a clear boundary with the agranular frontal cortex, and anteriorly extends round the frontal pole. There are eight areas in the human frontal region. Areas 44, 45 and 47 on the inferior frontal gyrus have similarities, and can be termed *subfrontal subregion*. The differences between the others are sometimes difficult to determine.

Area 8 – the *intermediate frontal area* – is a strip, wide superiorly and narrowing laterally which, like the agranular frontal area (6), crosses from the callosomarginal sulcus on the medial surface onto the lateral surface.

Area 9 – the *granular frontal area* – is similar to, but more extensive than, area 8. On the lateral surface it stops ventrally near the inferior frontal sulcus.

Area 10 – the *frontopolar area* – covers the frontal pole. Inferomedially it is demarcated by the superior rostral sulcus.

Area 11 – the *prefrontal area* – forms the anteroventral part of the frontal lobe on its orbital and medial surfaces, thus including most of the straight gyrus, the rostral gyrus and the anterior end of the superior frontal gyrus.

Area 44 – the *opercular area* – is a well-differentiated area of the inferior frontal gyrus – Broca's area. There is much variability of the sulci within it.

Area 45 – the *triangular area* – forms the triangular part of the inferior frontal gyrus. Its caudal border lies in the ascending ramus of the Sylvian fissure, its dorsal border in the inferior frontal sulcus and its rostral border near the radiate sulcus.

Area 47 – the *orbital area* – shares architectonic affinities with areas 44 and 45 and can be combined with them to form a *subfrontal subregion.*

Area 46 – the *middle frontal area* – is not clearly distinguishable from neighbouring areas. It includes about the middle third of the middle and the most anterior part of the inferior frontal gyri.

The parietal region coincides essentially with the parietal lobe, but inferiorly is difficult to differentiate from temporal and even occipital cortex; it is better distinguishable from the postcentral region at the postcentral sulcus.

Area 5 – the *preparietal area* – is delimited from neighbouring areas by the presence in layer V of large pyramidal cells almost the size of Betz giant cells, and a thick inner granular layer. Its thickness exceeds that of postcentral cortex. Its structure varies in individual cases, but its position is relatively constant. It begins in the caudal portion of the paracentral lobule, and narrows in the depths of the terminal branch of the callosomarginal sulcus on its rostral bank, extending to the lateral surface of the hemisphere.

Area 7 – the *superior parietal area* – extends from, medially the subparietal sulcus, laterally the intraparietal sulcus, posteriorly the parieto-occipital sulcus, and anteriorly the superior postcentral sulcus. One can distinguish a division into an anterior and posterior superior parietal area.

Area 40 – the *supramarginal area* – is ventral to the intraparietal sulcus around the posterior ramus of the Sylvian fissure, on the supramarginal gyrus. Anteriorly it borders the postcentral areas 2 and 43, separated from them by the inferior postcentral sulcus and the posterior subcentral sulcus. It has no sharp boundary with the temporal region (area 22).

Area 39 – the *angular area* – corresponds to the angular gyrus, widening around the posterior end of the superior temporal sulcus. Its boundaries with the occipital and temporal regions (areas 19 and 37) are ill-defined; its border with the parietal area is formed approximately by the intraparietal sulcus.

The occipital region includes the whole occipital lobe, and is divided into three structurally very different areas.

Area 17 – the *striate area* – is characterised by the calcarine type of cortex which is easily recognisable macroscopically. It is around the calcarine sulcus, and posteriorly extends a little round the occipital pole onto the lateral surface. Its individual borders are variable, with no relationships to "limiting sulci". The cuneus and the lingual gyrus form part of the striate area to variable extents, depending on the folding of the calcarine sulcus: usually the striate area extends further ventrally from the calcarine sulcus than dorsally. The dorsal striate area retreats entirely from the surface into the depths of the sulcus.

Area 18 – the *occipital area* – is a ring-like area that surrounds the striate area, more extensively laterally.

Area 19 – the *preoccipital area* – further surrounds occipital area 18, again especially laterally. Its boundaries are not related to sulci.

The temporal region is well delimited and homogeneous, stretching from the posterior margin of the insula over the whole vertical extent of the temporal lobe to the rhinal sulcus or the temporal incisura. It contains several clearly different areas of which certain, such as the transverse gyri, form important subregions of great functional importance.

Area 36 – the *ectorhinal area* – lies lateral to the rhinal sulcus and represents the first area of the neopallium adjacent to the archipallium, to which area 35 belongs. It is heterotypical with relatively few cells but a massive development of those of layers V and VI. It is the rostral extension of the lingual gyrus.

Area 37 – *occipitotemporal area.* – is a wide, but poorly circumscribed, transition zone between the adjacent occipital and temporal regions, distinct from preoccipital area 19 and temporal area 20.

Area 38 – the *temporopolar area* – forms the tip of the temporal lobe, without any clear external delimitation; it fuses laterally with areas 20, 21 and 22, and medially with area 36, and is characterised by its great depth.

Area 20 – the *inferior temporal area* – corresponds to the inferior temporal gyrus and blends rostrally and caudally with areas 37 and 38 without sharp borders.

Area 21 – the *middle temporal area* – is situated in the middle temporal gyrus, although not precisely.

Area 22 – the *superior temporal area* – is well differentiated from areas 20 and 21. Together with the cortex of the transverse gyri of Heschl (1878) (areas 41 and 42) it forms a homogeneous structural region. It was known that the transverse temporal gyri of Heschl had a different structure from most of the temporal lobe. Campbell (1905) differentiated a field within these gyri, his "audito-sensory area", contrasting it with the other temporal gyri, or "audito-psychic area". Elliot Smith (1907), in agreement with this, wrote: "The two transverse gyri of Heschl represent a sharply-defined anatomical area of this cortex", but gave no precise topographical description. The superior temporal area is superficial in the posterior two-thirds of the superior temporal gyrus, the deep part of which is partially occupied by areas 41, 42 and 52. Anteriorly it climbs onto the medial surface of the superior temporal gyrus; posteriorly it reaches the level of the vertical terminal branch of the Sylvian sulcus and blends with the supramarginal area.

Area 42 – the *lateral (posterior) transverse temporal area* – is medial to area 22, extending obliquely over the superior bank of the superior temporal gyrus, but partly on the free surface of the gyrus. It forms a well-demarcated crescent along the lateral edge of area 41.

Area 41 – the *medial (anterior) transverse temporal area* – corresponds to the anterior transverse gyrus. It is bordered medially by the parainsular area 52 from which it is sharply demarcated.

Area 52 – the *parainsular area* – forms a narrow band on the superior bank of the superior temporal gyrus along the posterior margin of the insula and represents a transitional area between the temporal cortex and the insula.

The insular region is distinguishable from surrounding regions by its easily recognisable laminar pattern, including the claustrum. It coincides approximately with the Sylvian fossa, but may encroach on the under surface of the frontal and temporal opercula. Brodmann divided the insula into two halves along the prolongation of the central sulcus, one posterior and granular, the other anterior and agranular, but without attributing numbers to them. Thus, like the central region, the insula is divisible according to the presence or absence of an inner granular layer.

The cingulate region: The crescent-shaped cingulate gyrus bordering the corpus callosum is divisible at the level of the central sulcus, like the insula, into a *postcingulate* and a *precingulate subregion*, the former with a distinct inner granular layer, while the latter (except area 32) does not have an inner granular layer. Thus the human cortical surface is divided at the level of the central sulcus into structurally different halves, an anterior agranular and a posterior granular, a trend that is also found in lower mammals.

Area 23 – the *ventral posterior cingulate area* – is in the ventral part of the posterior half of the cingulate gyrus and lies directly above the corpus callosum. It forms an arc around the splenium as far as the anterior bank of the parieto-occipital sulcus with which it gradually blends. Rostrally it fuses with the agranular precingulate subregion.

Area 31 – the *dorsal posterior cingulate area* – is in the dorsal portion of the posterior half of the cingulate gyrus and forms an arc around area 23 as far as the parieto-occipital sulcus. There is no clear outer border with area 23 or with the parietal cortex (area 7).

Area 24 – the *ventral anterior cingulate area* – in the ventral part of the anterior half of the cingulate gyrus next to the corpus callosum. Posteriorly it fuses gradually with a weakly granular transitional zone over the middle of the corpus callosum; anteriorly it extends as far as the rostrum. Its structure changes gradually from posterior to anterior.

Area 32 – the *dorsal anterior cingulate area* – forms a semicircle around the anterior end of the corpus callosum.

Area 33 – the *pregenual area* – is formed by a narrow strip of rudimentary cortex hidden in the callosal sulcus. Anteroinferiorly it stretches round the end of the rostrum of the corpus callosum, while posterosuperiorly it extends over the corpus callosum.

Area 25 – the *subgenual area* – is a small area of cortex inferior to the genu of the corpus callosum. Like the pregenual area it has a rudimentary (heterogenetic) laminar pattern.

The retrosplenial region consists of three crescent-shaped areas around the splenium of the corpus callosum. The retrosplenial cortex is partly heterogenetic.

Area 26 – the *ectosplenial area* – is apposed to the posterior end of the corpus callosum, hidden in the callosal sulcus. It has rudimentary lamination. Laterally it merges without a clear border with Area 29.

In **Area 29** – the *granular retrolimbic area* – there is a unique development of the inner granular layer (IV) and degeneration of layers II and III. It is a narrow semicircular area around the ectosplenial area and lies to a great extent in the depths of the callosal sulcus.

Area 30 – the *agranular retrolimbic area* – covers the edge of the isthmus of the cingulate gyrus, but also extends a short distance over the anterior bank of the calcarine sulcus. It forms a sort of arc around the other retrosplenial areas. The inner granular layer is degenerated while layers III and V are relatively well developed.

The hippocampal region includes the (heterogenetic) cortex between the hippocampal and rhinal sulci.

Area 27 – the *presubicular area* – lies lateral to the subiculum, separated by a sharp border, as a long, narrow zone along the hippocampal sulcus from the uncus to the tail of the hippocampus just under the corpus callosum.

Area 28 – the *entorhinal area* – is heterogenetic and lies medial to the rhinal sulcus and covers most of the head of the parahippocampal gyrus.

Area 34 – the *dorsal entorhinal area* lies mainly medial to the inferior rhinencephalic sulcus.

Area 35 – the *perirhinal area* – is a narrow band along the rhinal sulcus. The inner granular layer is missing. It forms the border between the archipallium and the neopallium, and it is difficult to decide whether it should be attributed to the one or the other.

Area 48 – the *retrosubicular area* – is at the caudal end of the perirhinal area (35) and lateral to the presubicular area (27).

1.1.5 Brodmann's Arealisation and Circuit Specialisation: The Pyramidal Cell

Pyramidal neurons are distinguished by their prominent apical dendrite and basal dendritic tree (Fig. 1.3). They comprise some 70–90 % of all neurons in cerebral neocortex (DeFelipe and Fariñas 1992). Pyramidal cells form rich plexuses of connections, often forming intrinsic lattices or patches, within cortical areas.

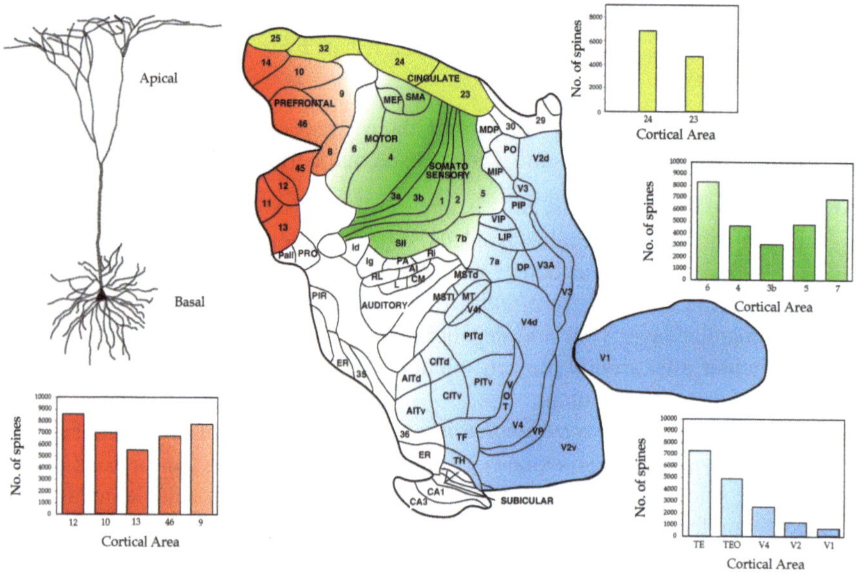

Fig. 1.3 *Top left*: a typical neocortical pyramidal cell. *Centre*: locations from which neurons were sampled in granular prefrontal (*red*), cingulate (*yellow*), sensorimotor (*green*) and visual (*blue*) cortex of the macaque monkey. *Right*: corresponding bar graphs of estimates of the total number of spines in the basal dendritic trees of layer III pyramidal cells in these areas. As each spine in mature cortex receives at least one excitatory input, differences in the estimates of the number of spines in different populations of cells may be taken as an approximation of numbers of excitatory inputs to these cells. Differences the complexity of the dendritic tree (size, number of branches) may also reflect differences in the integrative abilities of neurons (see Elston 2007 for a review). Note the trend towards more spinous cells with anterior progression from primary sensory areas such as Brodmann's areas 17 (visual) and 3 (somatosensory) into adjacent functionally related cortical areas. In visual cortex, in particular, there is a striking systematic increase in the number of spines on pyramidal cells with progression through a hierarchical series of visual areas, including Brodmann's area 18 (V2), into the temporal lobe. Note also the consistent trend for progressively more spinous cells with progression through somatosensory areas from the central sulcus to the angular gyrus (Data from Elston and Rosa 1998a; Elston et al. 1999a, 2001, 2005a, 2006, 2011; Elston 2000; Elston and Rockland 2002)

They form nearly all cortico-cortical connections, both ipsi- and contralateral, as well as most subcortical connections. Pyramidal cells contain the excitatory neurotransmitter glutamate: their discharge directly facilitates cortical activity, rather than inhibiting it. Arguably, pyramidal cells are the principal neurons of the cerebral cortex, generating nearly all cortically initiated excitation. Different subtypes of pyramidal cells have been reported based on aspects of their morphology, for example, inverted pyramidal cells, those whose axons project into the white matter, and those whose axons are restricted to the grey matter (see Feldman 1984; Nieuwenhuys 1994; Elston and DeFelipe 2002; Valverde 2002 for reviews). In addition, pyramidal cells have been subdivided according to their neurochemical content and receptor subunit profiles (Gabernet et al. 1999; González-Albo et al.

2001; Hof et al. 2001). It has also been proposed that pyramidal cells are not genetically fated to have their characteristic morphology. According to this theory, all variants of spiny neurons, from the typical pyramidal cell to the typical spiny stellate cell, are derived from a common precursor (Valverde 1988). Here we focus on the "typical" pyramidal cell of unknown neurochemical content observed in mature primate cortex. As a prelude to studying specialisations in pyramidal cell structure that have occurred during the evolution of different primate species we first set out how pyramidal cell structure varies among Brodmann's areas in a single species of macaque monkey (*Macaca fasicularis*), and outline how specialisation in cell structure may influence functional capabilities and, in turn, behavioural complexity.

1.1.6 Visual Cortex

The areas of the cerebral cortex that contain neurons involved in some form of visual processing are perhaps the most thoroughly explored in the macaque brain. Prior to the 1970s, most studies were restricted to Brodmann's area 17 (the striate, or primary (V1), visual cortex). Since then there has been an explosion in the number of studies in, and our understanding of, extrastriate visual cortex (Zeki 1969, 1978a) see (Kaas 1995; Rosa 1997; Kaas and Lyon 2001; Kaas and Preuss 2003; Zeki 2003; Rosa and Manger 2005) for reviews. Application of techniques such as electrophysiological mapping and imaging, and the development of specialised tracers, have revealed that visual processing is much more complex than previously thought, involving up to half the cortex and as many as 30 different areas (Fig. 1.3). Various theories have been presented for the existence of so many visual cortical areas, and how visual stimuli are processed by neurons in these areas (see Kaas 1987; Weller 1988; Felleman and Van Essen 1991; Rosa 1997; Kaas 2000 for reviews). In addition, many theories have been proposed regarding the recruitment and interaction of neurons in these different cortical areas during particular visual tasks. Two of the most popular models include the quasi-hierarchical model and the distributed processing model see (Mountcastle 1995) for a review. In the quasi-hierarchical model, visual inputs to cortex are processed through a series of cortical areas. These areas are not necessarily organised into a strict hierarchy, but there is some form of serial processing through select visual areas. In the distributed processing model, visual processing occurs in multiple cortical areas, but not necessarily in any form of hierarchy. That is not to say however, that the two theories are mutually exclusive. Mountcastle (1995), for example, highlighted how hierarchies may exist within a distributed system.

New methods of quantification (Elston and Rosa 1997; Elston 2001) have revealed marked, systematic differences in pyramidal cell structure (and cortical circuitry) in these different visual areas. Briefly, there is a trend for increasingly more complex pyramidal cells with progression from V1 to the second visual area (V2) and parietal visual areas (the lateral intraparietal area, LIP, and cytoarchitectonic area 7a), and temporal areas (V4, the middle temporal area, MT, cytoarchitectonic areas TEO and TE, and the superior temporal polysensory area, STP) (Elston and Rosa 1997, 2000; Elston et al. 1999a). The increase in the size of the dendritic tree, coupled with a concomitant increase in the number of dendritic branches and spine density, results in a progressive doubling in our estimates of the total number of spines in the basal dendritic trees of pyramidal cells through V1, V2, V4, TEO and TE. The functional implications of these specialisations in pyramidal cell structure in functionally related cortical areas are discussed in detail in the works of Elston (2002, 2007), Jacobs and Scheibel (2002), Spruston (2008) and DeFelipe (2011).

1.1.7 Somatosensory and Motor Cortex

Based on patterns of connectivity, neuronal response properties and, more recently, imaging studies, several theories have been put forward regarding normal function across, and cooperation between, Brodmann's sensorimotor areas (Mishkin 1979; Pons et al. 1987, 1992; Passingham 1997; Geyer et al. 2000). By injecting large numbers of pyramidal cells in some of these different cortical areas it has been possible to demonstrate marked and systematic differences in their size, branching complexity and spine density. More specifically, two trends of increasing morphological complexity have been revealed with progression from the central sulcus to adjacent cortical areas. There is a systematic increase in the size of pyramidal dendritic trees, their branching complexity, and spine density within their basal dendritic trees, with caudal progression from the primary somatosensory area on the posterior wall of the central sulcus (Brodmann's area 3; 3b) to the rostral bank of the intraparietal sulcus (Brodmann's area 5) and the exposed rostral portion of the inferior parietal lobule (Brodmann's area 7; 7b). There is also an increase in the size of the dendritic trees of pyramidal cells, their branching complexity, and the total number of spines within their basal dendritic trees, with rostral progression from the primary motor area on the anterior wall of the central sulcus (Brodmann's area 4) to the exposed lateral portion of the precentral gyrus (Brodmann's area 6 or premotor cortex) (Elston and Rockland 2002). These differences in size, branching complexity and spine density result in appreciable differences in our estimates of the total number of spines in the basal dendritic tree of the average cell in each cortical area (Fig. 1.3).

1.1.8 Cingulate Cortex

A study of the literature reveals little agreement regarding the functions performed in cingulate cortex. Some have attributed higher cognitive and emotional functions to the anterior cingulate cortex and vegetative functions to posterior cingulate cortex (Allman et al. 2001) whereas others have claimed the reverse (eg Baleydier and Mauguiere 1980). In a series of studies in which cortical activity was recorded in awake behaving monkeys by functional magnetic resonance imaging (fMRI), Dreher and colleagues revealed that neurons in anterior cingulate cortex, unlike those in posterior cingulate cortex, are often co-activated with granular prefrontal cortex (gPFC) during cognitive tasks (Dreher and Berman 2002; Dreher and Grafman 2003). Various other studies have also demonstrated "executive" or cognitive function in the anterior cingulate, including error detection and reward-based decision-making (Gemba et al. 1986; Carter et al. 1998; Bush et al. 2002; Shidara and Richmond 2002; Hadland et al. 2003).

While it is well known that these two regions of cingulate cortex can be distinguished by their cytoarchitecture, particularly the granular layer, relatively little is known of possible differences in their microcircuitry. In order to investigate this, layer III pyramidal cells were injected in the posterior cingulate gyrus (Brodmann's area 23) and their structure compared with that of cells injected in the anterior cingulate gyrus (Brodmann's area 24) (Elston et al. 2005a). These investigations revealed that pyramidal cells in area 24 were considerably larger, more branched and more spiny than those in area 23. Estimates of the total number of spines in the basal dendritic tree of the average cell reveal a two-fold difference between cells in areas 23 and 24 (Fig. 1.3). Moreover, pyramidal cell structure in area 24 more closely approximates that of cells in the gPFC, than does that of cells in area 23 (see below).

1.1.9 Prefrontal Cortex

Exactly what constitutes *prefrontal* cortex (PFC) has been interpreted in many ways. Some classify it as cortex that receives projections from the medial dorsal nucleus of the thalamus whereas others distinguish the PFC by cytoarchitecture (see Fuster 1997 for a review). Here, as intended by Brodmann (1913), the term is used to only include granular cortex anterior to the central sulcus (see Elston and Garey 2004). To avoid confusion here this region is referred to as the gPFC. Brodmann's original maps of the gPFC have been further refined by others by cytoarchitecture and patterns of corticocortical connectivity (Vogt and Vogt 1919; Walker 1940; Barbas and Pandya 1989; Petrides 1991; Preuss and Goldman-Rakic 1991a, b, 1991c; Petrides 1998; Pandya and Yeterian 2000; Petrides and Pandya 2001). Broadly speaking, these different areas have been grouped into the lateral, medial and orbital regions.

Prefrontal cortex has been the focus of intensive investigation in recent times because of its involvement in executive functions such as conceptual thinking, prioritising and planning (see Goldman-Rakic 1996; Fuster 1997; Barbas 2000; Petrides 2000; Miller and Cohen 2001 for reviews). Our understanding of the functions performed by neurons in the different regions within gPFC, and how other cortical regions participate in specific tasks, is growing rapidly with the advent of new methodologies (see Quintana and Fuster 1999; Passingham et al. 2000; Rolls 2000; Fuster 2001; Funahashi and Takeda 2002 for reviews). Orbital and medial gPFC are now believed to be involved in emotional behaviour and the processing of taste, reward, memory, affect and motivation. The lateral gPFC provides cognitive support to the temporal organisation of behaviour, speech and reasoning and is involved in executive control of voluntary motor movements. Caudo-rostral gradients characterised by different patterns of connectivity and functions have also been reported within each of these gross subdivisions (Goldman-Rakic 1987; Petrides 1987, 1991; Barbas 1992; Wilson et al. 1993; Barbas et al. 1999; Pandya and Yeterian 2000). What is clear from these studies is that because of the complexity of functions performed by neurons in gPFC, and the difficulty in quantifying neuronal responses to specific executive tasks, our understanding of prefrontal function is likely to lag behind that of sensory cortex for some time to come.

1.1.10 Synthesis and Speculation

One of the advantages of the cell injection approach is that it is not fraught with the methodological problems of anaesthetic state, training, attention or stimulus specificity associated with electrophysiological mapping and/or imaging studies. By injecting large numbers of pyramidal cells in different regions of gPFC, and relating these findings with those of pyramidal cell structure in other cortical regions, new information has emerged related to circuit specialization in different areas in the gPFC. The investigations of pyramidal cells in gPFC revealed that they are highly branched and spiny relative to those in many other cortical regions. For example, cells in the frontal eye field (FEF) were the most branched of all cells studied in the cerebral cortex of the macaque monkey (Elston and Rosa 1998b; Elston 2000; Elston et al. 2006). In addition, estimates of the number of spines in their dendritic trees revealed that they are more spiny than their counterparts in other cortical areas (Fig. 1.3). In particular, pyramidal cells in dorsolateral area 9d are considerably more spiny than those in visual, auditory, somatosensory, motor and cingulate cortex (Elston and Rosa 1997, 1998a; Elston et al. 1999a, 2005a, 2006, 2010; Elston and Rockland 2002). Pyramidal cells in the gPFC are, on average, up to 16-fold more spinous than those in V1 (Elston et al. 2001). That is not to say, however, that all pyramidal cells in gPFC are highly branched and spiny relative to those in other cortical regions. For example, those in prefrontal area 10 are less spiny than those in inferotemporal cortex (IT) and the anterior cingulate gyrus (Elston et al. 2005a,b, 2006). Some interpretations of the functional consequences of these regional differences in pyramidal cell structure are discussed below.

Multiple converging criteria, including cortical damage (both that inflicted by foreign objects and that which results from internal insult such as calcification or haemorrhage), experimental ablation studies, electrophysiological recording, imaging and theoretical studies, reveal that memories are stored across large expanses of cortex, including but not exclusive to neocortex (see Fuster 1995 for a review). While this remains a highly controversial field of investigation, there is widespread agreement that the hypothesis proposed by Hebb (1949) is fundamental to the storage of memory: that is, put simply, synapses are strengthened through use. Considerable advances have been made in our understanding of the molecular mechanisms involved at the synaptic level (see Kandel 2001 for a review). Here we turn our attention to how differences in circuit complexity may influence association in prefrontal cortex and cognitive functions.

During the last few years there has been a resurgence of interest in Hebbian type reinforcement at the circuit level (see Mel 2002; Chklovskii et al. 2004 for reviews). Of particular interest here is the distinction made between "weight-based" learning and "wiring-based" learning (Fig. 1.4). Put simply, weight-based learning refers to the synaptic reinforcement of a pre-existing synapse (in series) whereas wiring based learning refers to the establishment of a new synaptic contact by association (in parallel) (Chklovskii et al. 2004). How this bears relevance to the regional difference in pyramidal cell structure reported here becomes clear when we focus on the word "association". In particular, are cortical circuits composed of neurons with quantifiably different associative potential characterised by differing memory capacities? Intuitively, the answer is yes but let us consider how.

Typically, the associative potential of a neuron is measured by the number of inputs it can sample. As we have seen, pyramidal cells in V1 contain, on average, approximately 600 spines (putative excitatory inputs) in the basal dendritic trees. Those in prefrontal cortex contain, on average, more than 10,000 spines in their basal dendritic trees. Based on what is known of intrinsic connectivity in V1 and the gPFC (McGuire et al. 1991; Melchitzky 1998; Melchitzky et al. 2001), 20–30 % of inputs to pyramidal cells in both these cortical regions originate from neighbouring pyramidal cells (excitatory). By way of example, let us assume that cells in V1 are connected with 200 local pyramidal cells whereas those in the gPFC are connected with 3,000 neighbouring pyramidal cells. Application of these numbers to a wiring based model makes it clear that there is a greater potential for association in the gPFC. Extension of this logic to include multiple synapses across three or more levels reveals how the associative potential of a circuit is increased, particularly in the gPFC. The exponential increase in the associative potential in a multi-neuron circuit in the gPFC soon eclipses that in V1 as successive synaptic steps are added to the circuit. Applying this logic, based on what is known of the structure of neurons, reveals intermediate levels of association in inferotemporal and parietal association cortex, as well as cingulate cortex.

In addition, the functional capacity of individual neurons is influenced by the branching structure of their dendritic trees. In turn, the branching structure and distribution of ion channels throughout the dendritic trees of neurons will influence the association potential of the circuits they comprise. More specifically, the structure of the dendritic tree determines both the number of computational subunits, and their geometric interface with other neurons (Williams and Stuart 2002; Nolan et al. 2004; Krapp and Gabbiani 2005). Differences in the distribution of inputs throughout the dendritic trees of pyramidal cells, as exemplified by their

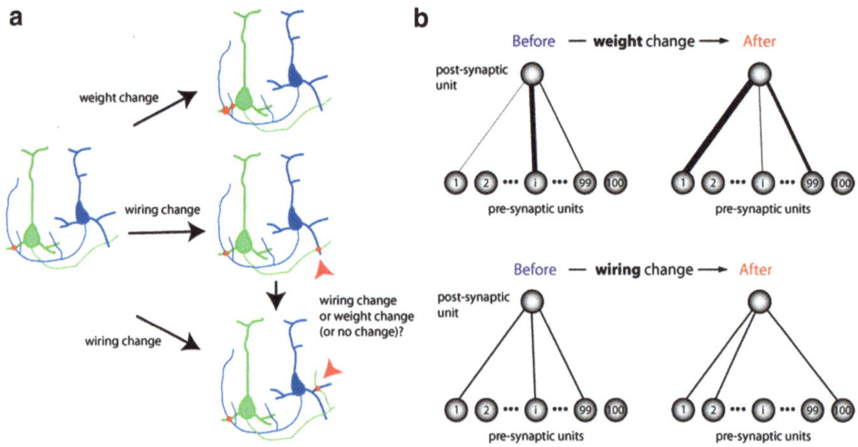

Fig. 1.4 Learning by changing weights versus wires. (**a**) Two neurons (*green*, *blue*), dendrites (*thick lines*), axons (*thin lines*) and synapses (*red circles*). Initial connectivity is from the *blue* to *green* neuron. *Top*: weight changes can occur through changes at existing connections (*larger red dot*), or by synapse formation (*new red dot*) or by elimination (*not shown*). *Middle* and *bottom*: a new connection is established from *green* to *blue* neuron, resulting in a wiring change. *Middle* and *bottom* cases are functionally equivalent if the postsynaptic integrative unit is the whole neuron, and are different if the postsynaptic unit is a single dendrite. (**b**) Storage capacity is a measure of total learning-related flexibility of the circuit, which may be represented as a graph of abstract units with weighted interconnections. If a network is sparsely connected, it is useful to distinguish weight versus wiring modes of plasticity. Consider a postsynaptic unit with $s = 10$ input connections (only three are shown), and a population of 100 potential pre-synaptic partners. Each connection has four possible stable values (0, 1, 2 and 3) denoted by *line thickness*. Assuming only weight changes, this gives $w = \log_2(4) = 2$ bits of capacity per synapse. However, if the wiring diagram can change during learning, the wiring-related capacity is $\log_2(100 \text{ choose } 10) = 46$ bits of storage, or 4.6 bits per synapse. The more axons that can serve as presynaptic partners for each postsynaptic site, the greater the in-principle wiring capacity advantage (From Chklovskii et al. 2004)

different branching structure, may result in a difference of up to two orders of magnitude in the memory storage capacity (Poirazi and Mel 2001).

1.1.11 Species Specialisations in Cortical Circuitry Among Brodmann's Areas

How can circuit specialisation among Brodmann's areas in a single species be interpreted among species? The cerebral cortex differs in volume by more than 10,000-fold in extant mammals (see Stephan et al. 1981; Hofman 1985; Kaas 1989; Krubitzer 2000; Kaas and Collins 2001; Northcutt 2002 for reviews). These differences in volume do not adhere strictly to any cladistic variable, as dramatic differences in the volume of the cerebral cortex can be found within any given order or family, even between subspecies (eg pygmy marmoset and common marmoset, greater and lesser galagos) (see Manger 2005 for a review). Not withstanding this variability, comparisons of the brain volume of large numbers of individuals suggest that the cerebrum has expanded in some species, remained relatively constant in others and possibly shrunk in yet other species. Narrowing the scope to primates, it is generally accepted that a common ancestor of extant primates had a relatively small cerebral cortex, which has become larger in different species (Tobias 1981). Presently available data yields no convincing evidence that there is a common or unifying principle regarding specialisation in pyramidal cell structure and brain size or volume. Instead, the data on pyramidal cell structure suggest that evolutionary and developmental features that act in concert to shape the mature neuronal phenotype are likely to vary in different cortical regions and species.

Based on data obtained from the occipital, parietal, temporal and frontal lobes of primates, two different trends have been revealed between pyramidal cell structure and cortical expansion: (1) expansion whereby mature cortex contains pyramidal neurons of similar structure or (2) expansion whereby mature cortex contains pyramidal neurons of increasingly complex structure (Elston et al. 2005b). Evidence can be gleaned for both outcomes. For example the total cortical volume occupied by V1 varies between primates, with as much as a five-fold difference reported between macaques and marmosets (Brodmann 1913), yet layer III pyramidal cells have a remarkably similar number of spines in these animals (Fig. 1.5). Evidence for expansion of cortex by the addition of progressively more complex pyramidal cells comes from the temporal lobe: there is a consistent and dramatic increase in the number of spines in the dendritic trees of pyramidal cells as a function of absolute distance from the occipital pole. Comparison of the gPFC data in galago, marmoset, vervet, macaque, baboon and man (Elston and Rosa 1998b; Elston et al. 2001, 2005b, 2006, 2011) suggest a general trend for increasingly more spiny pyramidal cells as this region as a whole has undergone cortical expansion (Fig. 1.5).

1.1.12 The Brain and Intelligence

Several attempts have been made in recent times to readdress the issue of brain size and intelligence by employing sophisticated statistical and imaging methodologies. Unfortunately, rather than clarify the issue these studies have created more confusion. For example, Bush and Allman (2004) based on a sample of 25 primate and 15 carnivore species, suggest that the frontal lobe "hyperscales" in primates. That is to say, the volume of the frontal lobe increases disproportionately with increasing total cortical volume in primates. However, Semendeferi and colleagues (Semendeferi et al. 2002) quantified the frontal lobes of five great apes and concluded that human frontal lobes were not disproportionately large in comparison to those of other great apes (as a function of total cerebral cortical volume). A reasonable conclusion from these combined data might be that the apparent

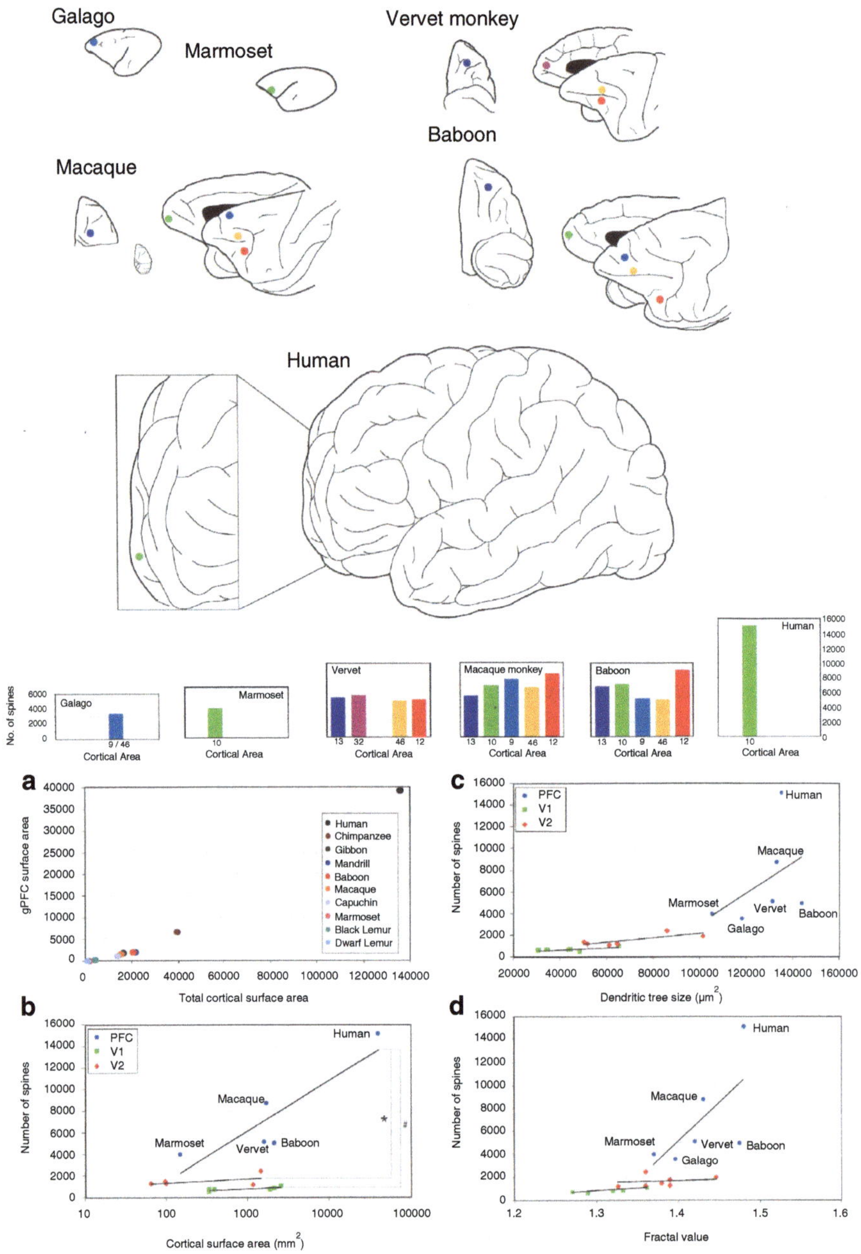

Fig. 1.5 *Top*: Schematics of brains of galago, marmoset, vervet, macaque, baboon and man in which pyramidal cells were injected in granular prefrontal cortex (*coloured dots*), revealing differences in the number of spines (*putative excitatory inputs*) in the basal dendritic trees among species (*colour bar graphs*). Note that pyramidal cells in human granular prefrontal cortex are considerably more spinous than those in prefrontal cortex of other primates. The number of

hyperscaling reported for primates by Bush and Allman (2004) results primarily from an increase in frontal lobe volume in the great apes collectively. However, frontal cortex hyperscaling is greater in lemurs and lorises compared with other primates. Thus, relative size of the frontal lobe to total cortical volume reveals little about cognitive abilities. Schoenemann and colleagues (2005) concluded that there is a disproportionately larger volume of white matter in human prefrontal cortex as compared to that in other primates, and suggested this reflects a substrate for human cognitive abilities. However, reanalysis of their data led Sherwood and colleagues (Schoenemann et al. 2005) to conclude that the white matter volume in the human brain is less that expected from the great ape data. Clearly, there is no agreement among these studies. Moreover, conspicuous in all of them is that the methodologies used do not allow the identification of the structure in question – prefrontal cortex. These authors were forced to make an approximation of prefrontal cortex. Furthermore, many of them ignore the most comprehensive data set available on prefrontal cortex, that of Brodmann (Brodmann 1912, 1913). Most of the confusion arises because inappropriate methodology has been applied to test the relationship between prefrontal size and "intelligence".

There is no denying that the human granular prefrontal cortex is unmatched in size in mammals. Some would conclude that this counts for little. Arguments are put forward to standardise these brain data with body weight and then further process the data by log transformation. Two important questions need to be considered when interpreting such an approach. (1) Does body weight have anything to do with intelligence and (2) is it appropriate to only accept the data if it holds up after being subjected to algorithms that homogenise them? A critical review of the fossil record and the comparative data leads to the conclusion that any relationship between brain or body weight and human intelligence is fallacious. The dramatic increase in human brain size and cognitive ability was not paralleled by a correlated increase in bodyweight: there was a three-fold increase in brain size

Fig. 1.5 (continued) spines in the basal dendritic tree of the "average" pyramidal cell is calculated by summing the product of the mean spine density and mean number of dendritic branches over successive 25 μm annuli across the entire dendritic tree of all neurons sampled in a given cortical area. *Bottom*: Plots of (**a**) Brodmann's (1913) data on the size of the granular prefrontal cortex (gPFC) versus the total cortical surface area in man, chimpanzee, gibbon, mandrill, baboon, macaque, capuchin monkey, marmoset, black lemur and dwarf lemur and the number of spines in the basal dendritic trees of pyramidal cells in the granular prefrontal cortex, primary (V1) and secondary (V2) visual areas and (**b**) the tangential surface area of each cortical area and (**c**) the size of the dendritic trees and (**d**) the number of spines versus the fractal value of the basal dendritic trees of pyramidal cells in gPFC, V1 and V2. Note in D the radical increase in the number of spines in gPFC as cells have increased their space-filling capacity. Moderated multiple regression revealed a significant difference between the slopes of regression lines of gPFC and V2 for comparisons between total number of spines in the dendritic trees of pyramidal cells versus cortical surface area (F_{change} (1,6) = 6.19, $p < 0.05$). Significance was approached for the comparison between the total number of spines in the dendritic trees of pyramidal cells versus cortical surface area for gPFC and V1 (F_{change} (1,7) = 5.20, $p < 0.057$) (Data from Elston 2001; Elston et al. 2001, 2005b, 2006, 2011)

during hominin evolution with relatively little change in body size (Tobias 1981). Moreover, the brain of modern *Homo sapiens* is apparently smaller than that of our Neanderthal cousins but our accomplishments suggest we are more intelligent. What is clear is that:

1. The human granular prefrontal cortex is larger than that in any other species;
2. The human brain is larger than that of its present day hominin cousins; and
3. The recent expansion of the human brain far exceeds brain or body weight trends of other mammals.

Moreover, let us not forget that there is still a glaring problem in this field of investigation, irrespective of the data set. All these studies focus on the gross structure of the brain. Any direct correlation between brain size and behavioural complexity (or intelligence) is founded on the premise that the structure is uniform in composition. The logic posits that a species with a larger brain (or part thereof), or higher than normal encephalisation quotient (brain to body weight ratio – EQ) is functionally superior (more intelligent). It is generally agreed that there is no relationship between brain size (or part thereof) and intelligence. Elephants are not more intelligent than humans. However, much credence is placed on the relationship between the brain and body weight (EQ) and intelligence. An increasing EQ is supposed to signify increasing intelligence (Jerison 1973). However, such an interpretation is fundamentally flawed. No brain structure is uniform across mammalia, let alone across vertebrates. The brain in a species of relatively high EQ may be less intelligent than the brain of a species with a lower EQ, due to differences in the functional capacity of their design. The size of the brain, cortex or prefrontal cortex must be considered with regard to the complexity of the circuits they are composed of. Anyone who has lived through the computer age is acutely aware that bigger does not necessarily mean better. On the contrary, the modern ethos is that smaller is better, and smaller increasingly means greater functional complexity within increasingly more complex circuit design. It follows that simply by being larger than that in other primates, the gPFC of human cannot satisfactorily explain human intelligence. Clearly one needs to look more closely ... to the microstructure.

Acknowledgements Supported by grants from the McDonnell Foundation, Hear and Say Australia.

References

Allman JM, Hakeem A, Erwin JM, Nimchinsky E, Hof P (2001) The anterior cingulate cortex. The evolution of an interface between emotion and cognition. Ann N Y Acad Sci 935:107–117
Baleydier C, Mauguiere F (1980) The duality of the cingulate gyrus in monkey: neuroanatomical study and functional hypothesis. Brain 103:525–554

Barbas H (1992) Architecture and cortical connections of the prefrontal cortex in the rhesus monkey. In: Chauvel P, Delgado-Escueta AV, Halgren E, Bancuad J (eds) Advances in neurology. Raven, New York, pp 91–115

Barbas H (2000) Connections underlying the synthesis of cognition, memory, and emotion in primate prefrontal cortices. Brain Res Bull 52:319–330

Barbas H, Pandya DN (1989) Architecture and intrinsic connections of the prefrontal cortex in the rhesus monkey. J Comp Neurol 286:353–375

Barbas H, Ghashghaei H, Dombrowski SM, Rempel-Clower NL (1999) Medial prefrontal corticies are unified by common connections with superior temporal corticies and distinguished by input from memory related areas in the rhesus monkey. J Comp Neurol 410:343–367

Berlin R (1858) Beitrag zur Structurlehre der Grosshirnwindungen. Inaugural dissertation. Junge, Erlangen

Betz W (1874) Anatomischer Nachweis zweier Gehirncentra. Centralblatt für die medicinischen Wissenschaften 12:578–580, 595–599

Bolton JS (1900) The exact histological localisation of the visual area of the human cerebral cortex. Philos Trans R Soc Lond B 193:165–222

von Bonin G (1953) Alfred Walter Campbell. In: Haymaker W (ed) The founders of neurology. Thomas, Springfield, pp 16–18

Brodmann K (1903a) Beiträge zur histologischen Lokalisation der Grosshirnrinde. Erste Mitteilung: Die Regio Rolandica. J Psychol Neurol 2:79–107

Brodmann K (1903b) Beiträge zur histologischen Lokalisation der Grosshirnrinde. Zweite Mitteilung: Der Calcarinatypus. J Psychol Neurol 2:133–159

Brodmann K (1905a) Beiträge zur histologischen Lokalisation der Grosshirnrinde. Dritte Mitteilung: Die Rindenfelder der niederen Affen. J Psychol Neurol 4:177–226

Brodmann K (1905b) Beiträge zur histologischen Lokalisation der Grosshirnrinde. Vierte Mitteilung: Die Riesenpyramidentypus und sein Verhalten zu den Furchen bei den Karnivoren. J Psychol Neurol 6:108–120

Brodmann K (1906) Beiträge zur histologischen Lokalisation der Grosshirnrinde. Fünfte Mitteilung: Über den allgemeinen Bauplan des Cortex pallii bei den Mammalieren und zwei homologe Rindenfelder im besonderen. Zugleich ein Beitrag zur Furchenlehre. J Psychol Neurol 6:275–400

Brodmann K (1908a) Beiträge zur histologischen Lokalisation der Grosshirnrinde. VI. Mitteilung: Die Cortexgliederung des Menschen. J Psychol Neurol 10:231–246

Brodmann K (1908b) Beiträge zur histologischen Lokalisation der Grosshirnrinde. VII. Mitteilung: Die cytoarchitektonische Cortexgliederung der Halbaffen (Lemuriden). J Psychol Neurol 10:287–334

Brodmann K (1909) Vergleichende Lokalisationslehre der Grosshirnrinde in ihren Prinzipien dargestellt auf Grund des Zellenbaues. JA Barth, Leipzig (see Garey 1994)

Brodmann K (1912) Neue Ergebnisse über die vergleichende histologische Lokalisation der Grosshirnrinde mit besonderer Berücksichtigung des Stirnhirns. Anat Anz 41:157–216

Brodmann K (1913) Neue Forschungsergebnisse der Grosshirnrindenanatomie mit besonderer Berücksichtigung anthropologischer Fragen. Gesellschaft deutscher Naturforscher und Ärtze 85:200–240 (see Elston and Garey 2004)

Bush EC, Allman JM (2004) The scaling of frontal cortex in primates and carnivores. Proc Natl Acad Sci USA 101:3962–3966

Bush G, Vogt BA, Holmes J, Dale AM, Greve D, Jenik MA, Rosen BR (2002) Dorsal anterior cingulate cortex: a role in reward-based decision making. Proc Natl Acad Sci USA 99:523–528

Cajal SR (1900–1906) Studien über die Hirnrinde des Menschen. JA Barth, Leipzig. Translated by J Bresler. Volume 1. Die Sehrinde, 1900. Volume 2. Die Bewegungsrinde, 1900. Volume 3. Die Hörrinde, 1902. Volume 4. Die Riechrinde beim Menschen und Säugetier, 1903. Volume 5. Vergleichende Strukturbeschreibung und Histogenesis der Hirnrinde, 1906

Campbell AW (1905) Histological studies on the localisation of cerebral function. Cambridge University Press, Cambridge

Carter CS, Braver TS, Botvinik MM, Noll D, Cohen JD (1998) Anterior cingulate cortex, error detection, and the online monitoring of performance. Science 280:747–749

Chklovskii DB, Mel BW, Svoboda K (2004) Cortical rewiring and information storage. Nature 431:782–788

Defelipe J (2011) The evolution of the brain, the human nature of cortical circuits, and intellectual creativity. Front Neuroanat 5:29

DeFelipe J, Fariñas I (1992) The pyramidal neuron of the cerebral cortex: morphological and chemical characteristics of the synaptic inputs. Prog Neurobiol 39:563–607

Dreher JC, Berman KF (2002) Fractionating the neural substrate of cognitive control processes. Proc Natl Acad Sci USA 99:14595–14600

Dreher JC, Grafman J (2003) Dissociating the roles of the rostral anterior cingulate and the lateral prefrontal cortices in performing two tasks simultaneously or successively. Cereb Cortex 13:329–339

Elston GN (2000) Pyramidal cells of the frontal lobe: all the more spinous to think with. J Neurosci 20:RC95 (1–4)

Elston GN (2001) Interlaminar differences in the pyramidal cell phenotype in cortical areas 7 m and STP (the superior temporal polysensory area) of the macaque monkey. Exp Brain Res 138:141–152

Elston GN (2002) Cortical heterogeneity: implications for visual processing and polysensory integration. J Neurocytol 31:317–335

Elston GN (2007) Specializations in pyramidal cell structure during primate evolution. In: Kaas JH, Preuss TM (eds) Evolution of nervous systems. Academic, Oxford, pp 191–242

Elston GN, Rosa MGP (1997) The occipitoparietal pathway of the macaque monkey: comparison of pyramidal cell morphology in layer III of functionally related cortical visual areas. Cereb Cortex 7:432–452

Elston GN, Rosa MGP (1998a) Morphological variation of layer III pyramidal neurones in the occipitotemporal pathway of the macaque monkey visual cortex. Cereb Cortex 8:278–294

Elston GN, Rosa MGP (1998b) Complex dendritic fields of pyramidal cells in the frontal eye field of the macaque monkey: comparison with parietal areas 7a and LIP. Neuroreport 9:127–131

Elston GN, Rosa MGP (2000) Pyramidal cells, patches, and cortical columns: a comparative study of infragranular neurons in TEO, TE, and the superior temporal polysensory area of the macaque monkey. J Neurosci 20:RC117 (1–5)

Elston GN, Rockland K (2002) The pyramidal cell of the sensorimotor cortex of the macaque monkey: phenotypic variation. Cereb Cortex 12:1071–1078

Elston GN, DeFelipe J (2002) Spine distribution in neocortical pyramidal cells: a common organizational principle across species. Prog Brain Res 136:109–133

Elston GN, Garey LJ (2004) New research findings on the anatomy of the cerebral cortex of special relevance to anthropological questions. University of Queensland Printery, Brisbane

Elston GN, Garey LJ (2009) Prefrontal cortex: Brodmann and Cajal revisited. In: LoGrasso L, Moretti G (eds) Prefrontal cortex: roles, interventions and traumas. Nova, New York, pp 245–259

Elston GN, Tweedale R, Rosa MGP (1999a) Cortical integration in the visual system of the macaque monkey: large scale morphological differences of pyramidal neurones in the occipital, parietal and temporal lobes. Proc R Soc Lond B 266:1367–1374

Elston GN, Benavides-Piccione R, DeFelipe J (2001) The pyramidal cell in cognition: a comparative study in human and monkey. J Neurosci 21:RC163 (1–5)

Elston GN, Benavides-Piccione R, DeFelipe J (2005a) A study of pyramidal cell structure in the cingulate cortex of the macaque monkey with comparative notes on inferotemporal and primary visual cortex. Cereb Cortex 15:64–73

Elston GN, Elston A, Casagrande VA, Kaas JH (2005b) Pyramidal neurons of granular prefrontal cortex the galago: complexity in the evolution of the psychic cell. Anat Rec 285:610–618

Elston GN, Okamoto T, Oga T, Dornan D, Fujita I (2010) Spinogenesis and pruning in the primary auditory cortex of the macaque monkey (Macaca fascicularis): an intracellular injection study of layer III pyramidal cells. Brain Res 1316:35–42
Elston GN, Benavides-Piccione R, Elston A, Zietsch B, DeFelipe J, Manger P, Casagrande V, Kaas JH (2006) Specializations of the granular prefrontal cortex of primates: implications for cognitive processing. Anat Rec 288A:26–35
Elston GN, Benavides-Piccione R, Elston A, Manger P, DeFelipe J (2011) Pyramidal cells in prefrontal cortex: comparative observations reveal unparalleled specializations in neuronal structure among primate species. Front Neuroanat 5:2
Feldman D (1984) Morphology of the neocortical pyramidal neuron. In: Peters A, Jones EG (eds) Cerebral cortex vol 1: cellular components of the cerebral cortex. Plenum, New York, pp 123–200
Felleman DJ, Van Essen DC (1991) Distributed hierarchical processing in primate cerebral cortex. Cereb Cortex 1:1–47
Funahashi S, Takeda K (2002) Information processes in the primate prefrontal cortex in relation to working memory processes. Rev Neurosci 13:313–345
Fuster JM (1995) Memory in the cerebral cortex. An empirical approach to neural networks in the human and nonhuman primate. MIT Press, Cambridge
Fuster JM (1997) The prefrontal cortex: anatomy, physiology, and neuropsychology of the frontal lobe. Lippincott-Raven, Philadelphia
Fuster JM (2001) The prefrontal cortex – an update; time is of the essence. Neuron 30:319–333
Gabernet L, Meskenaïte V, Hepp-Reymond MC (1999) Parcellation of the lateral premotor cortex of the macaque monkey based on staining with the neurofilament antibody SMI-32. Exp Brain Res 128:188–193
Garey LJ (1994) Brodmann's localisation in the cerebral cortex. Smith-Gordon, London. New edtion (2006) Springer, New York
Gemba H, Sasaki K, Brooks VB (1986) "Error" potentials in limbic cortex (anterior cingulate area 24) of monkeys during motor learning. Neurosci Lett 70:223–227
Geyer S, Matelli M, Luppino G, Zilles K (2000) Functional neuroanatomy of the primate isocortical motor system. Anat Embryol 202:443–474
Goldman-Rakic PS (1987) Circuitry of primate prefrontal cortex and regulation of behavior by representational memory. In: Plum F (ed) Handbook of physiology: the nervous system vol 5. American Physiological Society, Bethesda, pp 373–417
Goldman-Rakic PS (1996) The prefrontal landscape: Implications of functional architecture for understanding human mentation and the central executive. Phil Trans R Soc Lond B 351:1445–1453
González-Albo MC, Elston GN, DeFelipe J (2001) The human temporal cortex: characterization of neurons expressing nitric oxide synthase, neuropeptides and calcium binding proteins, and their glutamate receptor subunit profiles. Cereb Cortex 11:1170–1181
Hadland KA, Rushworth MFS, Gaffan D, Passingham RE (2003) The anterior cingulate and reward-guided selection of actions. J Neurophysiol 89:1161–1164
Hammarberg C (1895) Studien über Klinik und Pathologie der Idiotie, nebst Untersuchungen über die normale Anatomie der Hirnrinde. Berling, Akademische Buchdruckerie, Upsala
Hebb DO (1949) The organization of behaviour. Wiley, New York
Heschl RL (1878) Ueber die vordere quere Schläfenwindung des menschlichen Grosshirns. Braumüller, Vienna
His W (1904) Die Entwickelung des menschlichen Gehirns während der ersten Monate. Hirzel, Leipzig
Hof PR, Nimchinsky EA, Perl DP, Erwin JM (2001) An unusual population of pyramidal neurons in the anterior cingulate cortex of hominids contains the calcium-binding protein calretinin. Neurosci Lett 307:139–142
Hofman MA (1985) Size and shape of the cerebral cortex in mammals. I. The cortical surface. Brain Behav Evol 27:28–40

Jacobs B, Scheibel AB (2002) Regional dendritic variation in primate cortical pyramidal cells. In: Schüz A, Miller R (eds) Cortical areas: unity and diversity. Taylor and Francis, London, pp 111–131
Jerison HJ (1973) Evolution of the brain and intelligence. Academic, New York
Kaas JH (1987) Why does the brain have so many visual areas? J Cogn Neurosci 1:121–135
Kaas JH (1989) The evolution of complex sensory systems in mammals. J Exp Biol 146:165–176
Kaas JH (1995) The evolution of isocortex. Brain Behav Evol 46:187–196
Kaas JH (2000) Organizing principles of sensory representations. In: Bock GR, Cardew G (eds) Evolutionary developmental biology of the cerebral cortex, vol 228, Novartis foundation symposium. Wiley, New York, pp 188–205
Kaas JH, Collins C (2001) Evolving ideas of brain evolution. Nature 411:141–142
Kaas JH, Lyon DC (2001) Visual cortex organization in primates: theories of V3 and adjoining visual areas. Prog Brain Res 134:285–295
Kaas JH, Preuss TM (2003) Human brain evolution. In: Squire L, Bloom F, Landis SR, Zigmond M (eds) Fundamental neuroscience, 2nd edn. Academic, San Diego, pp 1147–1166
Kandel ER (2001) The molecular biology of memory storage: a dialog between genes and synapses. Biosci Rep 21:565–611
Krapp HG, Gabbiani F (2005) Spatial distribution of inputs and local receptive field properties of a wide-field, looming sensitive neuron. J Neurophysiol 93:2240–2253
Krubitzer LA (2000) How does evolution build a complex brain. In: Bock GR, Cardew G (eds) Evolutionary developmental biology of the cerebral cortex. Wiley, Chichester, pp 206–226
Lewis WB (1878) On the comparative structure of the cortex cerebri. Brain 1:79–96
Lewis WB (1881) Researches on the comparative structure of the cortex cerebri. Phil Trans R Soc 171:35–64
Lewis WB, Clarke H (1878) The cortical lamination of the motor area of the brain. Proc R Soc Lond 27:38–49
Manger PR (2005) Establishing order at the systems level in mammalian brain evolution. Brain Res Bull 66:282–289
Mauss T (1908) Die faserarchitektonische Gliederung der Grosshirnrinde bei den niederen Affen. J Psychol Neurol 13:263–325
McGuire BA, Gilbert CD, Rivlin PK, Wiesel TN (1991) Targets of horizontal connections in macaque primary visual cortex. J Comp Neurol 305:370–392
Mel BW (2002) Have we been Hebbing down the wrong path? Neuron 34:175–177
Melchitzky DS, Sesack SR, Pucak ML, Lewis DA (1998) Synaptic targets of pyramidal neurons providing intrinsic horizontal connections in monkey prefrontal cortex. J Comp Neurol 390:211–224
Melchitzky DS, González-Burgos G, Barrionuevo G, Lewis DA (2001) Synaptic targets of the intrinsic axon collaterals of supragranular pyramidal neurons in monkey prefrontal cortex. J Comp Neurol 430:209–221
Miller EK, Cohen JD (2001) An integrative theory of prefrontal cortex. Ann Rev Neurosci 24:167–202
Mishkin M (1979) Analagous neural models for tactual and visual learning. Neuropsychologia 17:139–149
Mountcastle VB (1995) The evolution of ideas concerning the function of neocortex. Cerebr Cortex 5:289–295
Nieuwenhuys R (1994) The Neocortex. An overview of its evolutionary development, structural organization and synaptology. Anat Embryol 190:307–337
Nolan MF, Malleret G, Dudman JT, Buhl DL, Santoro B, Gibbs E, Vronskaya S, Buzsáki G, Siegelbaum SA, Kandel A, Morozov A (2004) A behavioral role for dendritic integration: HCN1 channels constrain spatial memory and plasticity at inputs to distal dendrites of CA1 pyramidal neurons. Cell 24:719–732
Northcutt RG (2002) Understanding vertebrate brain evolution. Integr Comp Biol 42:743–756

Pandya DN, Yeterian EH (2000) Comparison of prefrontal architecture and connections. In: Roberts AC, Robbins TW, Weiskrantz L (eds) The prefrontal cortex. Oxford University Press, Oxford, pp 51–66
Passingham R (1997) Functional organisation of the motor system. In: Frackowiak RSJ, Friston KJ, Frith CD, Dolan RJ, Mazziotta JC (eds) Human brain function. Academic, San Diego, pp 243–274
Passingham RE, Toni I, Rushworth MFS (2000) Specialisation within the prefrontal cortex: the ventral prefrontal cortex and associative learning. Exp Brain Res 133:103–113
Petrides M (1987) Conditional learning and the primate prefrontal cortex. In: Perecman E (ed) The frontal lobes revisited. IRBN Press, New York, pp 91–108
Petrides M (1991) Functional specialization within the dorsolateral prefrontal cortex for serial order memory. Proc R Soc Lond B 246:299–306
Petrides M (1998) Specialised systems for the processing of mnemonic information within the primate frontal cortex. In: Roberts AC, Robbins TW, Weiskrantz L (eds) The prefrontal cortex. Oxford University Press, Oxford, pp 103–106
Petrides M (2000) Mapping prefrontal cortical systems for the control of cognition. In: Toga AW, Mazziotta JC (eds) Brain mapping: the systems. Academic, San Diego, pp 159–176
Petrides M, Pandya DN (2001) Comparative cytoarchitectonic analysis of the human and the macaque dorsolateral prefrontal cortex and corticocortical connection patterns in the monkey. Eur J Neurosci 16:291–310
Poirazi P, Mel BW (2001) Impact of active dendrites and structural plasticity on the storage capacity of neural tissue. Neuron 29:779–796
Pons TP, Garraghty PE, Mishkin M (1992) Serial and parallel processing of tactual information in somatosensory cortex of rhesus monkeys. J Neurophysiol 68:518–527
Pons TP, Garraghty PE, Friedman DP, Mishkin M (1987) Physiological evidence for serial processing in somatosensory cortex. Science 237:417–420
Preuss TM, Goldman-Rakic PS (1991a) Architectonics of the parietal and temporal association cortex in the strepsirhine primate *Galago* compared to the anthropoid Macaca. J Comp Neurol 310:475–506
Preuss TM, Goldman-Rakic PS (1991b) Ipsilateral cortical connections of granular frontal cortex in the strepsirhine primate *Galago*, with comparative comments on anthropoid primates. J Comp Neurol 310:507–549
Preuss TM, Goldman-Rakic PS (1991c) Myelo-and cytoarchitecture of the granular frontal cortex and surrounding regions in the strepsirhine primate *Galago* and the anthropoid primate *Macaca*. J Comp Neurol 310:429–474
Quintana J, Fuster JM (1999) From perception to action: temporal integrative functions of prefrontal and parietal neurons. Cereb Cortex 9:213–221
Rolls ET (2000) The orbitofrontal cortex. In: Roberts AC, Robbins TW, Weiskrantz L (eds) The prefrontal cortex. Oxford University Press, Oxford, pp 67–86
Rosa MGP (1997) Visuotopic organization of primate extrastriate cortex. In: Rockland K, Kaas JH, Peters A (eds) Cerebral cortex vol 12, extrastriate cortex in primates. Plenum Press, New York, pp 127–204
Rosa MG, Manger PR (2005) Homologies in the mammalian cerebral cortex: the case of the third visual area (V3). Clin Exp Pharmacol Physiol 32:327–339
Schoenemann PT, Sheehan MJ, Glotzer LD (2005) Prefrontal white matter volume is disproportionately larger in humans than in other primates. Nat Neurosci 8:242–252
Semendeferi K, Lu A, Schenker N, Damasio H (2002) Humans and great apes share a large frontal cortex. Nat Neurosci 5:272–276
Sengpiel F, Kind PC (2002) The role of activity in development of the visual system. Curr Biol 12:818–826
Shidara M, Richmond BJ (2002) Anterior cingulate: single neuron signals related to degree of reward expectancy. Science 296:1709–1711

Smith GE (1907b) A new topographical survey of the human cerebral cortex, being an account of the distribution of the anatomically distinct cortical areas and their relationship to the cerebral sulci. J Anat Physiol 41:237–254

Spruston N (2008) Pyramidal neurons: dendritic structure and synaptic integration. Nat Rev Neurosci 9:206–221

Stephan H, Frahm H, Baron G (1981) New and revised data on volume of brain structures in insectivores and primates. Folia Primatol 35:1–29

Tobias PV (1981) Evolution of the human brain, intellect and spirit, vol 1, Abbie memorial lecture. University of Adelaide, Adelaide, pp 5–70

Valverde F (1988) Competition for the sake of diversity. Behav Brain Sci 11:102–103

Valverde F (2002) Structure of the cerebral cortex. Intrinsic organization and comparative analysis of the neocortex. Rev Neurol 34:758–780

Vogt C, Vogt O (1919) Allgemeine Ergebnisse unserer Hirnforschung. J Psychol Neurol 25:279–462

Vogt O (1906) Der Wert der myelogenetischen Felder der Grosshirnrinde (Cortex pallii). Anatom Anz 29:273–287

Walker AE (1940) A cytoarchitectural study of the prefrontal areas of the macaque monkey. J Comp Neurol 73:59–86

Weller RE (1988) Two cortical visual systems in Old World and New World primates. Prog Brain Res 75:293–306

Williams SR, Stuart GJ (2002) Dependence of EPSP efficacy on synapse location in neocortical pyramidal neurons. Science 295:1907–1910

Wilson FAW, Ó Scalaidhe SP, Goldman-Rakic PS (1993) Dissociation of object and spatial processing domains in primate prefrontal cortex. Science 260:1955–1958

Zeki SM (1969) The secondary visual areas of the monkey. Brain Res 13:197–226

Zeki SM (1978a) The third visual complex of rhesus monkey prestriate cortex. J Physiol 277:245–272

Zeki SM (2003) Improbable areas in the visual brain. Trends Neurosci 26:23–26

Chapter 2
The Cytoarchitectonic Map of Constantin von Economo and Georg N. Koskinas

Lazaros C. Triarhou

Abstract In 1925 Constantin von Economo (1876–1931) and Georg N. Koskinas (1885–1975), working in the Psychiatric Clinic of Julius Wagner-Jauregg (1857–1940) at the University of Vienna, published their monumental *Atlas and Textbook of Cytoarchitectonics of the Adult Human Cerebral Cortex*, following in the footsteps of Theodor Meynert (1833–1892) and Korbinian Brodmann (1868–1918). Von Economo and Koskinas provided a much more detailed verbal and pictorial description of the variations in cellular structure (cytoarchitecture) of cerebral cortical layers, compared to Brodmann. By dissecting each gyrus and sulcus perpendicularly to its axis, von Economo and Koskinas successfully addressed the core problem of flattening out the convoluted polyhedral surface of the human cerebral mantle. They defined five structural cortical *types* (agranular, frontal, parietal, polar, and granulous) and 107 cytoarchitectonic area *modifications* (35 frontal, 13 limbic, 6 insular, 18 parietal, 7 occipital, 14 temporal, and 14 hippocampal). Their numerous discoveries include the *koniocortex*, i.e. the dusty appearance of sensory areas, and the identification, at the boundaries of koniocortex with ordinary isocortex in parietal, temporal and occipital areas, of thin bands with giant pyramidal cells, the so-called *parasensory zones*. Von Economo and Koskinas also provided the first comprehensive description of the distinct rod and corkscrew cells in cingulate and frontoinsular areas known today as "von Economo neurons" that are putatively involved in social behavior and the pathophysiology of neurodevelopmental and mental diseases. The cortical cytoarchitectonics system of von Economo and Koskinas may be especially meaningful in conjunction with modern studies on functional imaging in the human brain.

L.C. Triarhou (✉)
Economo-Koskinas Wing for Integrative and Evolutionary Neuroscience, University of Macedonia, Egnatia 156, Bldg. Z-312, 54006 Thessaloniki, Greece
e-mail: triarhou@uom.gr

S. Geyer and R. Turner (eds.), *Microstructural Parcellation of the Human Cerebral Cortex*, DOI 10.1007/978-3-642-37824-9_2, © Springer-Verlag Berlin Heidelberg 2013

2.1 Introduction

"The cortex is both chaos and order, and therein lies its strength." With these words the neuroanatomist Gerhardt von Bonin (1890–1979) summarized in his classical essay on the cerebral cortex (von Bonin 1950) the quintessence of the cerebral hemispheric mantle.

The inextricability of cerebral morphology and function was exemplified in the writings of the neurobiologist Christfried Jakob (1866–1956): "Form is stabilized function and function is change of form; the organism is a single entity that presents itself as form in the latent state and as function in the kinetic state... Form, structure and function are inseparable, if not identical, and only scholastic science has managed to separate them... Only a basis that is fundamentally biological, morphostructural and histophysiological at the same time, unified in an ample ontogenetic and phylogenetic context, can let us address in legitimate ways the fundamental questions of modern neuro- and psychobiopathology" (Jakob 1939, 1941; Triarhou 2010; Triarhou and del Cerro 2006).

One of the overarching grand challenges of neuroscience for the twenty-first century is how does the brain work and produce mental activity and how does physical activity in the brain give rise to behavior (Hougan and Altevogt 2008). It is argued that the field of understanding how the mind works may move forward to its full potential only when we gain a better insight into the physical instantiation of nervous systems by constructing connectional maps that integrate anatomy, neuronal activity and function.

In the early twentieth century, the holding tenet among neuroanatomists was that deciphering cortical cell architecture is a preamble to understanding the mind. Essential contributions to cortical histology by Félix Vicq d'Azyr (1748–1794), Theodor Meynert (1833–1892), Vladimir A. Betz (1834–1894), W. Bevan Lewis (1847–1929), Santiago Ramón y Cajal (1852–1934), Theodor Kaes (1852–1913), Christfried Jakob (1866–1956), Alfred Walter Campbell (1868–1937), Korbinian Brodmann (1868–1918), Oskar Vogt (1870–1959), Sir Grafton Elliot Smith (1871–1937), and Cécile Mugnier-Vogt (1875–1962) formed the basis upon which Baron Constantin von Economo (1876–1931) and Georg N. Koskinas (1885–1975), from patrician Greek families rooted in the Hellenic regions of Macedonia and Lacedaemonia, respectively (Fig. 2.1), produced their *magnum opus* on the adult human cerebral cortex (von Economo and Koskinas 1925). The historical merit and its modern perspective are discussed elsewhere (von Economo 2009; von Economo and Koskinas 2008; Zilles 2004; Zilles and Amunts 2012).

With his landmark monograph, Brodmann (1909) defined 44 cortical cytoarchitectonic areas in the human brain (and a total of 52 areas in the primate brain overall). He studied cortical cytoarchitecture in numerous mammals, from the hedgehog, with its unusually large archipallium, to primates and humans, and introduced the terms *homogenetic* and *heterogenetic formations* to denote two different basic cortical patterns with either the typical six layers or lacking the six-layer stage, respectively (Garey 2006; Zilles and Amunts 2010).

Fig. 2.1 Constantin von Economo (1876–1931), Professor of Psychiatry and Neurology at the University of Vienna, *left*, and Georg N. Koskinas (1885–1975), former Assistant at the Psychiatric and Neurological Clinic of the University of Athens, *right*. For his discovery of encephalitis lethargica, von Economo was nominated three times for the Nobel Prize in Physiology or Medicine between 1926 and 1932. Koskinas returned to Greece in 1927; after an unsuccessful application for the chair of Neurology at the University of Athens, he devoted himself to private practice in the suburb of Kifisia (Triarhou 2005) (Photo credits: Bildarchiv und Grafiksammlung der Österreichischen Nationalbibliothek, Vienna (Economo); Helios Encyclopedic Lexicon, Athens (Koskinas). Used by permission and protected by copyright law. Copying, redistribution or retransmission without the author's express written permission is prohibited)

Vogt and Vogt (1919) laid the foundations of myeloarchitectonics (the architecture of fiber pathways) and defined the structural features of *allocortex*, *proisocortex* and *isocortex*; they also analyzed the differences between *paleocortical*, *archicortical*, and *neocortical* regions (Vogt and Vogt 1919; Vogt 1927; Zilles 2004; Zilles and Amunts 2012).

Von Economo commenced his work on cortical cytoarchitectonics in 1912, and Koskinas joined him in 1919. Their Atlas and Text Volume were published in 1925, and included 150 new discoveries (Koskinas 1931). Von Economo and Koskinas (1925, 2008) defined 107 area *modifications*, and more than 60 area *transitions* (von Economo 2009), virtually raising the "resolution" of our cortical cytoarchitectonic register, compared to Brodmann's data, by a factor of four.

In subsequent decades, by combining cytoarchitectonics with myeloarchitectonics, Sanides (1962, 1964) placed emphasis on transitions or gradations that accompany "streams" of neocortical regions coming from paleocortical and archicortical sources (Pandya and Sanides 1973), while Vogt and Vogt (1919) had already spoken of "areal gradations".

Comprehensive tables correlating the 107 cortical areas defined by von Economo and Koskinas with the Brodmann areas can be found in a previous review article (Triarhou 2007b) and in the English edition of the Atlas (von Economo and Koskinas 2008).

The work of von Economo and Koskinas represents a gigantic intellectual and technical effort (van Bogaert and Théodoridès 1979). Their attempt to bring the existing knowledge into a more orderly pattern was emphatically acknowledged by von Bonin (1950) and Bailey and von Bonin (1951).

Spyridon Dontas (1878–1958), Professor of Physiology and Pharmacology at the University of Athens and President of the Academy of Athens, had remarked in 1926 upon meeting Koskinas: "The work of von Economo and Koskinas is monumental and constitutes a milestone of science, charting new paths for understanding the brain from an anatomical, physiological and pathological viewpoint. It stands as the first comprehensive reference on the architecture of the adult human cerebrum and will persevere as a perpetual scientific testimony" (Triarhou 2012).

The brain map and the systematic area naming by von Economo and Koskinas have regrettably not passed into widespread general use. However, it is clear that they brought together concepts and ideas of cortical organization and structure that had been developing over the preceding 30 years and which remain with us in the present era of cortical research; moreover, they introduced original terms and, by applying in a systematic manner nomenclatures derived from other authors and themselves, they codified the language that we use to describe the cortex to this day, essentially providing the first "ontology" of the cerebral cortex (Jones 2010).

2.2 Method

At the outset of their studies, von Economo and Koskinas devised an entire system of new methods to overcome the existing obstacles and difficulties, from the autopsy to the photographic documentation (Koskinas 1926, 1931; von Economo 2009; von Economo and Koskinas 2008). The following are some of the introduced innovations.

2.2.1 Sectioning

Instead of the widely adopted method of sectioning the whole brain serially, perpendicular to its fronto-occipital axis, von Economo and Koskinas obtained tissue sections always perpendicular to the axis of each gyrus or sulcus and in directions corresponding to their convoluted pattern (Figs. 2.2 and 2.3). They arrived at that idea by considering that, in order to be able to compare the various brain areas cytoarchitectonically, sections had to have a consistent orientation relative to the gyral surface, insofar as only then could the breadth of the entire cerebral cortex and of each cortical layer as well be represented in the sections in a precise way.

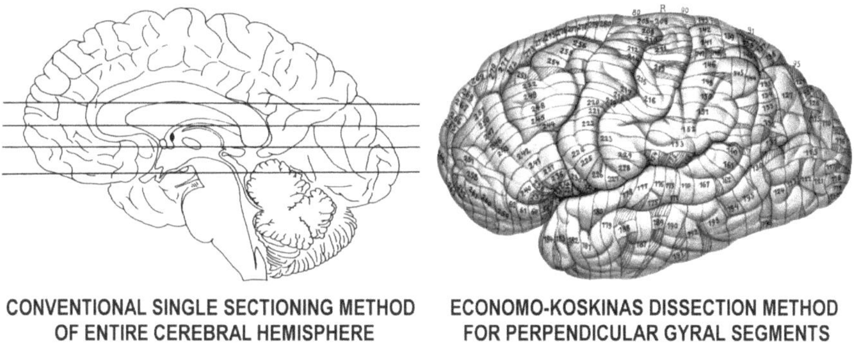

Fig. 2.2 The difference between the widely used method of obtaining whole single sections of the cerebral hemispheres, *left*, and the method devised by von Economo and Koskinas (1925, 2008) for dissecting each hemisphere into 250–350 tissue blocks, 4 mm in thickness, always perpendicular to the axis of each gyral or sulcal segment, *right*; hatched areas indicate "cancelled" tissue

2.2.2 *Staining*

The staining of the preparations was perfected such that a uniform tone was achieved not only of the single sections, but of all the series of sections into which each brain had been divided. That was mandated by the need, firstly, to define gradual differences of histological elements in neighboring areas of the cerebral cortex, and secondly, to achieve consistent photographic registrations.

2.2.3 *Specimen Depiction*

Most of the previous histological studies on cortical cytoarhitecture depicted their results schematically, and therefore subjectively. Instead of schematic drawings, and aiming at an exact documentation of the specimens, with all the relationships of the diverse neurons, von Economo and Koskinas used photography, which is the most objective testimony regarding form, size and arrangement, and turned to branches of science such as advanced optics and photochemistry.

The stained cortical sections were photographed using Carl Zeiss Planar lenses, which are special macro objectives with a considerably larger field than the common microscopy objectives, especially valuable for large area objects under relatively large magnifications. Planar lenses are used without an eyepiece. Additional details on technique can be found in my historical notes on Koskinas (Triarhou 2005) and von Economo (Triarhou 2006).

The depth of field that von Economo and Koskinas achieved in their photomicrographs, as well as the clarity and detail with which individual neurons can be visualized is remarkable. Their plates probably still represent the most comprehensive set of high resolution images of cortical histology ever assembled (Jones 2008).

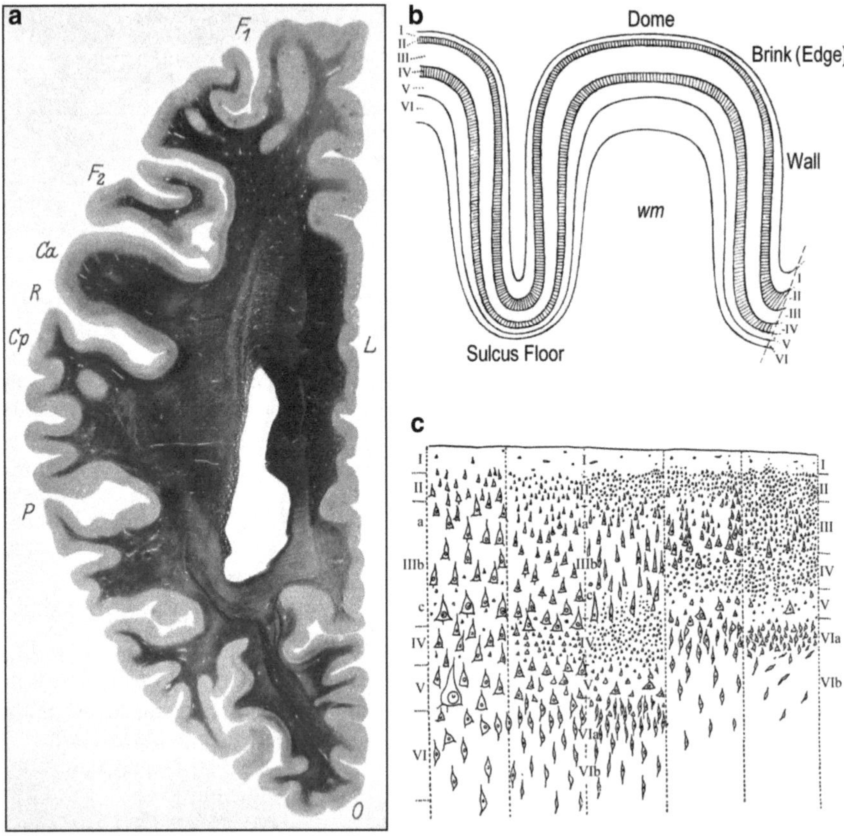

Fig. 2.3 (**a**) In obtaining sections of the entire cerebral hemisphere through conventional sectioning techniques, the real variations in layer thickness and cellular architecture cannot be studied consistently. The horizontal section through the left human cerebral hemisphere depicts such sizeable regional differences in cortical thickness and the random orientation of the gyri (von Economo and Koskinas 1925). Weigert method. F_1 and F_2, superior and middle frontal gyrus; *Ca*, precentral gyrus; *R*, central sulcus; *Cp*, postcentral gyrus, *P*, parietal lobe; *O*, occipital lobe; *L*, limbic gyrus. (**b**) A schematic drawing that depicts the varying thickness of the six cortical layers (I through VI) at the level of the *dome*, *brink (edge)*, *wall* and *valley (sulcus floor)* in a cortical gyrus. The two granular layers (external and internal) are hatched; *wm*, subcortical white matter (von Economo 2009). (**c**) The five fundamental structural types of isocortex: 1, agranular; 2, frontal; 3, parietal; 4, polar; 5, granulous or *koniocortex* (von Economo 1925, 1929, 2009; von Economo and Koskinas 1925, 2008)

2.3 General Part

The "General Part" of the Text Volume (von Economo and Koskinas 1925) covers introductory concepts on the cerebral cortex and its nerve cells, the structure and the development of the cortical layers, the composition and the meaning of the cortical laminar structure, the definition of cortical areas, and methodological issues (Fig. 2.4).

Brodmann (1909) grouped his 44 human cortical areas as 4 postcentral, 2 precentral, 8 frontal, 4 parietal, 3 occipital, 10 temporal, 6 cingulate, 3 retrosplenial, and 4 hippocampal.

Von Economo and Koskinas (1925, 2008) divided the cortex into seven lobes, which they denoted by their initials. The lobes were further subdivided into regions: the frontal lobe (F) into prerolandic, anterior (prefrontal) and orbitomedial (orbitomedial) regions; the superior limbic lobe (L) into anterior, posterior and retrosplenial regions; the parietal lobe (P) into postcentral (anterior parietal), superior, inferior and basal regions; and the temporal lobe (T) into supratemporal, temporal proper, fusiform and temporopolar regions. The insular (I) and occipital (O) lobes were not subdivided. The inferior limbic lobe consists of the hippocampus (H). For cytoarchitectonic area designations, they did not continue Brodmann's system of random numbers, but instead used letter codes, consisting of a Roman capital letter (the initial of the lobe), followed by a calligraphic capital to note the sequence of a gyrus within a lobe (e.g. F*B* means the second gyrus of the frontal lobe), and a Latin or Greek subscript for characteristic microscopic features (e.g. m = magnocellular, p = parvicellular, γ = gigantopyramidal).

Von Economo and Koskinas (1925, 2008) defined five fundamental "supercategories" of structural cortical types (agranular, frontal, parietal, polar and granulous) (Fig. 2.3c), further arranged into 54 *ground*, 76 *variant* and 107 cytoarchitectonic *modification* areas, plus more than 60 *transition* areas (von Economo 1925, 2009; von Economo and Horn 1930). Topographically, the 107 modification areas of von Economo and Koskinas are grouped into 35 frontal, 13 superior limbic, 6 insular, 18 parietal, 7 occipital, 14 temporal, and 14 inferior limbic or hippocampal (Figs. 2.5 and 2.6). Of the 107 modifications, 22 are allocortical, 22 heterotypic isocortical, and 63 homotypic isocortical. Von Economo and Koskinas (1925, 2008) separately analyzed the *dome*, *edge*, *wall* and *floor* of each cortical gyrus (Fig. 2.3b).

For certain cortical areas with a granular appearance of their cells in most layers, especially of gyral walls, associated primarily with sensory functions, von Economo and Koskinas (1923, 1925) introduced the term *koniocortex* to denote their dusty appearance.

Von Economo and Koskinas (1925, 2008) regularly saw a special type in a small band in sublayer IIIc at the boundary between any koniocortex (or sensory isocortex) and the ordinary surrounding isocortex in sensory parietal, occipital and temporal areas. Such zones contain giant pyramidal cells. They called these margin regions with magnocellular characteristics *parasensory zones*.

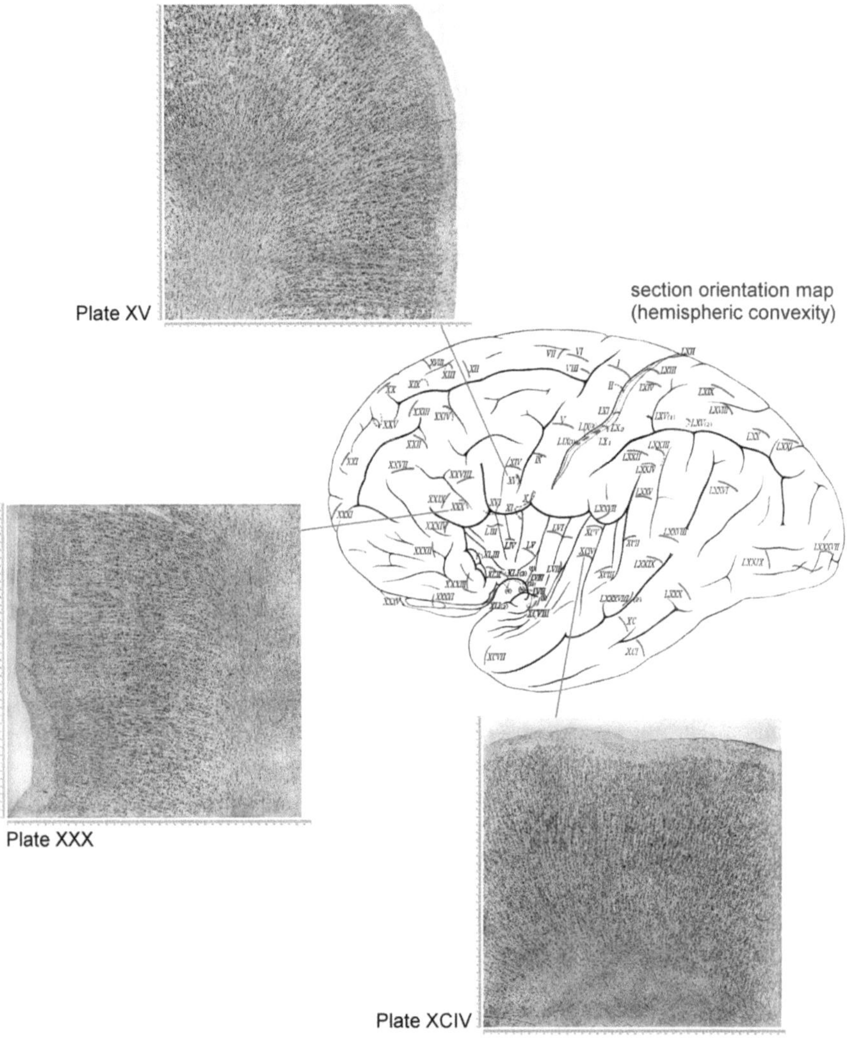

Fig. 2.4 Schematic map of the lateral (*convex*) facies of the hemispheric surface and three microphotographic plates from the Atlas of von Economo and Koskinas (2008), shown as examples: *Plate XV* – Magnocellular intermedio-agranular frontal area FCB_m (*Broca's area*) at the *foot* of the inferior frontal gyrus, anterior wall. *Plate XXX* – Triangular granular frontal area FD_γ in the inferior frontal gyrus, wall of the notch of pars triangularis *(incisura capi)*. *Plate XCIV* – Supratemporal area granulosa *TC* in first gyrus of Heschl (primary auditory cortex), middle, dome, with the typical "rain shower formation" *(Regenschauerformation)*. The detailed descriptions of the normal histological structure of the cerebral cortex depicted in the 112 microphotographic plates of the Atlas were explained in the accompanying Text Volume (von Economo and Koskinas 1925). The printing size of the original plates was 40 × 40 cm at a magnification of × 100, therefore covering a 4.0 × 4.0 mm true cortical area

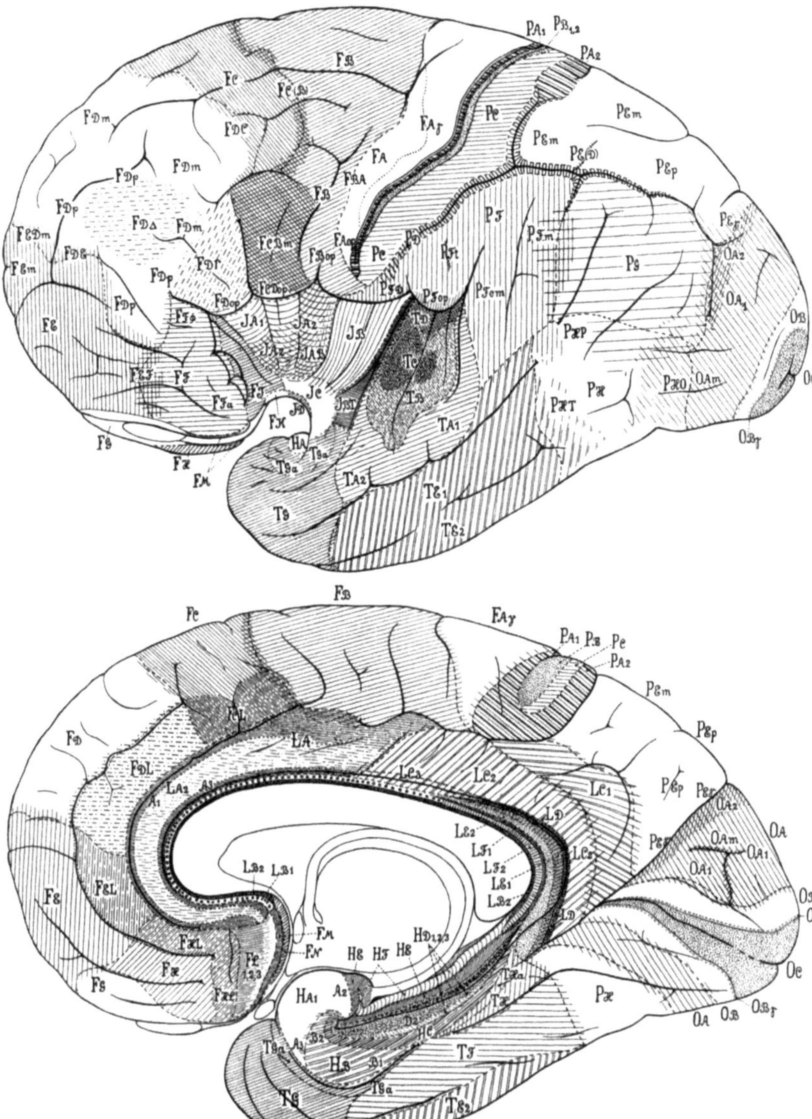

Fig. 2.5 Cytoarchitectonic maps of von Economo and Koskinas, showing cortical modification areas in the convex and median hemispheric facies of the human brain

Another crucial discovery was that of the large, spindle-shaped bipolar projection neurons in the inferior ganglionic layer (Vb) of the dome of the transverse insular gyrus, which are now called "von Economo neurons" (Watson et al. 2006) – although a more succinct term might be "von Economo-Koskinas neurons". The

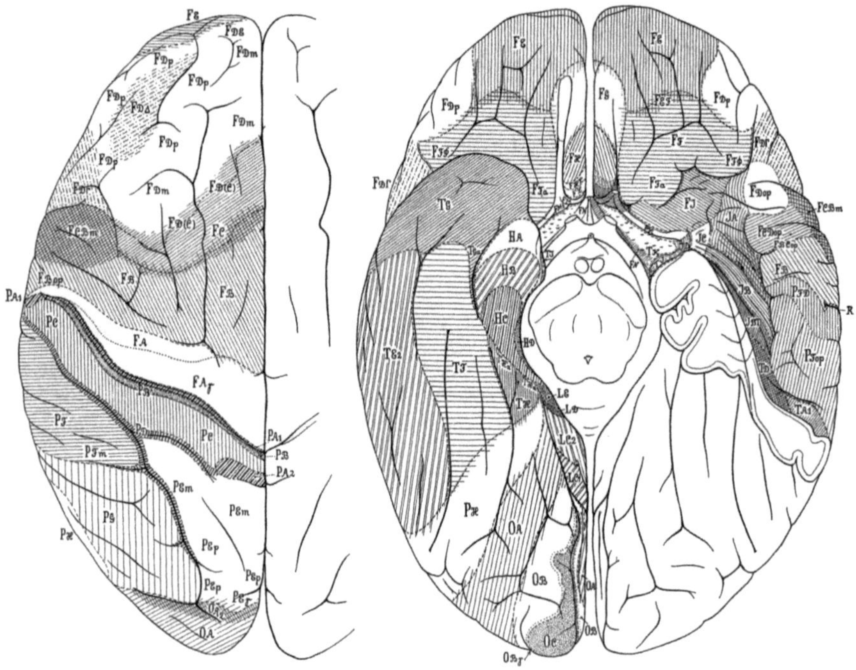

Fig. 2.6 Cytoarchitectonic maps of von Economo and Koskinas for the dorsal and ventral hemispheric surface

detailed morphology of these rod cells *(Stäbchenzellen)* and corkscrew cells *(Korkzieherzellen)* was documented by von Economo and Koskinas (1925) in cingulate (anterior) limbic and frontoinsular areas.

2.4 Special Part

The following text is a selection of some ideas discoursed in previous reviews (Triarhou 2007a, b) and in the new English editions of the Atlas (von Economo and Koskinas 2008) and of von Economo's shorter textbook of cortical cytoarchitectonics (von Economo 2009).

2.4.1 Frontal Lobe

Broca's motor speech area FCB_m in the inferior frontal gyrus was considered as a particular human characteristic by von Economo and Koskinas (1925), as well as by Brodmann (1909). The surface area of the pars opercularis of the inferior frontal gyrus is characterized by a distinct type of cortex, distinguishable from the posteriorly lying premotor cortex in area F*B* in the precentral gyrus; it continues rostrally as area FD_{γ} (Fig. 2.4).

Anteriorly one finds portions of areas F*D* and F*E*, which are rich in granule cells. Lesions in the prefrontal region result in disturbances of attention, psychomotor activity, will and emotivity. Von Economo (2009) termed such higher mental functions, localized in the frontal regions of the brain, "the active part of the psychic personality". Area FA_{γ} resembles the frontal core area more closely, with the consequence that a large part of area F*A* belongs to nonprimary motor cortex.

Area F*F* partly corresponds to the orbitofrontal proisocortex of the monkey that lies intercalated between the caudal orbitofrontal isocortex rostrally, and the orbitofrontal peripaleocortex caudally. Area FF_{a} in the human brain probably corresponds to the granular isocortex in the anterior part of the orbital surface of the frontal lobe in the macaque. Areas F*H* and F*HL* correspond to the paralimbic dysgranular isocortex on the ventromedial surface of the prefrontal cortex in the macaque, which lies intercalated between the frontopolar granular isocortex rostrally and the orbitomesial archicortical proisocortex of the straight gyrus caudally. Area F*J* appears to correspond to peripaleocortex in the inferior part of the transverse gyrus of the insula, and to the orbitofrontal peripaleocortex in the monkey (de Olmos 1990). The area F*F* lies rostrally and ventrally to area FD_{γ}.

Von Economo and Koskinas (1925) mark transitional types of cortex in their maps, beyond the 107 "standard" modifications; such transitions comprise the areas F*BA*, F*C(B)*, $FCD_{\rm op}$, F*DC*, F*DE*, FED_{m}, F*EF* and FE_{m}. Areas F*BA*, F*DC*, F*DE* and F*EF* denote transition forms (e.g. F*BA* marks the transition of area F*B* into F*A*, F*DC* the transition of F*D* into F*C*, and so on). The designation F*C(B)* implies a part of area F*C* with an admixture of the type of the neighboring area F*B*, whereas the subscript *m* in the areas FED_{m} and FE_{m} signifies cellular variations with magnocellular features. Area $FCD_{\rm op}$ is a transitional opercular variant between areas $FC_{\rm op}$ and $FD_{\rm op}$.

2.4.2 Parietal Lobe

Brodmann (1909) defined four areas (1, 2, 3, 43) in the postcentral region, whereas von Economo and Koskinas six (PA_1, PA_2, PB_1, PB_2, P*C* and P*D*). In the parietal region, Brodmann defined four areas (5, 7, 39, 40), and von Economo and Koskinas nine (P*ED*, PE_{m}, PE_{p}, PE_{γ}, P*F*, PF_{t}, $PF_{\rm op}$, PF_{cm} and P*G*). The basal parietal region P*H* most likely belongs to the visual cortex and includes the functionally defined areas V4 and V5 (Zilles and Palomero-Gallagher 2001). The proposed subdivisions of the anterior parietal cortex by von Economo and Koskinas are still in use.

Area P*A* is located in the depths of the central sulcus. Area P*B* is "sensory koniocortex", located on the caudal bank of the central sulcus. Concerning the primary somatosensory cortical areas and the subdivision of the posterior parietal lobe into a superior and an inferior lobule, the accepted terminology of von

Economo and Koskinas forms the basis for modern cytoarchitectonic analyses and experiments in primates (Zilles 2004). Brodmann did not delineate any transition zones in the posterior parietal lobe, whereas von Economo and Koskinas marked such transition zones between the areas P*E*, P*F*, P*G*, P*H* and O*A*, in agreement with the observations of Eidelberg and Galaburda (1984).

Beyond the standard modifications, transition parietal areas are PC_γ, P*E(D)*, P*FD* and PF_m. Areas PC_γ and PF_m denote cellular variations containing giant pyramidal and magnocellular neurons, respectively. Area P*E(D)* is a variant of area P*E* with an admixture of the neighboring cortical type P*D*. The functionally defined secondary somatosensory cortex (SII) is located in the parietal operculum, hidden within the Sylvian fissure. Brodmann areas 40 and 43 extend into the parietal operculum and are candidates for SII on topographic grounds; they partially correspond to the opercular modification PF_{op} and to the subcentral area P*FD*, respectively. In the supramarginal gyrus of the rostral inferior parietal cortex, von Economo and Koskinas subdivide Brodmann area 40 into the five areas P*F*, PF_{cm}, PF_m, PF_{op} and PF_t, confirmed in general lines by Caspers et al. (2006). In the caudal inferior parietal cortex, Caspers et al. (2006) distinguish a caudal region termed P*G*p and a rostral region termed P*G*a, this latter fitting to area P*G* in the angular gyrus (roughly Brodmann area 39).

2.4.3 Temporal Lobe and Insula

Based on pathological and physiological considerations, von Economo (1927, 2009) localized the understanding of word *speech* in area TA_1 of the left hemisphere, the understanding of word *sense* in the caudal transitional region of area TA_1 towards area P*F*, and the understanding of *music* in area TA_2 and the temporal pole; the appreciation of higher tones in parts in the bottom of the Sylvian fissure, while that of lower tones more towards outer portions. Area T*C* is koniocortex, i.e. sensory cortex representing *primary audition*, and receiving fibers from the medial geniculate body.

Von Economo and Horn (1930) investigated the cytoarchitectonics of the auditory cortex further in the adult and juvenile human brain. They found the superior temporal surface and the length of the Sylvian fissure larger on the left side. Initial attempts at investigating the cytoarchitectonics of the auditory cortex by Campbell (1905), Rosenberg (1907) and Brodmann (1909), who had identified it with Brodmann area 41, had missed the most characteristic feature that this area shares with all other "sensory" cortices, i.e. the "granularity" (Meyer 1977); that was first described by von Economo and Koskinas (1925). Von Economo and Horn (1930) attribute the striking variations in size among individuals and between the two hemispheres possibly to handedness or differences in musicality.

The koniocortex of the human temporal lobe encompasses areas T*C* and T*D* and is located on Heschl's gyrus (transverse temporal gyrus); area T*A* contains Wernicke's speech area, while the cerebral "belt" areas most likely correspond to areas T*A* and T*B* (Chiry et al. 2003; Webster and Garey 1990). Within the area T*C*, which closely corresponds to the "core" region of Hackett et al. (2001), von Economo and Horn (1930) describe 11 distinct types of granular cortex. The "belt" field of the human auditory cortex, on the other hand, seems to correspond to the medial portion of the koniocortical T*D* sector of von Economo and Koskinas (1925).

In sections cut perpendicularly to the radial orientation of layer III apical dendrites, the small pyramidal cells are arranged in short radial columns that partially extend into layers II and IV (Hackett et al. 2001); such a feature seems to correspond to what von Economo and Koskinas called the "rain shower formation" (Fig. 2.4). With regard to the columnar organization of the belt region, layer III pyramidal cells are arranged in organized vertical columns, which von Economo and Koskinas called the "organ pipe formation".

Von Economo and Koskinas (1925) and von Economo and Horn (1930) were among the first investigators to notice individual differences of the auditory fields and marked asymmetries between the two hemispheres: Heschl's gyrus is generally single and longer on the left side and double and shorter on the right side; the *planum temporale* (located caudally to Heschl's gyrus, in area T*B*) is larger on the left side (Webster and Garey 1990). Such asymmetries may underscore the modern idea of a functional differentiation of the two cerebral hemispheres and the predilection of the left hemisphere (right ear) for verbal tests, and that of the right hemisphere (left ear) for music recognition (Brodal 1981).

In contrast to Brodmann, von Economo and Koskinas divide the medial temporal lobe into a rostral area T*G* and two caudal areas, T*H* and T*F*, with area T*G* further subdivided into a medial area TG_{α} and a larger lateral area T*G* (Suzuki and Amaral 2003). Like Elliot Smith (1907), von Economo and Koskinas also illustrate the temporal polar cortex as being continuous with the anteroventral portion of the medial temporal lobe. The nomenclature and cortical demarcations of Brodmann (1909) regarding the medial temporal lobe in primates is somewhat vague and varying across species, whereas the analyses of von Economo and Koskinas are more detailed (Suzuki and Amaral 2003). Areas T*F* and T*H* belong to the posterior part of the parahippocampal gyrus; the anterior part of the parahippocampal gyrus comprises mainly the entorhinal cortex and the associated perirhinal cortex (Amaral and Insausti 1990). Area T*J* seems to be homologous to the hyperchromic, coarse-cell temporopolar peripaleocortex in the macaque (de Olmos 1990). Besides area T*J* the peripaleocortical agranular claustral region (Brodmann area 16 in the *Cercopithecus*) is also homologous to a certain extent to the human area I*D* (Zilles 2004).

The insula includes areas IA_1, IA_2, I*B*, I*C* and I*D*. The gradual transition of area I*A* backwards over the central sulcus of the insula to area I*B* is denoted by von Economo and Koskinas as area I*AB*, which is characterized by a condensation of the granular layers and a reduction of pyramidal cell size.

2.4.4 Occipital Lobe

The primary visual area or striate cortex is area *OC*, the parastriate cortex is area *OB*, and the peristriate cortex is area *OA*. A borderzone at the boundaries of Brodmann areas 17 and 18, containing giant pyramidal cells in the lower part of layer III, is area OB_γ *(limes parastriatus gigantopyramidalis)*. The total surface of koniocortex in the visual sensory sphere (area *OC*) in both hemispheres was estimated at about 50 cm^2 and the total number of cells at about 1.4×10^9, i.e. 10 % of the total number of neurons of the entire cerebral cortex. Thus, the area striata appears four times richer in cells than any other cortical region (Koskinas 1969; von Economo 1927; von Economo and Koskinas 2008).

2.4.5 Superior Limbic and Inferior Limbic (Hippocampal) Gyrus

One concern in the localization of functions in the human cerebral hemispheres is the boundary between the retrosplenial/cingulate and the parahippocampal cortices. Brodmann (1909) depicted the retrosplenial cortex as fully surrounding the posterior and ventral edge of the splenium of the corpus callosum. Von Economo (1927, 2009) provided the first subregional map of the posterior cingulate gyrus and showed a termination of the retrosplenial areas *LE* and *LD* at a plane caudal but not ventral to the splenium (Vogt et al. 2001).

Every section through the retrosplenial cortex includes a segment of allocortical hippocampus and ectosplenial area *LF*. At allocortical-isocortical transition points in the primate telencephalon, modern anatomists recognize the concept of a "dysgranular" cytoarchitecture (a weakly defined layer IV); such points are found in orbitofrontal, insular, and anterior and posterior cingulate cortices (Vogt et al. 2001). Ngowyang (1934) had described a "dysgranular region" in the frontal lobe, associated with areas *FC* and *FCL*. Going forward, the granular layer appears sporadically, making this area "hypogranular" or "dysgranular"; forward of Brodmann area 6, the prefrontal cortex, and continuing through the frontal pole, the cortex is "eugranular" (DeMyer 1988).

The area *LD* is dysgranular rather than agranular, as it was originally thought (Vogt et al. 2001); its layer IV has a variable thickness, interrupted by large SMI-32 immunopositive neurons in the sublayers IIIc and Va. Brodmann (1909) referred to area 30 as agranular. Von Economo (1927, 2009) was quite explicit that area *LD* is not merely agranular, but that the "granulous" layer of area *LE* is not continuous with the isocortical layer of area LC_2. Von Economo vacillated on the presence of a layer IV in area *LD* and showed a layer III(IV) below layer III. A dysgranular layer IV has a variable thickness and may even disappear as the neurons of the sublayers IIIc and Va intermingle. The dysgranular concept for a cortical architecture was obviously not defined during the early years of cortical cytoarchitectonics in terms of the chemical signature of neurons, since histochemical methods were not available. In a series of studies spanning over 30 years, Vogt et al. (2001) have described in the primate brain the dysgranular nature of area *LD* and its profound differences with the cytoarchitecture of the granular Brodmann area 23a.

The inferior limbic lobe comprises the hippocampal gyrus from the isthmus until near the temporal pole and contains the entire uncinate gyrus, the subiculum, the dentate gyrus and Ammon's horn. Above the splenium, the hippocampal rudiment, the indusium griseum, or areas LB_2 and *HF*, there is a single layer of densely packed SMI-32 (nonphosphorylated neurofilament) immunopositive neurons. Adjacent to the indusium griseum is the subicular rudiment or area *HE*, which has fewer and more dispersed neurons. These two areas together form the fasciolate gyrus on the dorsal surface of the corpus callosum (Vogt et al. 2001).

2.5 Discussion

Brodmann maps are commonly used to either designate cytoarchitectonic areas as such, or as a "shorthand system" to designate some region on the cerebral *surface* (DeMyer 1988). Macroscopic extrapolation of Brodmann projection maps are effected on the atlas of Talairach and Tournoux (1988), rather than being based on real microscopic cytoarchitectonics. Such specifications of Brodmann areas may lead to erroneous results in delineating cortical regions, something that may in turn lead to erroneous hypotheses regarding the involvement of specific brain systems in normal or pathological situations (Uylings et al. 2005).

Von Economo (1927, 2009) was the first to use subregional maps, which are invaluable in resolving difficult topological problems (Fig. 2.7). Talairach and Tournoux (1988) emphasize the shortcoming of Brodmann's reconstruction technique in not distinguishing areas on the gyral surfaces from areas in the sulcal depths, something may lead to miscalculations of the depth of the callosal sulcus and related areas, and placing e.g. Brodmann areas 29 and 30 on gyral surfaces. Because the architecture of each cortical area cannot yet be determined by the current imaging modalities, it is imperative that standardized atlases seeking to localize specific areas rely heavily on neuroanatomical observations, rather than Brodmann's reconstructions onto the convoluted human brain surface (Vogt et al. 2001).

On the other hand, the *perpendicular* sectioning method of von Economo and Koskinas (1925, 2008), which was consistently used to analyze the dome, wall and floor of each cortical gyrus, practically solves the generalized mapmaker's problem of flattening nonconvex polyhedral surfaces (Schwartz et al. 1989), which also constitutes a core problem in cortical research.

Microscopically-defined borders usually differ from gross anatomical landmarks; cytoarchitectonics reflect the inner organization of cortical areas and their morphofunctional correlates (Zilles 2004). Despite the integration of multifactorial descriptors such as chemoarchitecture, angioarchitecture, neurotransmitter,

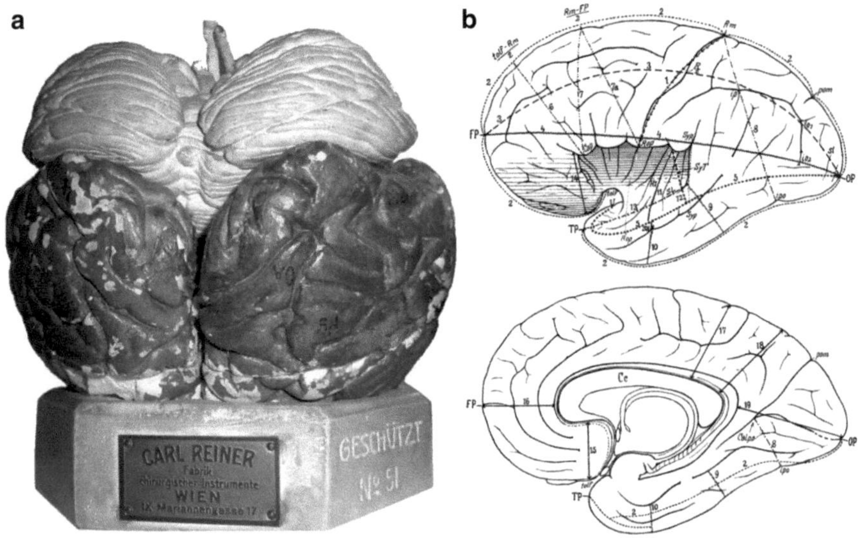

Fig. 2.7 (**a**) A plaster model of the human brain made in the 1920s, with cytoarchitectonics marked according to the system of von Economo and Koskinas (1925, 2008), used by von Economo (2009) for his lectures (courtesy of Fabrikation Chirurgischer Instrumente Carl Reiner GmbH, Vienna). (**b**) Schematic drawing of 26 encephalometric constants in the lateral and medial cerebral hemispheric facies, suggested by von Economo (1929) on the basis of macroscopic and cytoarchitectonic criteria as reference points, for future studies to determined variations among individuals, gender and talent differences, and alterations associated with nervous and mental diseases

receptor and gene expression patterns, as well as white matter tracts, it is clear that the knowledge of the classical anatomy remains fundamental (Toga and Thompson 2007). The structure of the cerebral cortical layers incorporates, and reflects, the form of their constitutive cells and functional connections; the underpinnings of neuronal connectivity at the microscopic level are paramount to interpreting any macroscopic clue yielded by neuroimaging studies.

The century-long endurance of the cytoarchitectonic analyses of Brodmann (1909) and of von Economo and Koskinas (1925) is in part due to the fact that these brain scientists did not hypothesize much about function; their only supposition was that anatomical subdivisions reflect functional variations, and that future functional and clinical studies would validate their anatomical subdivisions. In fact, there are examples of such cytoarchitectonic subdivisions in the motor, parietal and striate cortex that reflect functional differentiation to an unexpected degree (Bartels and Zeki 2005).

In a similar line of reasing, Koskinas (1926) argued: "The mind has its organic locus, its seat, its altar in the cerebral cortex. That is why one may be justified in claiming that the anatomical and the physiological exploration of that noblest of the organs deserves the utmost attention of science." And later wrote: "Provided that, as a general principle, each physiological function presupposes a corresponding anatomical basis, one understands how important the study of brain structure becomes. From a precise knowledge of the structure of the cerebral cortex, we may expect to shed light upon issues of the utmost importance, such as the anatomical basis of mental phenomena and the relationship between certain attributes and brain structure. But can we not hypothesize that the limitations of anatomically tracing deficits of the mind is simply a matter of the sophistication of our methods and the acuity of our foresight?" (Koskinas 1931).

The underlying concept is that cytoarchitectonic differences indicate functional differences (the "neural hardware" that includes cell types, connectivity, synaptic interactions and molecular events) and that functional differences necessitate cytoarchitectonic differences; by being "blind" with respect to function, the cytoarchitectonics approach ensures a degree of objectivity and data longevity, since observers document mere facts (Bartels and Zeki 2005).

Modern "probabilistic" atlases of the human brain bridge high-resolution in vivo data with neurocytology, and spatially normalize them to a common reference space; thus, they provide the means for moving from a descriptive to a hypothesis-driven science (Mazziotta et al. 1995). Nonetheless, in hypothesis-driven neuroimaging research, the interpretation of findings may vary depending on the specific paradigm, and attributing a function to a given area rarely goes unchallenged (Bartels and Zeki 2005).

In the fad of "cognitive brain mapping" and its purported representations in the human brain, color images generated by software can be adjusted to denote so-called "activations" with much ambiguity, and occasionally lead to fallacious findings unworthy of attempted replication. "Functional segregation", i.e. the common notion that mental functions are localized in cell clusters at specific cortical sites, is based on the old, hard-dying conception that a particular conscious process must have a delineated seat in the brain (Smith 2010), as "modern phrenologists, equipped with the powerful tools of functional MRI, seek to relate tiny pseudo-colored patches of slightly enhanced cortical activity associated with some limited cognitive function to an underlying structural correlate" (Jones 2008).

Functional MRI, as one technique that allows a correlation between structure and function, has limitations insofar as the measurements are not in real time and the spatial resolution only recently reached the mm level. Even the hypothetical development of a technique, which would noninvasively image neural activity at a spatial resolution of 1 mm and a temporal resolution of 1 msec, would still appear coarse relative to the size of the neuronal soma (5–100 μm) or the synaptic gap (20–40 nm) (Hougan and Altevogt 2008).

A key element in defining cortical areas is connectivity, and the guiding principle of neurohistologists that cortical areas form parts of connectional networks is now being adopted by the neuroimaging community; besides the streams of intrinsic cortico-cortical connections, no cortical area is without re-entrant projections from the thalamus, while each cortical area is undoubtedly governed, like the thalamocortical connections, by ontogenetic mechanisms (Jones 2008).

Particular emphasis on the U-shaped fibers of the frontal lobe and its connections with subcortical nuclei of the thalamus and the medial temporal lobe was placed by Christfried Jakob in his hodological approach of the early 1900s (Théodoridou and Triarhou 2012). Those pathways are currently emphasized in imaging studies (Catani et al. 2012). Jakob had identified all the major tracts of the limbic circuitry early on, preceding James W. Papez (1883–1958) by almost three decades (Triarhou 2008, 2009). Bearing a direct relevance to the clinic, Jakob's network approach provided a prescient anatomical framework for the concept of "diaschisis" – as elaborated by Constantin von Monakow (1853–1930) in 1914 to highlight the possible recovery of dysfunctional distant regions connected to destroyed areas – and for what would eventually become an intense area of research on the neural underpinnings of memory, emotion and behavioral disorders associated with frontal lobe damage (Catani and Stuss 2012).

As the necessity emerges to move from brain localization to connectivity imaging, methods such as high-resolution two-photon imaging are used to visualize functionally-defined afferent inputs on cortical dendritic spines in vivo with single-synapse resolution (Chen et al. 2011), and the relationship between structure and function in cortical synaptic circuits is studied by combining in vivo physiology with network anatomy. For example, a functional property of specific cortical neurons can be characterized by two-photon calcium imaging and a portion of these neurons' local interconnections can be traced with large-scale electron microscopy of serial thin sections (Bock et al. 2011). Thus, it is becoming possible to address hitherto intractable neurobiological questions through the technological advances that permit the combination of functional imaging and neuronal wiring (Briggman et al. 2011) through a high-speed reconstruction of neurite connectivity while performing reliable analyses of large neuroanatomical datasets (Helmstaedter et al. 2011).

The novel approaches for analyzing brain imaging data aim at providing levels of specificity with narrower confidence intervals in determining the dynamics of local neural population responses to their native temporal resolution (Tyler and Likova 2011). Furthermore, to better understand the anatomical organization of structures that form the basis of cognitive information processing, morphological data may be distilled and synthesized into a single interactive visualization that represents a fundamental blueprint upon which cognitive functions must be implemented (Solari and Stoner 2011). In such a framework, functional circuits corresponding to memory, cognition and cortical information flow are described in terms of distinguishable neuronal groups and cortical systems in order to elucidate the basis of distinct homotypical cognitive architecture in multiple independent visualizations that constitute an annotated view of "neuroanatomical consilience" (Solari and Stoner 2011).

Acknowledgments The author gratefully acknowledges the support of Karger Publishers, the Bodossakis Foundation, the Academy of Athens, the Hellenic Ministry of National Education, the Hellenic Neurological Society, and the University of Macedonia in his effort to revive the neuroanatomical works of von Economo and Koskinas, as well as the invaluable encouragement by the family members of the two eminent scientists.

References

Amaral DG, Insausti R (1990) Hippocampal formation. In: Paxinos G (ed) The human nervous system. Academic, San Diego, pp 711–755

Bailey P, von Bonin G (1951) The isocortex of man. University of Illinois Press, Urbana

Bartels A, Zeki S (2005) The chronoarchitecture of the cerebral cortex. Philos Trans R Soc Lond [Biol Sci] 360:733–750

Bock DD, Allen Lee W-C, Kerlin AM, Andermann ML, Hood G, Wetzel AW, Yurgenson S, Soucy ER, Kim HS, Reid RC (2011) Network anatomy and in vivo physiology of visual cortical neurons. Nature 471:177–182

Briggman KL, Helmstaedter M, Denk W (2011) Wiring specificity in the direction-selectivity circuit of the retina. Nature 471:183–188

Brodal A (1981) Neurological anatomy in relation to clinical medicine. Oxford University Press, Oxford

Brodmann K (1909) Vergleichende Lokalisationslehre der Großhirnrinde. J.A. Barth, Leipzig

Campbell AW (1905) Histological studies on the localization of cerebral function. Cambridge University Press, Cambridge

Caspers S, Geyer S, Schleicher A, Mohlberg H, Amunts K, Zilles K (2006) The human inferior parietal cortex: cytoarchitectonic parcellation and interindividual variability. Neuroimage 33:430–448

Catani M, Stuss DR (2012) At the forefront of clinical neuroscience. Cortex 48:1–6

Catani M, Dell'Acqua F, Vergani F, Malik F, Hodge H, Roy P, Valabregue R, Thiebaut de Schotten M (2012) Short frontal lobe connections of the human brain. Cortex 48:273–291

Chen X, Leischner U, Rochefort NL, Nelken I, Konnerth A (2011) Functional mapping of single spines in cortical neurons in vivo. Nature 475:501–505

Chiry O, Tardif E, Magistretti PJ, Clarke S (2003) Patterns of calcium-binding proteins support parallel and hierarchical organization of human auditory areas. Eur J Neurosci 17:397–410

de Olmos J (1990) Amygdaloid nuclear gray complex. In: Paxinos G (ed) The human nervous system. Academic, San Diego, pp 583–710

DeMyer W (1988) Neuroanatomy. Harwal Publishing, Malvern

Eidelberg D, Galaburda AM (1984) Inferior parietal lobule: divergent architectonic asymmetries in the human brain. Arch Neurol 41:843–852

Elliot Smith G (1907) A new topographical survey of the human cerebral cortex, being an account of the distribution of the anatomically distinct cortical areas and their relationship to the cerebral sulci. J Anat Physiol (Lond) 41:237–254

Garey LJ (2006) Brodmann's localisation in the cerebral cortex. Springer, New York

Hackett TA, Preuss TM, Kaas JH (2001) Architectonic identification of the core region in auditory cortex of macaques, chimpanzees, and humans. J Comp Neurol 441:197–222

Helmstaedter M, Briggman KL, Denk W (2011) High-accuracy neurite reconstruction for high-throughput neuroanatomy. Nat Neurosci 14:1081–1088

Hougan M, Altevogt B (2008) From molecules to minds—challenges for the 21st century: workshop summary. National Academies, Washington, DC

Jakob C (1939) Folia neurobiológica Argentina, Atlas I: El cerebro humano—su anatomía sistemática y topográfica. Aniceto López Editor, Buenos Aires

Jakob C (1941) Folia neurobiológica Argentina, Tomo I: neurobiología general. Aniceto López Editor, Buenos Aires

Jones EG (2008) Cortical maps and modern phrenology. Brain 131:2227–2233

Jones EG (2010) Cellular structure of the human cerebral cortex. Brain 133:945–946

Koskinas GN (1926) Cytoarchitectonics of the human cerebral cortex. Proc Athens Med Soc 92:44–48

Koskinas GN (1931) Scientific works published in German. Pyrsus, Athens

Koskinas GN (1969) Lobus occipitalis: makroskopische und mikroskopische Beschreibung. Privately Published, Athens

Mazziotta JC, Toga AW, Evans AC, Fox PT, Lancaster JL (1995) Digital brain atlases. Trends Neurosci 18:210–211
Meyer A (1977) The search for a morphological substrate in the brains of eminent persons including musicians: a historical review. In: Critchley M, Henson RA (eds) Music and the brain: studies in the neurology of music. William Heinemann, London, pp 255–281
Ngowyang G (1934) Die Cytoarchitektonik des menschlichen Stirnhirns I. Cytoarchitektonische Felderung der Regio granularis und Regio dysgranularis. Monogr Natl Res Inst Psychol Acad Sin (Shanghai) 7:1–68
Pandya DN, Sanides F (1973) Architectonic parcellation of the temporal operculum in rhesus monkey and its projection pattern. Z Anat Entwickl-Geschichte 139:127–161
Rosenberg L (1907) Histologische Untersuchung über die Cytoarchitektonik der Heschl'schen Windungen. Neurol Cbl 14:685–686
Sanides F (1962) Die Architektonik des menschlichen Stirnhirns. Springer, Berlin/Göttingen/Heidelberg
Sanides F (1964) The cyto-myeloarchitecture of the human frontal lobe and its relation to phylogenetic differentiation of the cerebral cortex. J Hirnforsch 47:269–282
Schwartz EL, Shaw A, Wolfson E (1989) A numerical solution to the generalized mapmaker's problem: flattening nonconvex polyhedral surfaces. IEEE Trans Pattern Anal Mach Intell 11:1005–1008
Smith DF (2010) Cognitive brain mapping for better or worse. Perspect Biol Med 53:321–329
Solari SVH, Stoner R (2011) Cognitive consilience: primate non-primary neuroanatomical circuits underlying cognition. Front Neuroanat 5:1–23. doi:10.3389/fnana.2011.00065 (article 65)
Suzuki WA, Amaral DG (2003) Where are the perirhinal and parahippocampal cortices? A historical overview of the nomenclature and boundaries applied to the primate medial temporal lobe. Neuroscience 120:893–906
Talairach J, Tournoux P (1988) Co-planar stereotaxic atlas of the human brain. 3-dimensional proportional system: an approach to cerebral imaging (trans: Rayport M). G. Thieme Verlag, Stuttgart
Théodoridou ZD, Triarhou LC (2012) Challenging the supremacy of the frontal lobe: early views (1906–1909) of Christfried Jakob on the human cerebral cortex. Cortex 48:15–25
Toga AW, Thompson PM (2007) What is where and why it is important. Neuroimage 37:1045–1049
Triarhou LC (2005) Georg N. Koskinas (1885–1975) and his scientific contributions to the normal and pathological anatomy of the human brain. Brain Res Bull 68:121–139
Triarhou LC (2006) The signalling contributions of Constantin von Economo to basic, clinical and evolutionary neuroscience. Brain Res Bull 69:223–243
Triarhou LC (2007a) The Economo-Koskinas atlas revisited: cytoarchitectonics and functional context. Stereotact Funct Neurosurg 85:195–203
Triarhou LC (2007b) A proposed number system for the 107 cortical areas of Economo and Koskinas, and Brodmann area correlations. Stereot Funct Neurosurg 85:204–215
Triarhou LC (2008) Centenary of Christfried Jakob's discovery of the visceral brain: an unheeded precedence in affective neuroscience. Neurosci Biobehav Rev 32:984–1000
Triarhou LC (2009) Tripartite concepts of mind and brain, with special emphasis on the neuroevolutionary postulates of Christfried Jakob and Paul MacLean. In: Weingarten SP, Penat HO (eds) Cognitive psychology research developments. Nova, Hauppauge, pp 183–208
Triarhou LC (2010) Revisiting Christfried Jakob's concept of the dual onto-phylogenetic origin and ubiquitous function of the cerebral cortex: a century of progress. Brain Struct Funct 214:319–338
Triarhou LC (2012) Cytoarchitectonics of the human cerebral cortex: the 1926 presentation by Georg N. Koskinas (1885–1975) to the Athens Medical Society. In: Bright P (ed) Neuroimaging—cognitive and clinical neuroscience. InTech Publishers, Rijeka/Vienna, pp 1–16

Triarhou LC, del Cerro M (2006) Semicentennial tribute to the ingenious neurobiologist Christfried Jakob (1866–1956)—2. Publications from the second Argentina period, 1913–1949. Eur Neurol 56:189–198
Tyler CW, Likova LT (2011) Estimating neural signal dynamics in the human brain. Front Syst Neurosci 5:1–17. doi:10.3389/fnsys.2011.00033 (article 33)
Uylings HBM, Rajkowska G, Sanz-Arigita E, Amunts K, Zilles K (2005) Consequences of large interindividual variability for human brain atlases: converging macroscopical imaging and microscopical neuroanatomy. Anat Embryol 210:423–431
van Bogaert L, Théodoridès J (1979) Constantin von Economo: the man and the scientist. Verlag der Österreichischen Akademie der Wissenschaften, Vienna
Vogt O (1927) Architektonik der menschlichen Hirnrinde. Zbl Gesamte Neurol Psychiatrie 45:510–512
Vogt C, Vogt O (1919) Allgemeinere Ergebnisse unserer Hirnforschung. J Psychol Neurol (Leipz) 25:279–461
Vogt BA, Vogt LJ, Perl DP, Hof PR (2001) Cytology of human caudomedial cingulate, retrosplenial, and caudal parahippocampal cortices. J Comp Neurol 438:353–376
von Bonin G (1950) Essay on the cerebral cortex. Charles C. Thomas, Springfield
von Economo C (1925) Die fünf Bautypen der Grosshirnrinde. Schweiz Arch Neurol Psychiatrie 16:260–269
von Economo C (1927) Zellaufbau der Grosshirnrinde des Menschen: Zehn Vorlesungen. Julius Springer, Berlin
von Economo C (1929) Wie sollen wir Elitegehirne verarbeiten? Z Gesamte Neurol Psychiatrie 121:323–409
von Economo C (2009) Cellular structure of the human cerebral cortex (trans, ed: Triarhou LC). Karger, Basel
von Economo C, Horn L (1930) Über Windungsrelief, Maße und Rindenarchitektonik der Supratemporalfläche, ihre individuellen und ihre Seitenunterschiede. Z Gesamte Neurol Psychiatrie 130:678–757
von Economo C, Koskinas GN (1923) Die sensiblen Zonen des Großhirns. Klin Wochenschr 2:905
von Economo C, Koskinas GN (1925) Die Cytoarchitektonik der Hirnrinde des Erwachsenen Menschen: Textband und Atlas mit 112 Mikrophotographischen Tafeln. Springer, Vienna
von Economo C, Koskinas GN (2008) Atlas of cytoarchitectonics of the adult human cerebral cortex (trans, rev, ed: Triarhou LC). Karger, Basel
Watson KK, Jones TK, Allman JM (2006) Dendritic architecture of the von Economo neurons. Neuroscience 141:1107–1112
Webster WR, Garey LJ (1990) Auditory system. In: Paxinos G (ed) The human nervous system. Academic, San Diego, pp 889–944
Zilles K (2004) Architecture of the human cerebral cortex: regional and laminar organization. In: Paxinos G, Mai JK (eds) The human nervous system, 2nd edn. Elsevier/Academic, Amsterdam/London, pp 997–1055
Zilles K, Amunts K (2010) Centenary of Brodmann's map: conception and fate. Nat Rev Neurosci 11:139–145
Zilles K, Amunts K (2012) Architecture of the cerebral cortex. In: Mai JK, Paxinos G (eds) The human nervous system, 3rd edn. Elsevier/Academic, Amsterdam/London, pp 836–895
Zilles K, Palomero-Gallagher N (2001) Cyto-, myelo-, and receptor architectonics of the human parietal cortex. Neuroimage 14:S8–S20

Chapter 3
The Myeloarchitectonic Studies on the Human Cerebral Cortex of the Vogt-Vogt School, and Their Significance for the Interpretation of Functional Neuroimaging Data

Rudolf Nieuwenhuys

Abstract The human cerebral cortex contains numerous myelinated fibres, many of which are concentrated in tangentially organized layers and radially oriented bundles. The spatial organization of these fibres is by no means homogeneous throughout the cortex. Local differences in the thickness and compactness of the fibre layers, and in the length and strength of the radial bundles, render it possible to recognize areas with a different *myeloarchitecture*. The neuroanatomical subdiscipline aimed at the identification and delineation of such areas is known as *myeloarchitectonics*. There is another, closely related neuroanatomical subdiscipline, named *cytoarchitectonics*. The aims and scope of this subdiscipline are the same as those of myeloarchitectonics, viz. parcellation. However, this subdiscipline focuses, as its name implies, on the size, shape and arrangement of the neuronal cell bodies in the cortex, rather than on the myelinated fibres.

At the beginning of the twentieth century, two young investigators, Oskar and Cécile Vogt founded a centre for brain research, aimed to be devoted to the study of the (cyto+myelo) architecture of the cerebral cortex. The study of the cytoarchitecture was entrusted to their collaborator Korbinian Brodmann, who gained great fame with the creation of a cytoarchitectonic map of the human cerebral cortex. The significance of this work is highlighted in the contribution of Laurence Garey to this volume. Here, we focus on the myeloarchitectonic studies on the cerebral cortex of the Vogt-Vogt school, because these studies are nearly forgotten in the present

---

"This is an updated version of a review article that has originally been published in the journal Brain Structure and Function 2012 (DOI 10.1007/s00429-012-0460-z)"

R. Nieuwenhuys (✉)
The Netherlands Institute for Neuroscience, Royal Netherlands Academy of Arts and Sciences, Meibergdreef 47, 1105 BA Amsterdam, The Netherlands

(Private address): The Netherlands Institute for Neuroscience, Royal Netherlands Academy of Arts and Sciences, Papehof 25, 1391 BD Abcoude, The Netherlands
e-mail: RudolfN@planet.nl

S. Geyer and R. Turner (eds.), *Microstructural Parcellation of the Human Cerebral Cortex*, DOI 10.1007/978-3-642-37824-9_3, © Springer-Verlag Berlin Heidelberg 2013

attempts to localize functional activiations and to interprete findings in modern neuroimaging studies.

Following introductory sections on the principles of myeloarchitectonics, and on the achievements of three myeloarchitectonic pioneers who did not belong to the Vogt-Vogt school, the pertinent literature is reviewed in some detail. These studies allow the conclusion that the human neocortex contains about 185 myeloarchitectonic areas, 70 frontal, 6 insular, 30 parietal, 19 occipital, and 60 temporal. It is emphasized that the data available, render it possible to compose a myeloarchitectonic map of the human neocortex, which is at least as reliable as any of the classic architectonic maps.

During the realization of their myeloarchitectonic research program, in which numerous able collaborators were involved, the Vogts gradually developed a general concept of the organization of the cerebral cortex. The essence of this concept is, that this structure is composed of about 200 distinct, juxtaposed 'Rindenfelder' or 'topistische Einheiten', which represent fundamental structural as well as functional entities. The second main part of this article is devoted to a discussion and evaluation of this 'Vogt-Vogt concept'. It is concluded that there is converging quantitative cytoarchitectonic, receptor architectonic, myeloarchitectonic, hodological, and functional evidence, indicating that this concept is essentially correct.

The third, and final part of this article deals with the problem of relating particular cortical functions, as determined with neuroimaging techniques, to particular cortical structures. At present, these 'translation'- operations are generally based on adapted, three-dimensional versions of Brodmann's famous map. However, it has become increasingly clear that these maps do not provide the neuroanatomical precision to match the considerable degree of functional segregation, suggested by neuroimaging studies. Therefore, we strongly recommend an attempt at combining and synthesizing the results of Brodmann's cytoarchitectonic analysis, with those of the detailed myeloarchitectonic studies of the Vogt-Vogt school. These studies may also be of interest for the tnterpretation of the myeloarchitectonic features, visualized in modern in vivo mappings of the human cortex.

3.1 Introduction

In 1898, the young couple Oskar and Cécile Vogt (born in 1870 and 1875, respectively) took a remarkable initiative. They founded in Berlin a private centre, to be devoted entirely to the study of the structure and function of the central nervous system. This 'Neurologische Zentralstation' grew out to one of the largest and most prestigious research institutes of Germany. In 1902, it was incorporated under the name 'Neurologisches Zentralstation' into Berlin University and in 1931 it was transformed into the 'Kaiser Wilhelm Institut für Hirnforschung', comprising a large research centre with 30 scientific and 70 technical collaborators, as well as a research clinic (Forschungsklinik) with 60 beds. This 'KWIH' was established in the Berlin suburb of Buch. In 1937 Oscar Vogt was forced to give up his directorship of the institute for political reasons. However, generous financial support of the steel tycoon and armaments manufacturer E. A. Krupp and the Rockefeller foundation enabled the Vogts to establish a new research facility in the small city of Neustadt in the Black Forest. In this Institute, which was named 'Institut für Hirnforschung und allgemeine Biologie', they remained scientifically active almost to their death. Oscar Vogt died on July 31th, 1959, and Cécile followed him on May 4th, 1962. Adolf Hopf succeeded Oscar Vogt as director of the Institute in Neustadt. In 1965, it was relocated to Düsseldorf, where it flourishes, under the name 'C. & O. Vogt – Institut für Hirnforschung der Universität Düsseldorf', up to the present as a centre for brain research. The historical data just briefly reviewed are derived from Hopf (1970a) and Klatzo (2002), to which the reader is referred for details.

The research program of the newly founded laboratory in Berlin was comprehensive and ambitious. It was decided that the emphasis would be laid on the 'higher' and more complex centres of the brain, particularly the cerebral cortex (Vogt 1903). As regards the latter structure, the Vogts had not to start from the very first beginning. At the turn of the twentieth century, it was known already that the cerebral cortex is composed of cells of many different kinds (Fig. 3.1a), that the somata of these cells are arranged in layers, and that the size, shape, density and arrangement of the somata in these layers could display considerable local differences (Fig. 3.1b, c). It was also known that the cerebral cortex contains numerous myelinated fibres, forming tangentially organized plexuses and radially arranged bundles, which, just like the neuronal somata, could show marked local differences (Fig. 3.1b, c). Finally, there was clinical and experimental evidence, indicating that the cortex harbours centres with clearly different functions (Fig. 3.2).

The approach chosen by the Vogts was relatively simple and remained essentially unchanged throughout their long scientific career (Vogt 1903, 1943; Vogt and Vogt 1919, 1936, 1942, 1954, 1956). It included a systematic analysis of those structural features of the cerebral cortex that can be readily recognized with relatively weak magnifications, with a view at identifying and delimiting fundamental morphological units within that organ, assuming that the units or areas of distinct structure thus identified, would also proven to be organs of special function. The results of these architectonic studies were expected to provide an adequate basis for clinicopathological studies, as well as for the study of the brains of geniuses ('Elitegehirne'), and feeble-minded people.

At the time that the Vogts began their studies there were two staining procedures for brain tissue, which yielded reliable and reproducible results, *i. e.* the Nissl stain for neuronal cell bodies, and the Weigert stain and its variants for myelinated nerve fibres. Thus, it became routine in the Neurologisches Zentralstation to prepare serial sections of human and animal brains, and to stain these series according to the two procedures mentioned. The systematic study of the material thus prepared led to the emergence of two new neuroanatomical subdisciplines, which were designated by O. Vogt (1903) as *cytoarchitectonics* and *myeloarchitectonics*.

The cytoarchitectonics of the cerebral cortex became the specialism of Korbinian Brodmann, who joined the Vogts in 1901 and remained attached to their laboratory until 1909. Brodmann studied the cellular structure of the cortex in a considerable number of mammals, including the hedgehog, the flying fox, the lemur, the guenon, and the human, resulting in an impressive series of publications (Brodmann 1903a, b, 1905a, b, 1906, 1908a, b), and in a summarizing monograph

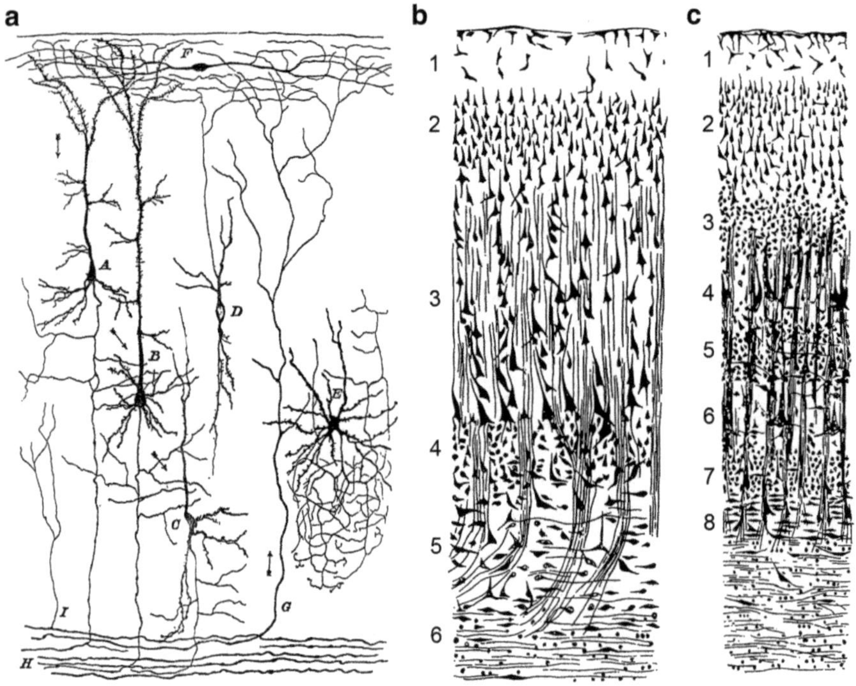

Fig. 3.1 The organization of the cerebral cortex. (**a**) The principal cellular types of the cerebral cortex of mammals, according to Cajal (1894). (**b**, **c**) Meynert's (1884) illustration of his six-layered precentral cortex (**b**), and his eight-layered occipital cortex (**c**)

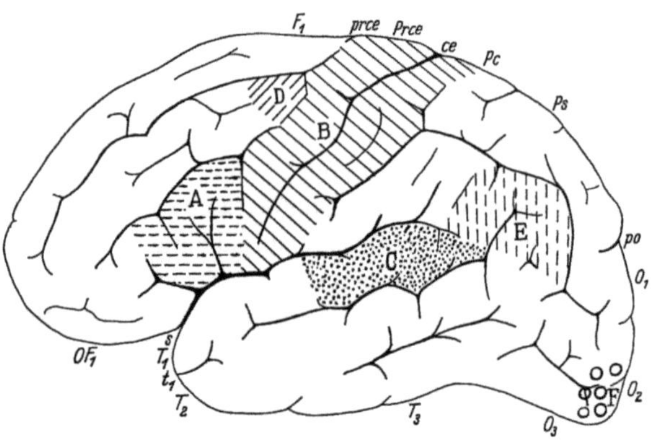

Fig. 3.2 Lateral view of the human telencephalon. Areas, the functions of which were known at the end of the nineteenth century are marked. *A*, motor speech centre of Broca; *B*, the somatomotor and somatosensory areas; *C*, auditory area; *D*, center concerned with writing; *E*, Wernicke's sensory speech centre; *F*, visual area (Modified from Vogt and Vogt 1954)

(Brodmann 1909). His famous map of the cytoarchitecture of the human cerebral cortex was first published in Brodmann (1908a), and its final version appeared in Brodmann (1914). It is of note that the practising of the cytoarchitectonics of the human cerebral cortex remained by no means confined to the institute of the Vogts. Von Economo and Koskinas (1925, Vienna), Bailey and Von Bonin (1951, Urbana, Illinois) and Sarkissov et al. (1955, Moscow) all published comprehensive works on the human cortex, resulting in complete cytoarchitectonic parcellations of that organ (see Braak 1980 and Zilles and Amunts 2010, for synopsis and further developments).

Study of the myeloarchitectonics of the cerebral cortex, on the other hand, remained largely concentrated in the laboratories of the Vogts, from its beginning, marked by the appearance of O. Vogt's (1910a) preliminary note on the human frontal cortex, to its end, marked by the appearance of Hopf's (1970b) study of the human parietal cortex. The comprehensive program included analyses of the myeloarchitecture of all parts of the human cortex, as well as the cortices of a number of mammals, including the hedgehog (Flores 1911), the mangabey (Mauss 1908), the orang-utan (Mauss 1911) and the chimpanzee (Beck 1929; Strasburger 1937b; Gerhardt 1938).

3.2 The Principles of Myeloarchitectonics

Preparations stained with the Weigert method reveal that the neocortex contains numerous myelinated fibres, which show two principal orientations, tangential and radial. The tangential fibres tend to form local concentrations or bands, the most conspicuous of which can be clearly observed with the naked eye in unstained sections. The radial fibres are concentrated in bundles or radii. O. Vogt (1903) designed a basic plan ('Grundschema') of the myeloarchitectonic organization of the neocortex, which formed the point of departure of all myeloarchitectonic studies produced in his laboratory. This scheme is shown in Fig. 3.3, together with a comparable basic scheme of the cytoarchitectonic organization. In both schemes the neocortex is subdivided into six layers. The cytoarchitectonic layers are: (I) the cell-poor zonal layer, (II) the external granular layer, (III) the external pyramidal layer, (IV) the internal granular layer, (V) the internal pyramidal layer, and (VI) the multiform layer. The corresponding myeloarchitectonic layers are, in order to avoid confusion, designated with Arabic, rather than with Roman numerals.

1. The *zonal layer* is differentiated into four sublayers, the narrow sublayer 1°, which contains only very few fibres, and the external, intermediate and deep sublayers 1a, 1b and 1c, of which 1a contains clearly more fibres than 1b and 1c.
2. The *dysfibrous layer* which contains, just like sublayer 1°, only very few fibres.
3. The *suprastriate layer* has three sublayers, of which the superficial sublayer $3a^1$ is more rich in fibres than the remaining sublayers $3a^2$ and 3b. In certain cortical regions, sublayer $3a^1$ show a distinct concentration of fibres, known as the *stripe of Kaes-Bechterew*. Sublayer 3b is characterized by the presence of the end-segments of the radial bundles.

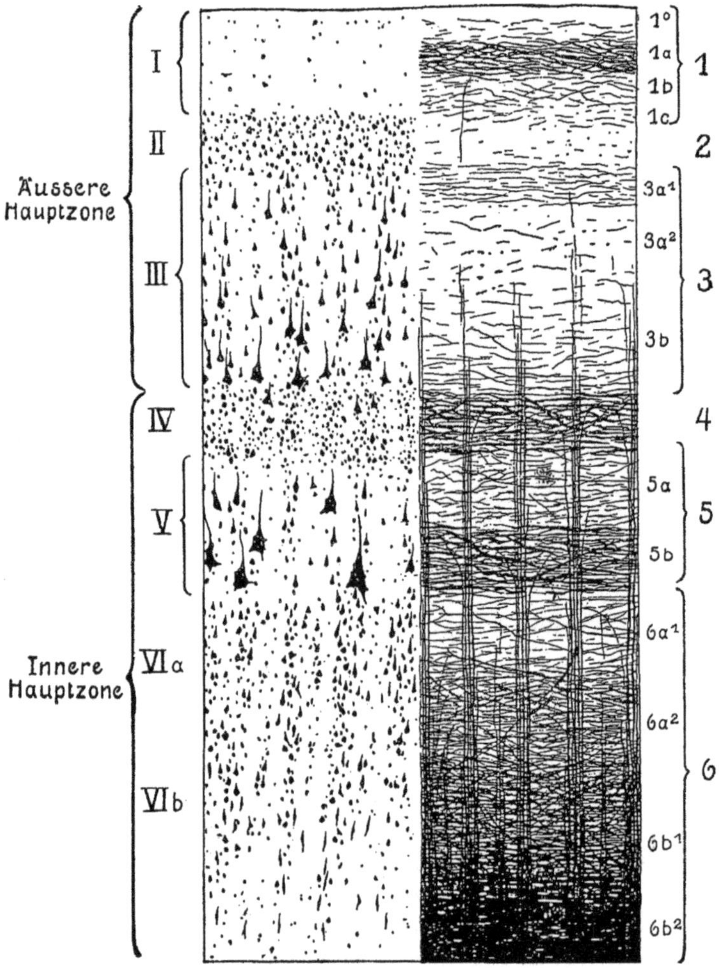

Fig. 3.3 O. Vogt's (1903) basic schemes of the cytoarchitectonic layers (designated with *Roman numbers*), and the myeloarchitectonic layers (designated with *Arabic numbers*)

4. The *external stria* or *outer stripe of Baillarger* forms a dark band of tightly packed, tangential fibres.
5a. The *intrastriate layer* is generally relatively poor in tangential fibres, thus contrasting with the bordering stripes of Baillarger.
5b. The *internal stria* or *inner stripe of Baillarger* is again a dense plexus of tightly packed tangentially oriented fibres.
6. This layer is subdivided into the pale substriate lamina $6a^1$ and laminae $6a^2$, $6b^1$ and $6b^2$, which show an increasing wealth of tangentially oriented fibres. Sublayer $6b^2$ forms the zone of transition to the subcortical white matter.

Variations in the number and density of the tangential and radial fibres define the boundaries of te myeloarchitectonic areas. With regard to the lamination of the

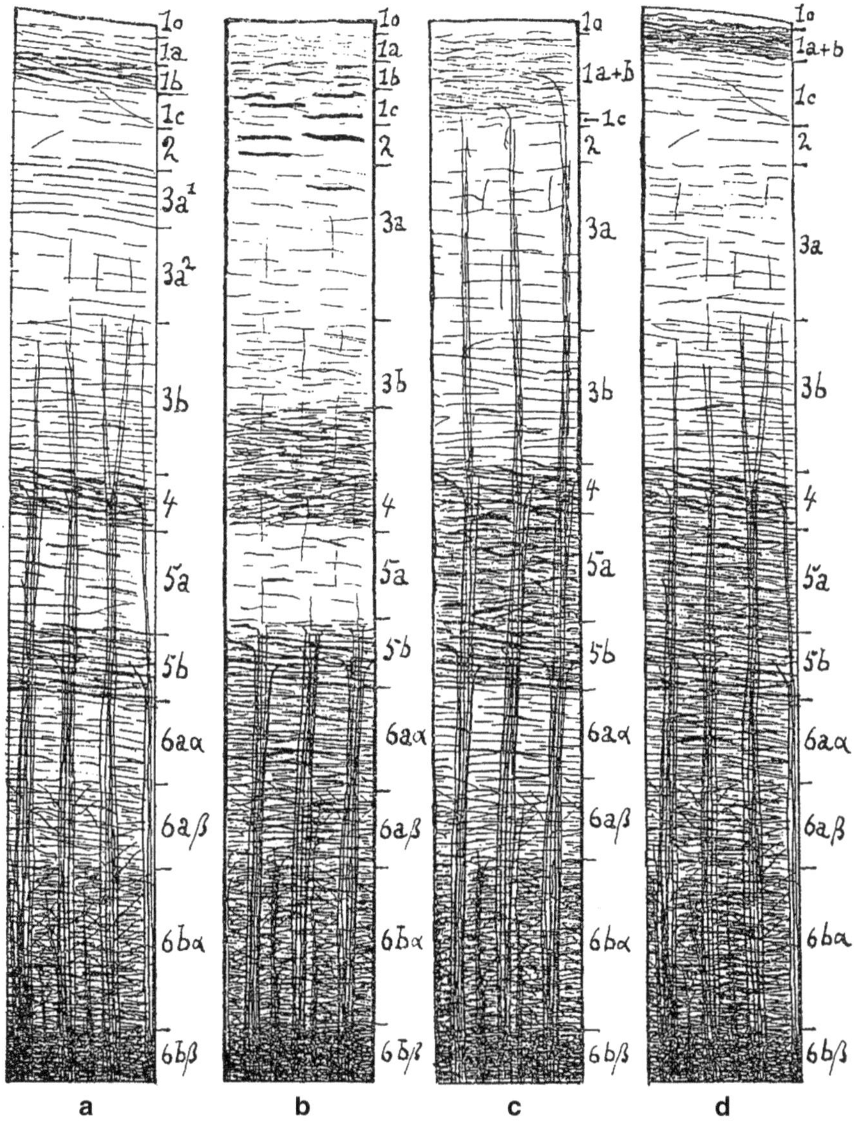

Fig. 3.4 The four principal types of myeloarchitectonic layering, according to O. Vogt (1910a). (**a**), bistrate type; (**b**), unistriate type; (**c**), unitostriate type; (**d**), astriate type

tangential fibres, O. Vogt (1910a, b, 1911) distinguished four principal types (Fig. 3.4).

(a) A *bistriate* type, characterized by the presence of two distinct and clearly separated bands of Baillarger (Fig. 3.4a). It should be added that the situation in which the density of fibres in the outer stripe equals that in the inner stripe is designated as *typus equodensus*, and that the situations in which the fibres in the outer stripe are more densely or less densely arranged than those in the inner stripe, are designated as *typus externodensior* and *internodensior*, respectively.

(b) A *unistriate* type, in which only the external stripe of Baillarger can be distinguished as a separate entity. The inner stripe, though present, cannot be delineated because of the high fibre content of the substriate lamina 6a1 (Fig. 3.4b, in which the substriate lamina is labeled 6aα).

(c) A *unitostriate* type, in which the fibre-poor interstriate layer is lacking and the two stripes of Baillarger form a single plexus (Fig. 3.4c).

(d) An *astriate* type, in which, due to the presence of an unusually large number of tangential fibres in intrastriate layer 5a and substriate lamina $6a^1$, layers 4–6 form a single dark and undivided fibre zone (Fig. 3.4d).

As regard the disposition of the bundles of radial fibre bundles, O. Vogt distinguished three types, *euradiate*, *supraradiate* and *infraradiate*. In the *euradiate type*, the radii do not extend beyond the level of the suprastriate layer (Fig. 3.4a, d); in the *supraradiate type*, the bundles traverse almost the entire width of the cortex and reach the zonal layer (Fig. 3.4c), whereas in the *infraradiate type*, the radii are very short and terminate already in the fifth layer at the level of the inner stripe of Baillarger (Fig. 3.4b).

The radii do not vary only in length, but also in breadth, number and calibre of their fibres. On the basis of differences in the breadth of radii, *latoradiate*, *medioradiate* and *tenuiradiate types* were distinguished. Differences in the number of radii led to distinction of *densoradiate*, *modicoradiate* and *sparsoradiate types*, whereas differences in the size of the fibres forming the radii found their expression in *grossoradiate* and *finoradiate types*.

O. Vogt codified many other myeloarchitectonic variations as types. Thus, he referred to an overall wealth of myelinated fibres in a given area as *typus dives*, whereas an overall scarcity of such fibres was characterized as *typus pauper*. The various tangential layers generally consist of a plexus of thin fibres of about equal size ('Grundfasern'), in which individual fibres of larger size ('Einzelfasern') are embedded. O. Vogt designated layers, in which the coarse individual fibres are scarce, as representing a *typus tenuifibrosus*, and layers in which these fibres are numerous as representing a *typus grossofibrosus*.

It should be appreciated that the myeloarchitectonics of the cortex, as developed by O. Vogt, in spite of its extremely detailed and intricate typology, remained a purely descriptive and qualitative neuroanatomical subdiscipline. Several attempts at the development of a more quantitative and more objective myeloarchitectonics have been made. Thus, Braitenberg (1962) devised a method in which the light absorption, being directly proportional to the fibre density, is systematically measured in narrow strips of Weigert-stained sections of the cerebral cortex, passing from the pial surface to the white matter. With this method, he recorded the fibre density in sections taken from 14 different cortical regions. Representation of the results in graphic form yielded quite characteristic curves for most of the regions studied (Fig. 3.5a, b). Braitenberg also produced an interesting diagram, explaining the relationships between the basic functional wiring and the

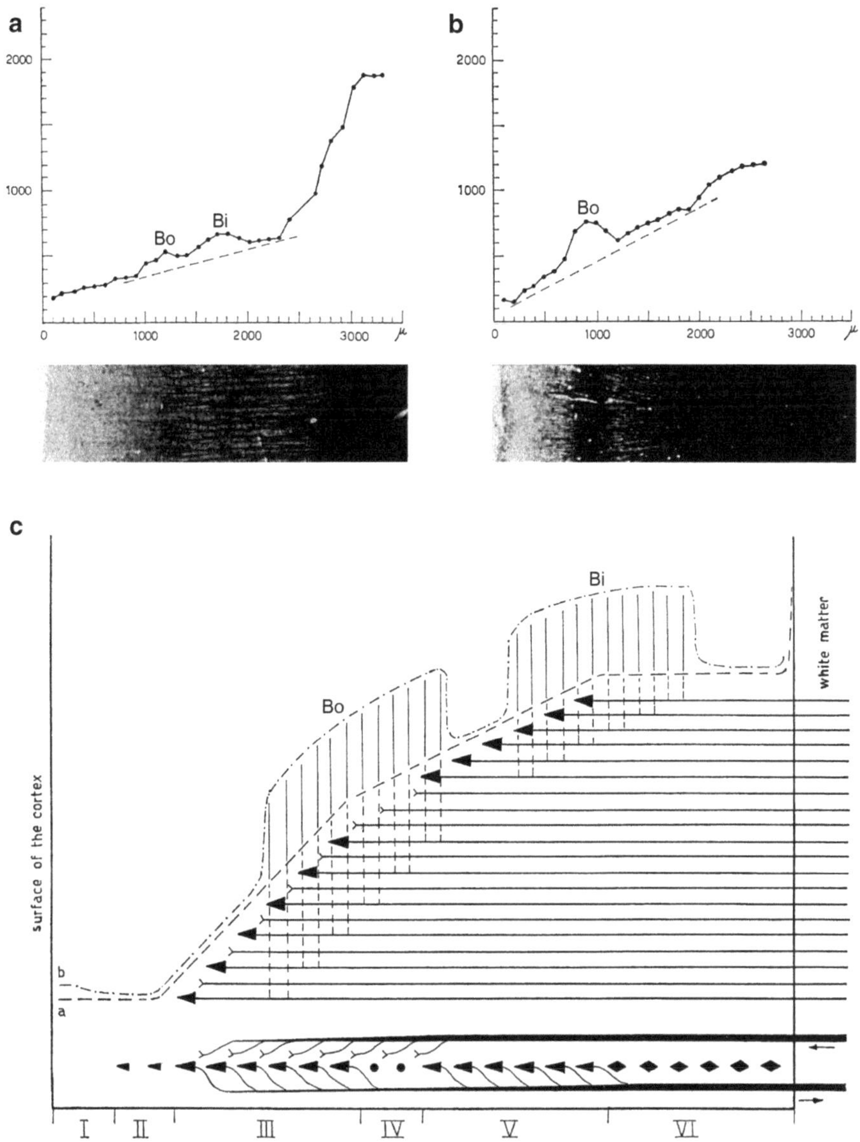

Fig. 3.5 Braitenberg's (1962) photometric analysis of the myeloarchitecture of the human cerebral cortex. (**a**, **b**) Graphic representations of photometric analyses of strips of myelin-stained sections of the cortex, with distance from the surface of the cortex (in mμ) on the abscissa, and fibre-density (in arbitrary units) on the ordinate. The tangents of the curves show different slopes. The outer and inner stripes of Baillarger (Bo, Bi) produce local prominences over the tangents. (**a**) Cortex of middle frontal convolution, showing the presence of both stripes. (**b**) Area striata with the pronounced outer stripe of Baillarger (or line of Gennari, or line of Vicq d'Azyr), after which it is named. (**c**) Diagram, clarifying the relationships between the overall course of the photometric curves, the myeloarchitecture, and the basic functional wiring of the neocortex. The following hodological features are taken into account: (1) The number of specific afferent fibres in the cortex

myeloarchitecture of the cortex, and the overall course of his photometric curves (Fig. 3.5c). Hopf (1966) developed a photometric method for determining the extent of myelinization in the cortex, which closely resembled that of Braitenberg. With the aid of this technique, he successfully explored the myeloarchitecture of the human frontal, temporal and parietal cortices (Hopf 1968a, b, 1969, 1970b).

In what follows, the literature on the myeloarchitecture of the human cerebral cortex, as produced by the Vogts and their numerous disciples, will be reviewed first. Next, a general concept, concerning the organization of the cerebral cortex, which has been developed from the myelarchitectonic studies reviewed, will be discussed, and finally, some remarks will be made on the functional parcellation of neocortex. It is felt appropriate, however, to preface this long story with a few remarks on the work of three myeloarchitectonic pioneers, who did not belong to the Vogt-Vogt-school.

3.3 Notes on the Work of Three Myeloarchitectonic Pioneers

3.3.1 Alfred Walter Campbell (1868–1937)

The pathologist Campbell published in 1905 a monograph, entitled: “Histological studies on the localisation of cerebral functions”. The opening paragraph, which gives a clear view on the perspective in which the author placed his work, may be quoted in full: “The process leading to the accomplishment of functional localisation in the cerebral cortex is such a complicated one, and involves so many side issues, that perfection cannot be attained or even hoped for until the fruits of investigation in a number of departments are thoroughly weighed, sifted and assorted. It is anticipated that the observations set forth in this research will help to establish the value of histological work as an auxiliary force in the final settlement of that functional subdivision of the cerebral cortex at which we aim” (l.c. p. XV).

The normal human material, on which the work is based, consisted of three cerebral hemispheres completely examined for both nerve cells and nerve fibres, and three additional hemispheres examined for fibres only. Campbell distinguished

---

Fig. 3.5 (continued) decreases if we ascend from the white matter; most of these fibres form their terminal ramifications in layers III and IV. (2) The number of efferent fibres increases if we descend to the white matter; practically all of these fibres are produced by pyramidal neurons, situated principally in layers III and V. These two features explain the steady increase in myelinization if we pass in the cortex from superficial to deep, and therewith the overall course of curve a. (3) The axons of the pyramidal neurons produce long, tangentially running collaterals, which tend to assemble at two different cortical levels. It is these concentrations of pyramidal collaterals that, forming the stripes of Baillarger, produce the local increases in fibre density shown in curve (**b**)

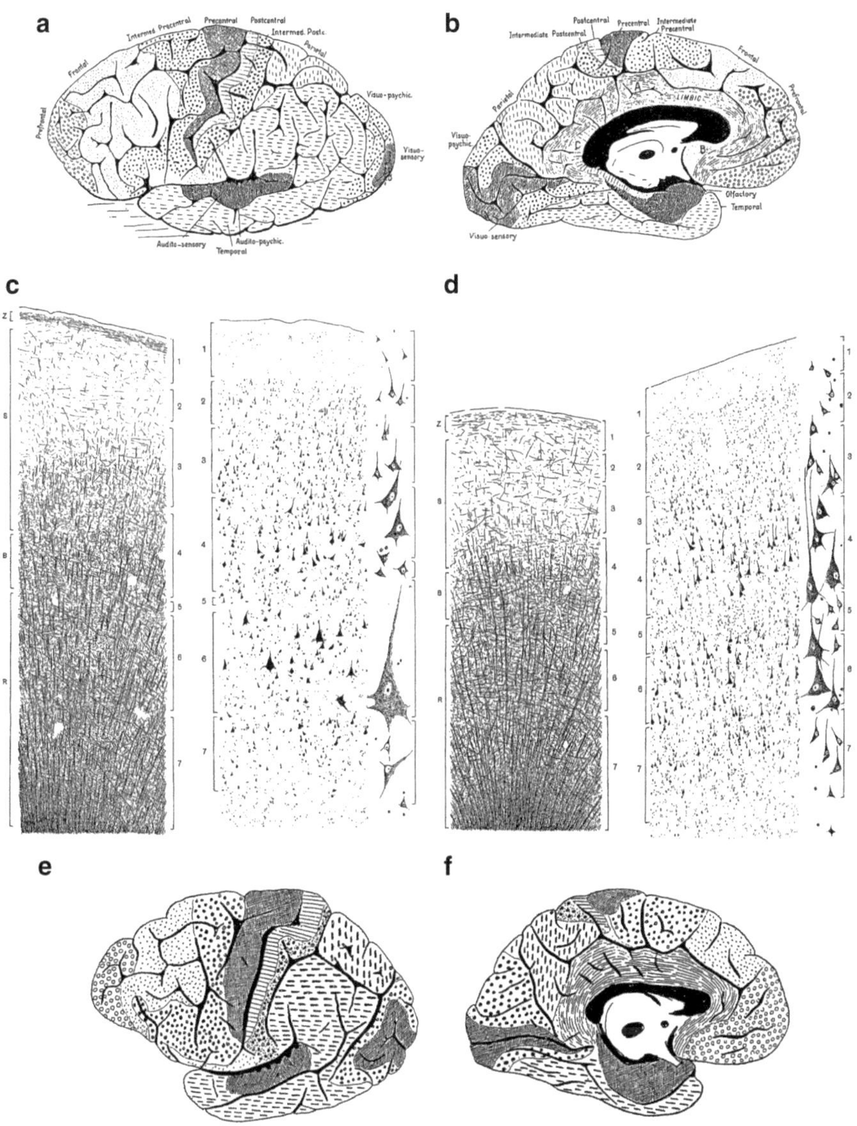

Fig. 3.6 Some illustrations from Campbell's (1905) monograph: "Histological Studies on the Localisation of Cerebral Function". (**a**, **b**) Architectonic map of the human cerebral cortex; (**a**), lateral view; (**b**), medial view. (**c**, **d**) Drawings showing the structure of the precentral or motor area (**c**), and the postcentral or somatosensory area of the human cerebral cortex (**d**). In each of these figures the pattern of myelinated fibres is shown to the *left*, and the arrangement of the neuronal cell bodies to the *right*. Representative cell types are shown at a higher magnification to the far *right*. (**e**, **f**) Architectonic map of the cerebral cortex of the chimpanzee

16 cortical areas, which he designated either with topographical (such as: frontal, postcentral, and temporal), or with (provisional) functional names (such as: olfactory, visuo-sensory and visuo-psychic). All of these areas were described in some detail, and their cytoarchitecture and myeloarchitecture were recorded in beautiful drawings (Fig. 3.6c, d). The resultant map, *which represents the only complete, combined cyto- and myeloarchitectonic map of the human cerebral cortex, produced thus far*, is shown in Fig. 3.6a, b. It will be seen that in Campbell's parcellation, the frontal cortex is unusually large; that the intermediate-central cortex, which roughly corresponds to Brodmann's area 6, is not demarcated from Broca's motor speech region (areas 44 and 45 of Brodmann), and that the temporal area does not only occupy the inferior and middle temporal convolutions, but rather extends over vast regions of the parietal lobe.

Campbell did not confine himself to the human brain, but also studied and mapped the cortex of several mammals, including the cat, the orang-utan and the chimpanzee (Fig. 3.6e, f).

3.3.2 Grafton Elliot Smith (1871–1936)

The anatomist Elliot Smith published in 1907 a detailed map of the human cerebral cortex, which, surprisingly, was exclusively based on macroscopic observations (Fig. 3.7a, b). He made fresh sections at many locations of the hemisphere, using differences in the width and distinctness of the stripes of Baillarger as the main criteria for his parcellation. In unstained preparations, these stripes can be recognized as whitish bands, contrasting with the darker hue of the cortical grey matter (Fig. 3.7c). With the aid of this simple technique, Elliot Smith was able to distinguish about 50 different cortical areas. He noticed that most of these areas have precise relations to various stable sulci. As regards the nature of the interareal boundaries, Elliot Smith (1907, p. 240) took a firm stand: "There is a very widespread belief that the characters of an area merge gradually and imperceptably into those of the neighboring areas, but this is entirely mistaken. The changes in structure occur with the utmost abruptness, so that it is possible to determine with absolute precision the exact boundaries of each area,"

3.3.3 Theodor Kaes (1852–1913)

Kaes was psychiatrist and prosector at the asylum Friedrichsburg near Hamburg. Between 1891 and 1904, he published a series of papers on the various techniques, used for the staining of myelinated fibres, and on the myeloarchitecture of the human cerebral cortex. He summarized his findings and views in his opus magnum (Kaes 1907), entitled: "Die Grosshirnrinde des Menschen in ihren Massen und ihrem Fasergehalt. Ein gehirnanatomischer Atlas mit erläuterndem Text", comprising, apart from a concise text, an atlas consisting of 90 large colour plates, showing the myeloarchitecture of 12 selected cortical regions in 45 individuals, and

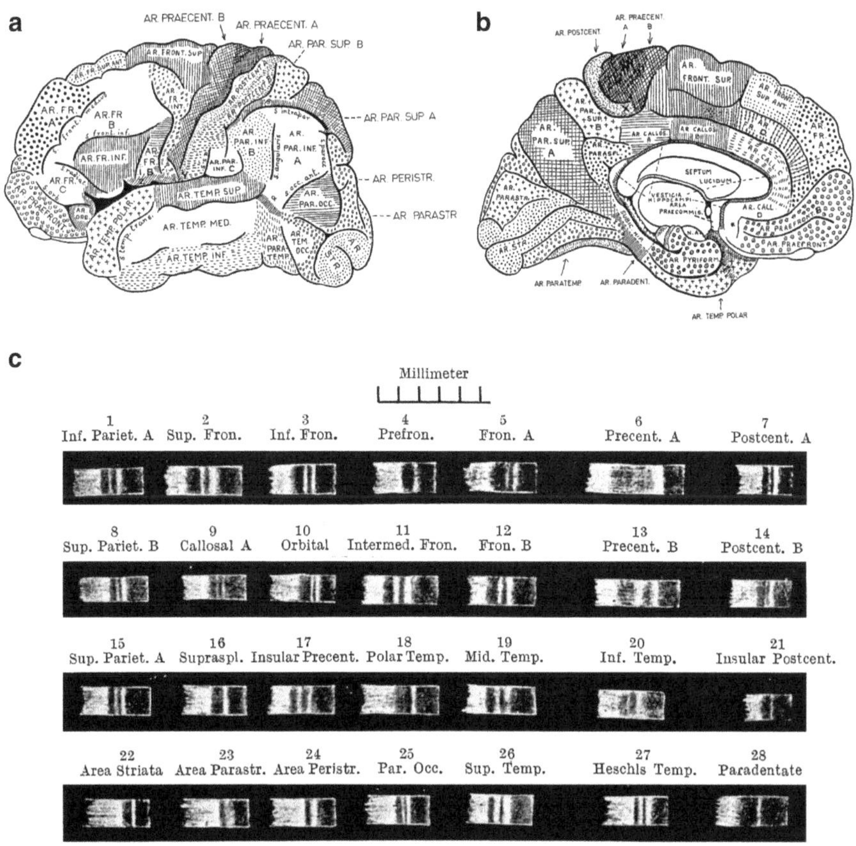

Fig. 3.7 Elliot Smith's (1907) 'myeloarchitectonic' analysis of the human cerebral cortex. (**a**, **b**) Architectonic map, based entirely on the study of fresh, unstained macroscopic sections of the cerebral cortex. (**c**) Pictures showing 28 of the about 50 areas, distinguished by Elliot Smith

numerous tables and curves, documenting thousands of measurements of the total width of the cortex, and of its various layers and zones.

Kaes collected 45 human brains, ranging from three months postnatal to 97 years, among which several mentally retarded and criminal individuals. He selected 12 regions in both hemispheres of these brains for further analysis. These regions were designated as: (1) vordere Stirne, (2) hintere Stirne, (3) vordere Zentralwindung, (4) hintere Zentralwindung, (5) Operculum, (6) Insel, (7) vordere Schläfe, (8) hintere Schläfe, (9) oberer Scheitel, (10) unterer Scheitel, (11) Sehrinde, and (12) Gyrus fornicatus. Most unfortunately, any further indication concerning the exact location of the regions selected, is lacking in the work. Sections, taken from these 12 regions, were stained according to the Weigert-Wolters technique. The best-stained sections were carefully drawn and included in the atlas. Thus, in the atlas 24 sections of the cortex of each individual investigated were included, 12 from the left, and 12 from the right hemisphere (Fig. 3.8).

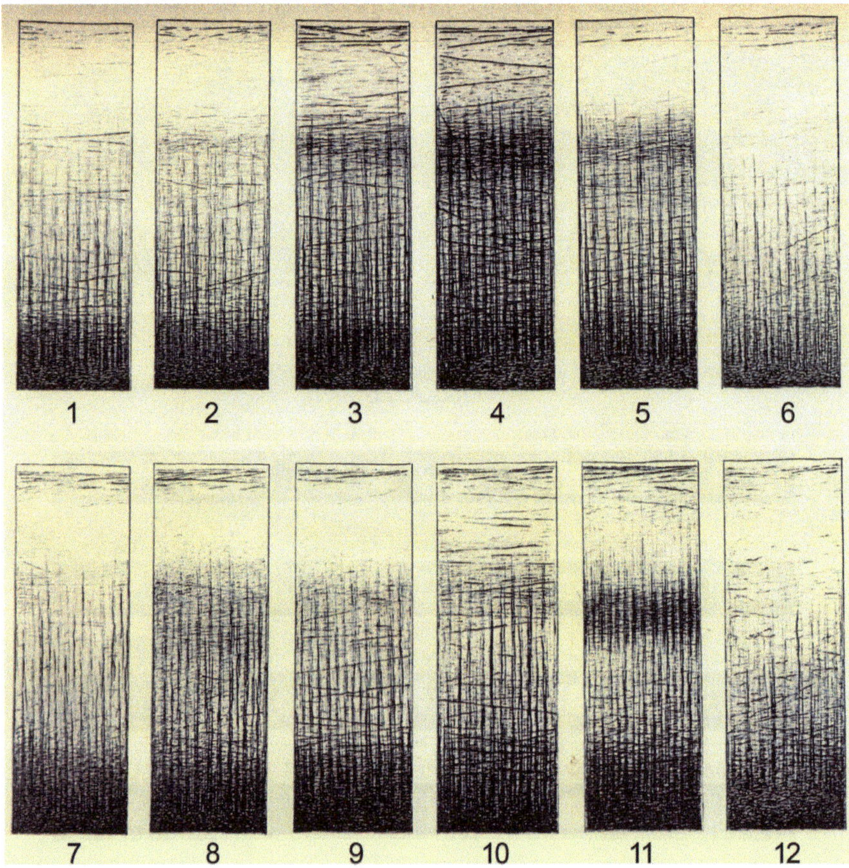

Fig. 3.8 One of the 90 coloured plates, with which Kaes (1907) illustrated his great work on the human cerebral cortex, reproduced at half of the original size. The plate shows the structure of the cortex, as seen in preparations stained with Wolters' variant of the Weigert technique, in 12 representative cortical areas. The numbers are specified in the text

Kaes used the material, thus selected, for a qualitative and quantitative analysis of the postnatal development of the cortex under normal and abnormal circumstances. In his quantitative studies, he divided the cortex into an external principal zone ('äußere Hauptzone'), encompassing layers I-III, and an internal principal zone ('innere Hauptzone'), consisting of layers IV-VI (Fig. 3.3). A detailed discussion of the results of Kaes, falls outside the scope of the present review. Hence, I confine myself to some of his main conclusions: (1) Cortices, which are poor in myelinated fibres, are in general wider than cortices containing numerous fibres. (2) The internal principal zone attains the peak of its development around the 19th year of life; the external principal layer continues developing until the 45th year of life and beyond. (3) The findings on brains of mentally retarded individuals ('Idiotengehirne'), confirm the rule, mentioned above under 1. (4) The brains of criminals show generally an abnormally low weight, and an infantile level of development.

3.4 Myeloarchitectonic Parcellations of the Human Neocortex

3.4.1 The Frontal Lobe

The literature on the myeloarchitecture of the frontal lobes is voluminous and encompasses studies of O. Vogt (1910a, b), C. and O. Vogt (1919), Strasburger (1937a, b, 1938), Braitenberg (1956), Hopf (1956, 1968a), and Sanides (1962, 1964). The myeloarchitectonic parcellation of this lobe, presented by O. Vogt (1910a), is complex (Fig. 3.9). He distinguished six *regions*, which were designated with Roman numerals. Each of these regions was subdivided into several (two to four) *subregions*, and these were, in their turn, further subdivided into *divisions*, and locally even still further into *subdivisions*. Finally, one or several *areas* were delineated within each of the (sub) divisions. In total, 66 myeloarchitectonic areas, designated with Arabic numerals, were distinguished within the frontal lobe. These numerals have nothing to do with the – also Arabic – numerals, used by Brodmann (1909) for his cytoarchitectonic areas. Each of the entities distinguished was designated with a full Latin name, referring to particular myeloarchitectural features characterizing that particular entity. A survey of O. Vogt's nomenclature for region III and its subdivisions is presented in Table 3.1. O. Vogt does not specify the histological material he used in this study. Strasburger (1937b) mentioned that it was principally based on serial sections of a single hemisphere, designated as A 18r. Vogt's (1910a) paper ends abruptly after the description of the last area (Fig. 3.10).

In a subsequent publication, O. Vogt (1910b) comments briefly on the findings just reviewed, in relation to the results of the cytoarchitectonic analyses of the cortex, published shortly before by Brodmann (1909). He points out that in general, the myeloarchitectonic approach is superior to the cytoarchitectonic one, because the number of cortical areas that can be delineated with the aid of the former, far exceeds that delineable with the aid of the latter. Unfortunately, he does not address the specific relationship between the results of his myeloarchitectonic, and Brodmann's cytoarchitectonic parcellation of the frontal lobe. O. Vogt (1910b) also points out that the relation between the sulci, separating the various convolutions, and the myeloarchitectonic boundaries, is variable in the frontal lobe. Some of the sulci coincide with such boundaries, but many others do not.

C. and O. Vogt (1919) published pictures of the myeloarchitecture of a number of frontal areas, some of which are reproduced in Fig. 3.11.

Before turning to a discussion of the publications of Strasburger (1937a, b, 1938), brief attention should be paid to Mauss' (1908, 1911) studies on the cortex of some non-human primates. This author analyzed the myeloarchitecture of the

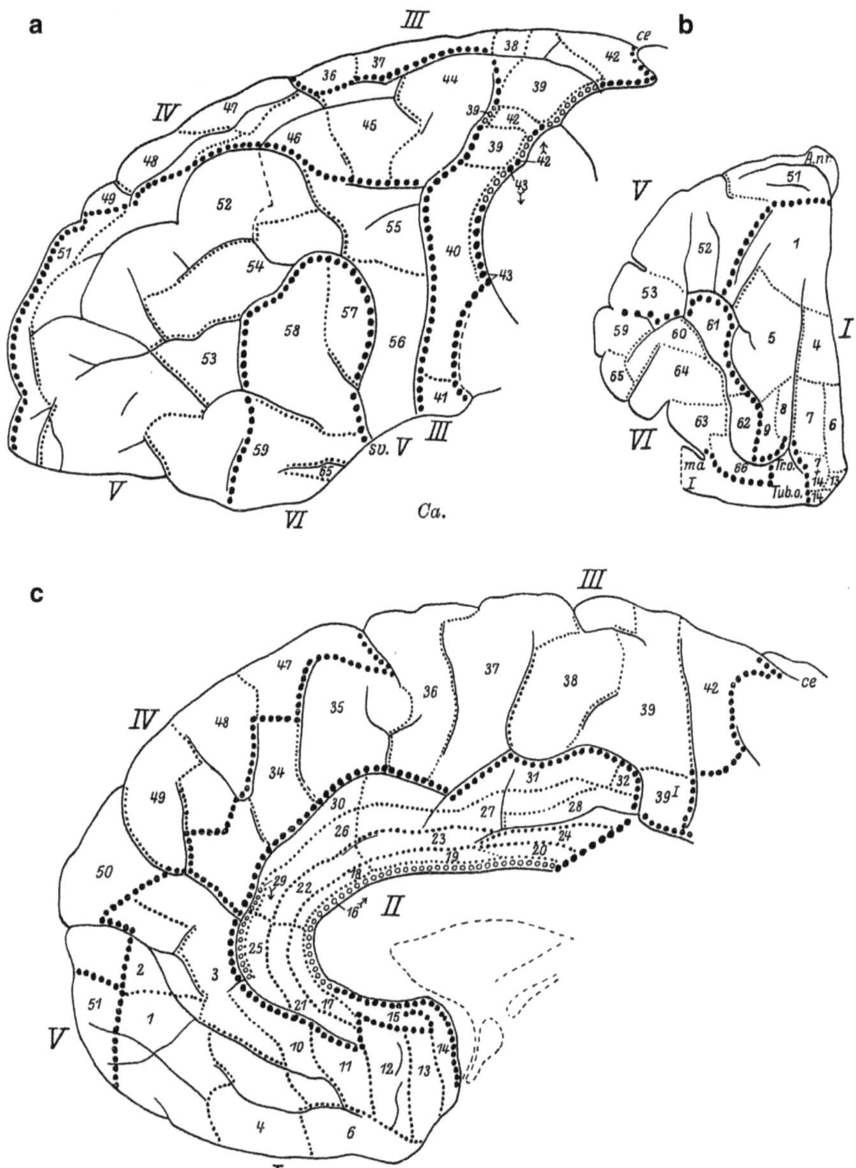

Fig. 3.9 Lateral (**a**), basal (**b**), and medial views (**c**) of the human frontal lobe, showing the myeloarchitectonic parcellation of O. Vogt (1910a, b)

cortex in two 'lower' monkeys, the rhesus macaque *Macaca mulatta* and the mangabey *Cercocebus fulginosus* (Mauss 1908), and in two anthropomorph monkeys, the gibbon *Hylobatus spec.* and the orang-utan *Pongo pygmaeus* (Mauss 1911). His principal results can be summarized as follows: (1) The results

Table 3.1 O. Vogt's (1910a) myeloarchitectonic parcellation of a part of the human frontal lobe

III. Regio unistriata euradiata grossofibrosa (areae 33–43)		
α Subregio unistriata progrediens		
	αα Divisio pauper	
		33 area latofasciculata
		34 area sublatofasciculata
		35 area aequofasciculata
	αβ Divisio dives	
		36 area subtenuistriata
		37 area aequostriata
β Subregio propeastriata		
		38 area subunistriata
		39 area propeastriata
γ Subregio unistrata degrediens		
		40 area dives
		41 area pauper
δ Subregio astriata		
		42 area typica
		43 area atypica

Fig. 3.10 The end of O. Vogt's (1910a) preliminary study on the myeloarchitectonic parcellation of the human frontal lobe

β) **Subregio unitostriata.**

5a kaum abgehoben.

βα) **Divisio trizonalis.**

1c breiter.

63. *Area pauper.*

Sehr arm an Grundfasern, so daß in *4—5b* die Einzelfasern vollständig vorherrschen.

64. *Area intermedia.*

Auch Vermehrung der Einzelfasern.

65. *Area dives.*

Auch noch weitere Zunahme der Einzelfasern.

Enthält gegenüber 58 und damit natürlich gegenüber 57 und 59 dickere und mehr Einzelfasern in *1a + b* und *4—6*, aber weniger Grundfasern, wenigstens in *1—3*.

ββ) **Divisio propebizonalis.**

1c schmäler.

66. *Area propebizonalis.*

of the myeloarchtectonic analysis of the cortex of the mangabey (Fig. 3.12) closely resemble those of Brodmann's (1905a) cytoarchitectonic analysis in the same species. (2) In the frontal lobe of the mangabey, 11 different myeloarchitectonic areas can be distinguished, nine of which (areae 4, 6, 8, 9, 10, 11, 12, and 24) can be readily equated to similarly numbered cytoarchitectonic areas. (3) The total number of delineable myeloarchitectonic cortical areas in the orang-utan (Fig. 3.13) is larger than that in the mangabey: 41 versus 31. (4) Using similarity in structure and similarity in position as criteria, 26 of the 31 myeloarchitectonic areas, present in the cortex of the mangabey, can be readily homologized with (similarly numbered) areas in the cortex of the orang-utan. (5) Although the areas numbered 8 in the maps of the mangabey and the orang-utan differ considerably in extent, they should, nevertheless, be considered as homologous. (6) The cortex of the orang-utan

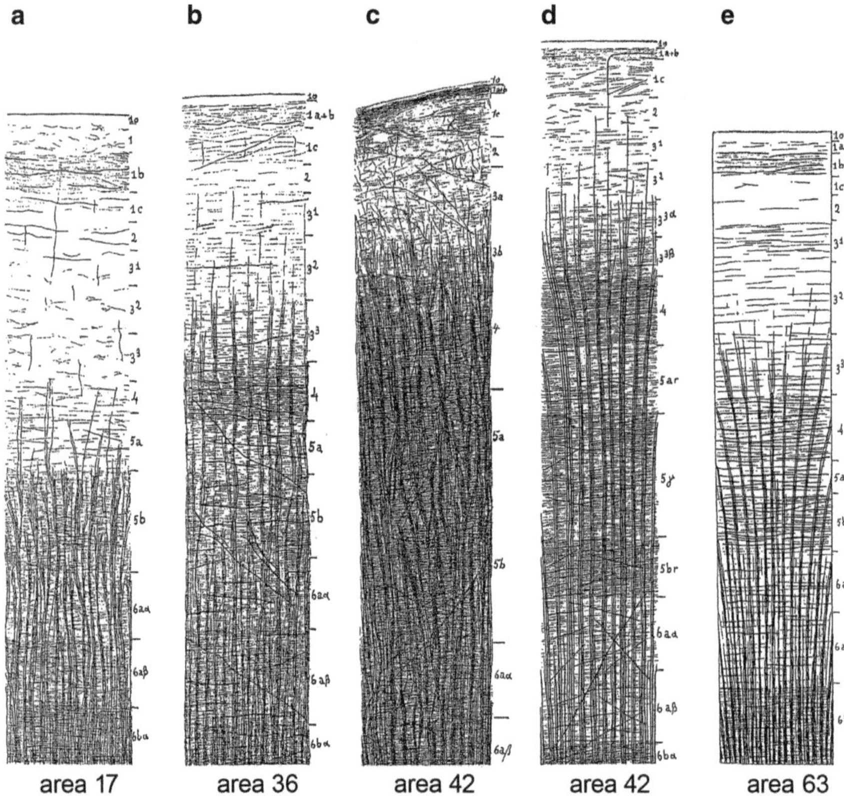

Fig. 3.11 The myeloarchitecture of some frontal areas, as depicted by C. and O. Vogt (1919). (**a**) Area 17, which is situated in the rostrobasal part of the cingulate gyrus (Fig. 3.9c), is of the *infraradiate* type, because the radii terminate principally already at the level of the transition of layers 5a and 5b. Layer 1 is subdivided, just as in the 'Grundschema' (Fig. 3.3), into four sublayers, one of which (1b) contains numerous 'Einzelfasern'. There is a paucity of both radial and tangential fibres in layers 2–4, which contrasts with the much higher fibre-density seen in layers 5b and 6. Layer 5a occupies, with regard to the density of its fibres, an intermediate position between these two sets of layers. (**b**) Area 36 forms part of the *regio unistriata euradiata grossofibrosa* (Fig. 3.9a, c: *III*). Because of its rich endowment of fibres it is characterized as *area dives*. (**c**) Area 42 also forms part of *regio III* (Fig. 3.9a, c), and corresponds to a part of the cytoarchitectonic *area gigantocellularis*. The radii contain numerous coarse '*Einzelfasern*'. O. Vogt (1910a, p. 230) characterized this area as "absolut astriär". C. and O. Vogt (1919) emphasized that the marking of the layers 4- 6aβ in the figure is only based on the tracing of these various cortical levels into adjacent cortical areas with a more distinct lamination. (**d**) Another picture of the myeloarchitecture of area 42, based on a preparation subjected to a further differentiation ('stärkere Entfärbung'). It will be seen that distinct local differences in the density of the horizontal fibres become manifest after this procedure. (**e**) Area 63, forms part of the orbitofrontal sector of the *regio unitostriata* (Fig. 3.9b). Although the term 'unitostriate' refers to the situation in which both stripes of Baillarger form together a single broad band (Fig. 3.4c), in area 63 these two stripes, forming layers 4 and 5b, are nevertheless discernable as separate formations. Area 63 is designated as *area pauper* (Fig. 3.9b), because of its extreme poverty in 'Grundfasern'

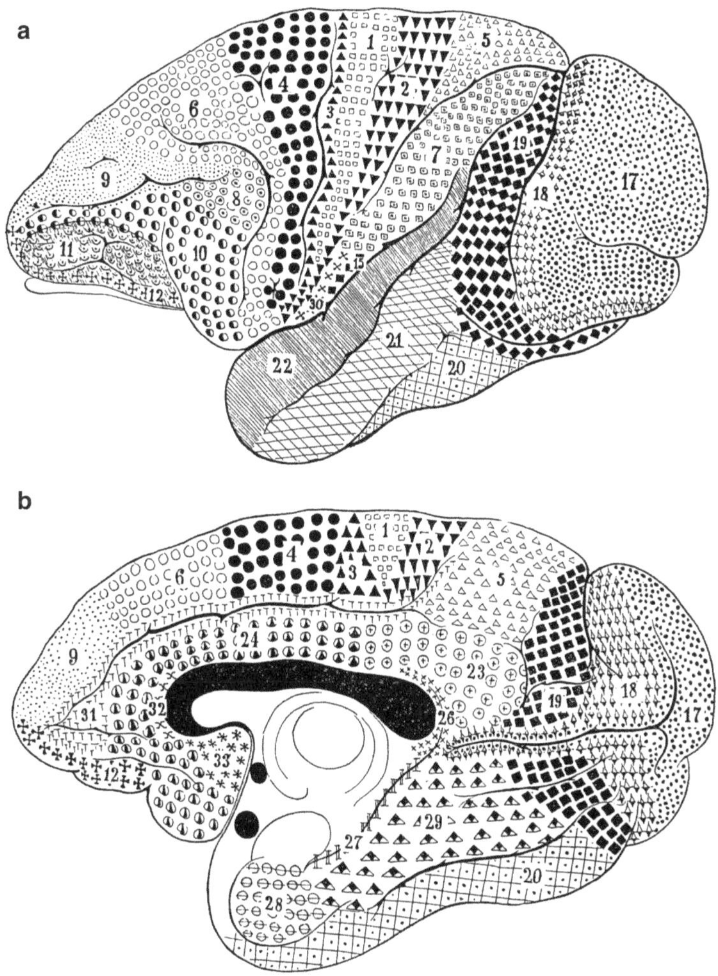

Fig. 3.12 Mauss' (1908) myeloarchitectonic map of the cortex of the mangabey *Cercocebus fulginosus*; (**a**), lateral view; (**b**), medial view. The numbers, indicating the various areas, correspond largely to those used by Brodmann (1909), in his map of the cytoarchitecture of the cortex of the same species

contains 10 myeloarchitectonic areas, which could not be identified in the mangabey. (7) All of the 11 myeloarchitectonic areas, present in the frontal lobe of the mangabey, i.e. areas 4, 6, 8, 9, 10, 11, 12, 24, 31, 32 and 33, have (similarly numbered) homologues in the frontal lobes of the orang-utan. (8) The frontal lobe of the orang-utan contains a single new myeloarchitectonic area, which is numbered 37 (Figs. 3.12 and 3.13).

Strasburger (1937a, b) thoroughly analyzed the myeloarchitecture of the frontal lobe in a human hemisphere (A 39r), and in hemispheres of two different chimpanzees (A 117l, A 118l). He compared the results of his analysis of the

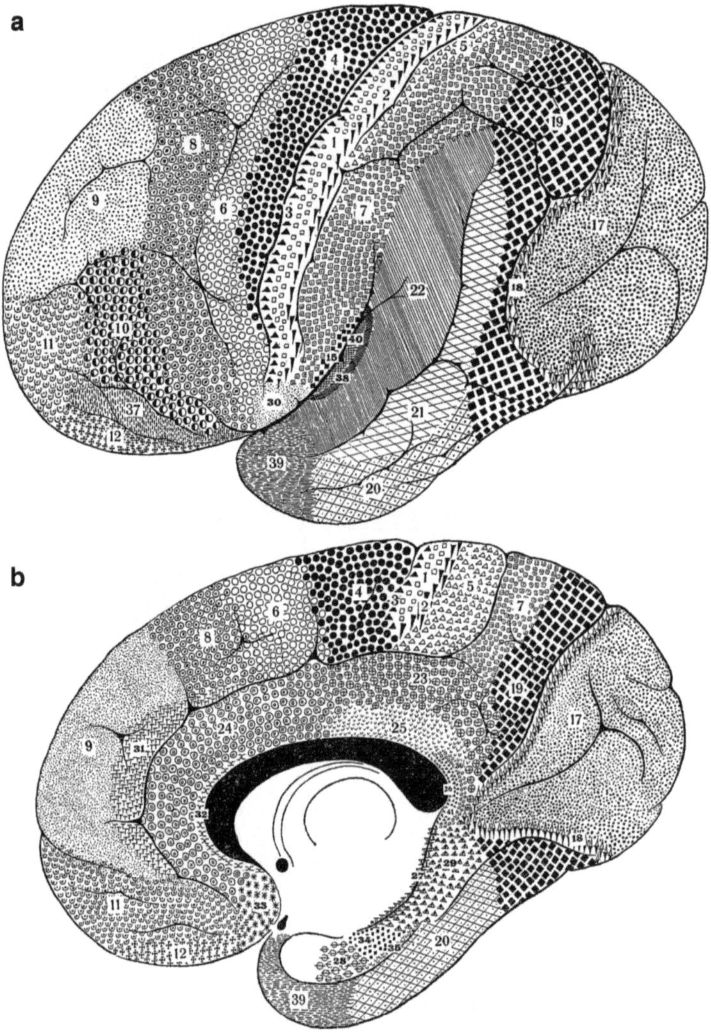

Fig. 3.13 Mauss' (1911) myeloarchitectonic map of the cortex of the orang-utan *Pongo pygmaues*; (**a**), lateral view; (**b**), medial view

human frontal lobe (Fig. 3.14), with those of O. Vogt (1910a; Fig. 3.9), adopting the numbering system for the various areas of the latter; moreover, he compared the frontal lobes of the two chimpanzees studied (Fig. 3.16) with each other, as well as with that of the human (Fig. 3.14). Contrary to O. Vogt (1910a), Strasburger illustrated his descriptions of the various areas, with numerous large, very detailed drawings, some of which are reproduced (at half of their original size) in Fig. 3.15. He not only depicted individual areas, but laudably also complexes of two adjacent areas (e. g. Fig. 3.15b, c), enabling the reader to visualize the structural differences between these areas. The criteria for the establishment of homologies, used by

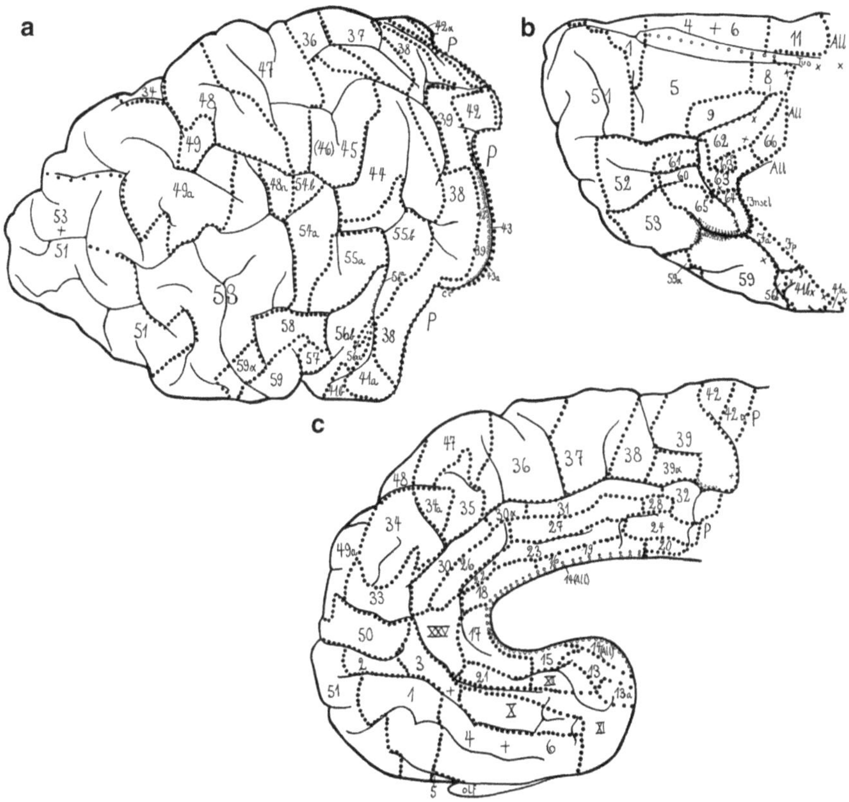

Fig. 3.14 Lateral (**a**), basal (**b**), and medial views (**c**) of the human frontal lobe, showing the myeloarchitectonic parcellation of Strasburger (1937a). *All*, allocortex; *cc*, central sulcus; *Insel*, island of Reil; *olf*, olfactory bulb; *P*, parietal lobe

Strasburger, were the same as those of Mauss, namely, (a) similarity in structure, and (b) similarity in position. If he was certain that a particular area in the human frontal lobe was homologous to an area delineated by O. Vogt (1910a), he adopted the number given to that area by the latter. However, if he considered the homology between a particular 'Strasburger' area and a particular 'Vogt' area as highly probable, but not entirely certain, he designated the area in question with the *Roman equivalent* of Vogt's (Arabic) number of that area (cf. areas X, XI, XII and XXV in Fig. 3.14c, with areas 10, 11, 12 and 25 in Fig. 3.9c). A similar procedure was followed in the comparison of the parcellations of the frontal cortex of the two chimpanzees studied, with that of the human frontal cortex (see the numerous Roman numerals in Fig. 3.16b, d). In his characterization of the myeloarchitecture of the various areas, Strasburger paid particular attention to: (a) the disposition of the stripes of Baillarger (as regards the areas depicted in Fig. 3.15, he considered areas 51 and 61 as bistriate; areas 2, 21, 26, 30 and 50 as

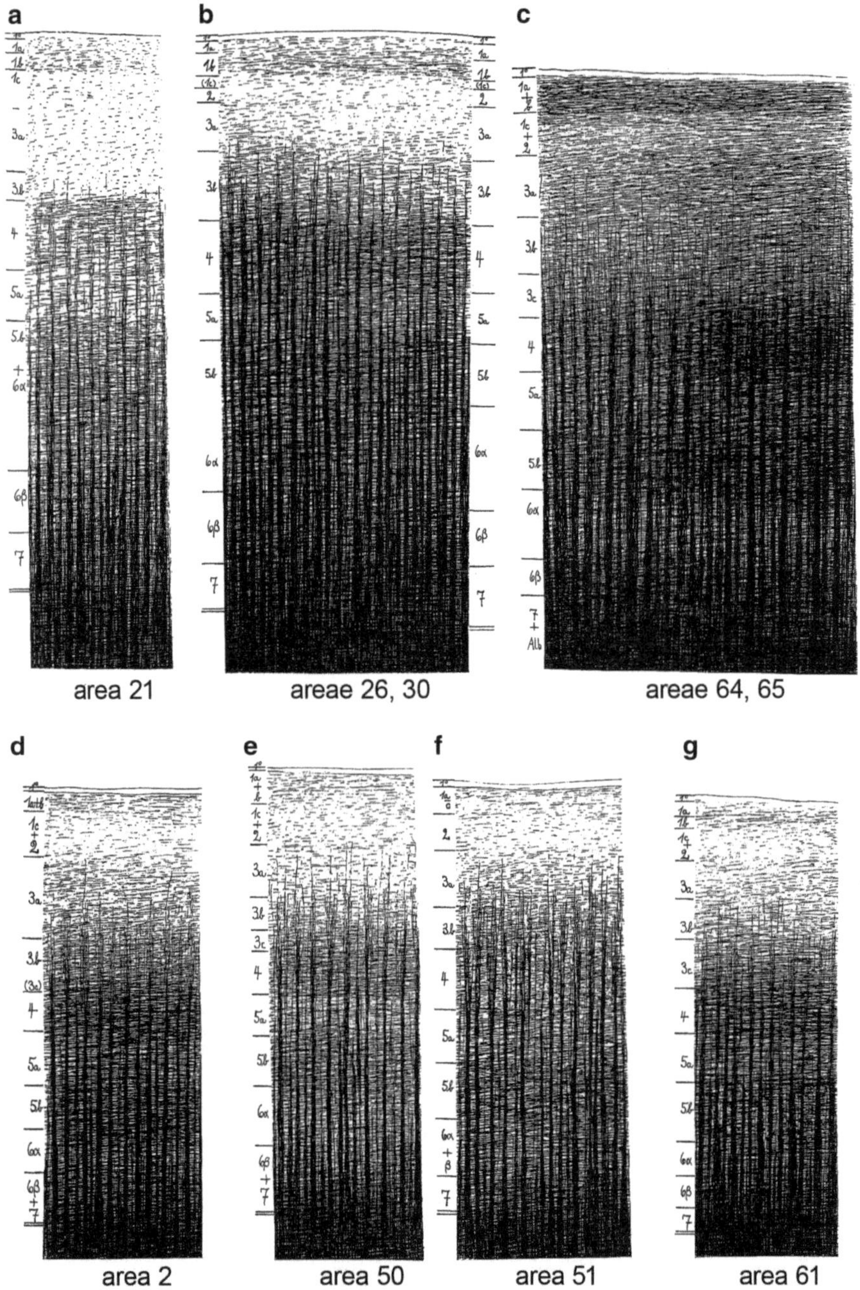

Fig. 3.15 The myeloarchitecture of some areas in the human frontal lobes, as depicted by Strasburger (1937a)

unistriate, and areas 64 and 65 as unitostriate); (b) the distinctness of the various sublayers in layer 1, and (c) the size and length of the radii.

The number of myeloarchitectonic areas, delineated by Strasburger in the human frontal cortex exceeds that of O. Vogt: 78 versus 66. If we disregard the slight doubts, expressed in the Roman numerals (see above), 58 of the areas distinguished by Strasburger appeared to be directly comparable to 'Vogt-areas'. Strasburger remained unable to identify homologues of 'Vogt-areas' 7, 40 and 46. He split each of the five 'Vogt-areas' 41, 54, 55, 56 and 65, into two separate areas: 41a, 41b, 54a, 54b and so on. Finally, Strasburger distinguished 10 new areas in the frontal cortex of the human specimen studied. He did not designate these areas with new numbers. Rather, he took the number of another, adjacent area, and added the letter a, or α to it (see e. g. 49a and 59α in Fig. 3.14a).

As regards the comparison between the frontal lobes of the two chimpanzees (A 117l, A 118l), and the human studied (A 39r), Strasburger (1937b) reported that, in general, the fibre layers in the human (Fig. 3.15) are more distinct than in the chimpanzee (Fig. 3.16e–h), and that the number of areas in the human is larger: 78 versus 60 (in A 118l) or 65 (in A 117l). Some areas, including 14, 22, 26, 30, 30α, 44 and 45, present in the human frontal cortex (Fig. 3.14), could not be identified in the chimpanzee. Moreover, several sets of separate areas in the human frontal cortex, appeared to be represented by a single area ('Sammelfeld') in the chimpanzee (see, for example, 1+4 in Fig. 3.16b, d; 37+38 in Fig. 3.16a, c, and 48+49 in Fig. 3.16a, c, h).

Strasburger (1937b, Table 6, p. 603) compared the results of his myeloarchitectonic parcellation of the frontal cortex of the chimpanzee (50–55 areas; Fig. 3.16a–d), with those of Mauss (1911), obtained from a similar study in the orang-utan (12 areas; Fig. 3.13). He considered areas 4, 6 and 12 of Mauss directly comparable to fields 42, 39 and 6, respectively, of his parcellation. The remaining nine areas of Mauss were homologized with smaller or larger sets of areas delineated by him. To give a single example: area 11 of Mauss corresponds, according to Strasburger, with 'his' areas 2+3, 1+4+8, 5, 9 and 61.

In a subsequent study, Strasburger (1938) presented a detailed myeloarchitectonic analysis of frontal areas 56–66, as distinguished by O. Vogt (1910a; Fig. 3.9a, c), in both hemispheres of six human brains (A 20, A 22, A 27, A 34, A 38, A 39). With a single exception (area 63 in A 201), all of the 12 areas could be identified in all of the 12 hemispheres investigated. The 12 fields showed only very slight left-right or interspeciminal structural and positional differences. The size (volume) of some areas showed considerable differences, however.

Braitenberg's (1956) myeloarchitectonic analysis of the human frontal cortex differs from that of O. Vogt (1910a) and Strasburger (1937a, b, 1938), in that he based his parcellation exclusively on structural differences that could be clearly observed in his Weigert-material, either with the naked eye, or with the magnifying glass. He also indicates to have included only those structural entities in his map (Fig. 3.17), that were clearly distinguishable in all of the series studied, though he does not mention on how many series his study was actually based. Another difference with the approach of O. Vogt and Strasburger is that he took the gross anatomy, i. e. the gyrification of the frontal lobe as point of departure. Thus, his

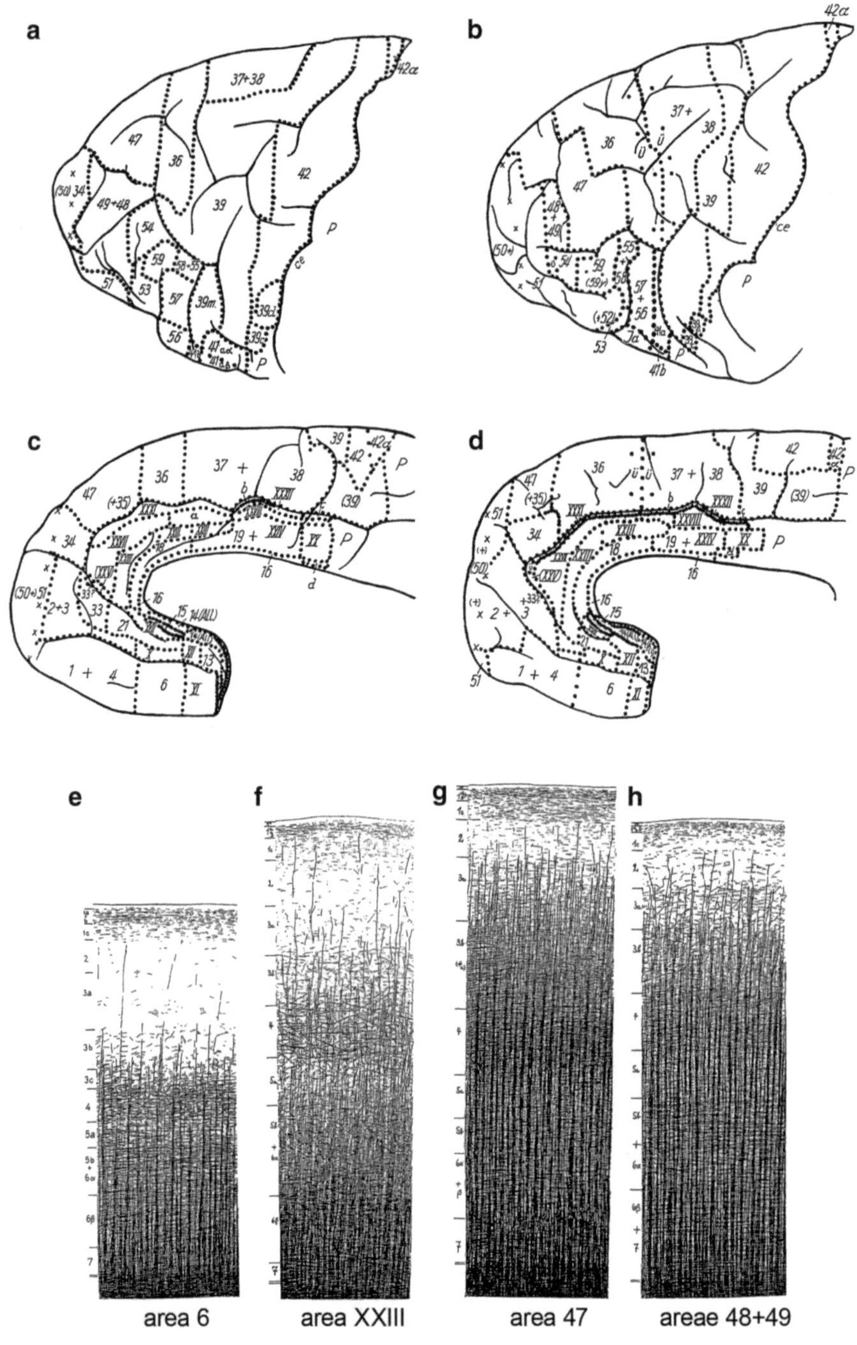

Fig. 3.16 Myeloarchitectonic parcellation of the frontal cortex of the chimpanzee, according to Strasburger (1937b). (**a, b**) Lateral (**a**) and medial views (**b**) of the frontal lobe of specimen A 117l. (**c, d**) Lateral (**c**) and medial views (**d**) of the frontal lobe of specimen A 118l. (**d–f**) The myeloarchitecture of some fields in the frontal cortex of specimen A 117l. For abbrevaiations, see Fig. 3.11 (Reproduced from Strasburger 1937b)

frontal regions F1, F2, and F3 correspond (partly) to the superior, middle, and inferior frontal gyri, respectively, whereas his region F1 recta is centered around the straight gyrus (gyrus rectus), which forms the medial part of the basal surface of the frontal lobe. Braitenberg delineated the following myeloarchitectonic entities within the frontal lobe (Fig. 3.17):

1. An unnamed elongated, astriate region, situated directly in front of the central sulcus. This region corresponds, according to Braitenberg, with area 4 of Brodmann, and with areas 42 and 43 of O. Vogt (in what follows, all area numbers correspond, unless otherwise stated, to those of the latter).
2. The unistriate Regio Frontalis 1 (*r. F1*), which includes the rostral part of the precentral gyrus, most of the superior frontal gyrus, and the medial part of the basal surface of the frontal lobe. It can be divided into a dorsal part (*r. F1 dors.*), and a 'straight' part (*r. F1 recta*). *r. F1 dors.* includes parts of O. Vogts' regions III and IV, and encompasses areas 33–41 (cf. Fig. 3.6). *r. F1 recta* corresponds roughly with O. Vogts' region I, and with the stretch of cortex occupied by areas 1–14.
3. The bistriate Regio Frontalis 2 (*r. F2*), which corresponds to region V, and to areas 46–55 of O. Vogt. This region is divisible into to dorsal and ventral moieties, designated as *r. F2 dors.* and *r. F2 vent.* or – *polaris*. The boundary between these two divisions of *r. F2* is not sharp.
4. A small unistriate area on the medial surface of the frontal lobe, corresponding to area 2, manifesting itself, on the basis of its extraordinary wealth of fibres, clearly as a separate entity (Fig. 3.14a).
5. The, mostly unitostriate, Regio Frontalis 3 (*r. F3*), which corresponds to region VI of O. Vogt, and which is clearly further differentiated into separate fields than *r. F1* and *r. F2*. This region contains, according to O. Vogt, 11 different areas, which he numbered 56–65 (Fig. 3.9a, b). Reference to Fig. 3.17b, c shows, that Braitenberg was able to identify all of these areas, except for 61, in his material. It is of note that Strasburger (1938), as we have seen, delineated the same areas in both hemispheres of six different brains.
6. The Regio cinguli anterior (*r. cingul. ant.*), which corresponds to region II of O Vogt, and which is characterized by very feebly developed stripes of Baillarger, and very short radii. Braitenberg indicates that the presence of subtle myeloarchitectonic differences, recognizable only with the aid of higher magnifications, render it possible to recognize the equivalents of areas 15–32 within the confines of this region.

Braitenberg emphasizes that the boundaries between all of the six myeloarchitectonic entities just discussed, are distinct, except for that between *r. F2 dors.*, and *r. F2 vent.* However, he also points out to have observed the following changes, in passing from caudodorsal to rostroventral in *r. F1* and *r. F2*: (i) A *gradual decrease* in the number of fibres; (ii) a *gradual decrease* in the size (calibre) of the individual fibres, and (iii) a *gradual decrease* in the width of the cortex. These three phenomena appeared to be correlated with (iv), a cytoarchitectonic change, viz. *a gradual increase* in the number of granule cells in layer IV of the cortex. These observations are noteworthy, in relation to the fact that, within the orthodox Vogt-Vogt-school

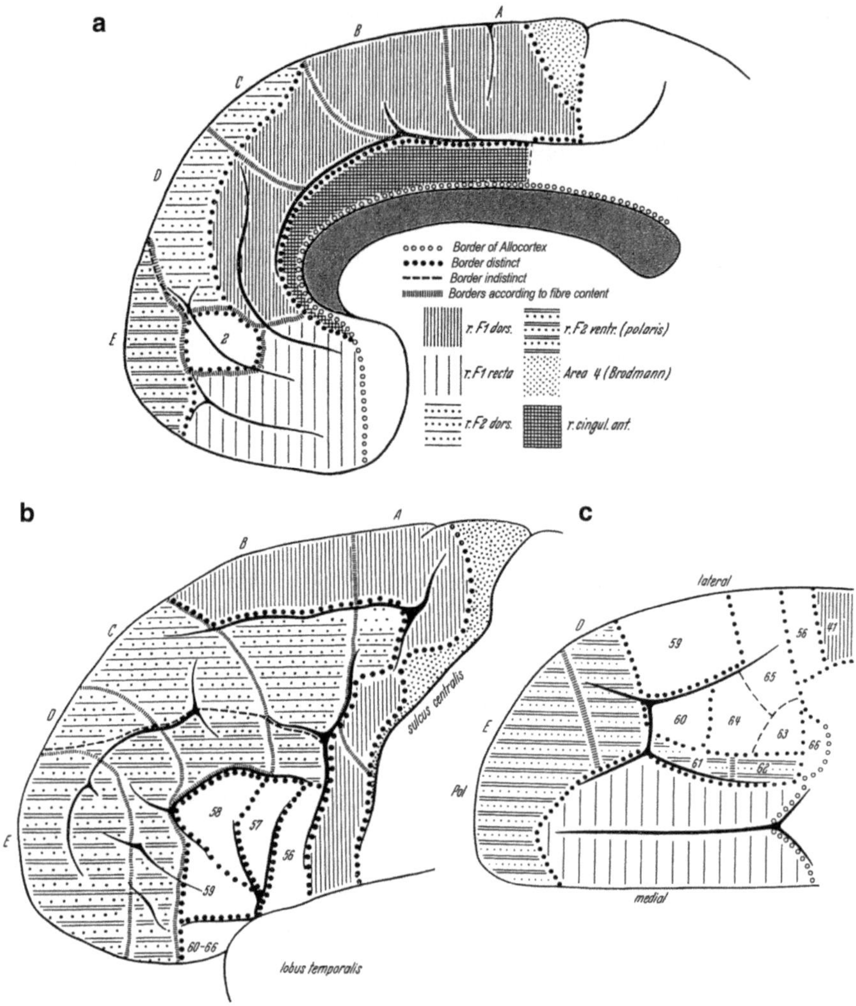

Fig. 3.17 Lateral (a), medial (b), and basal views (c) of the human frontal lobe, showing the myeloarchitectonic parcellation of Braitenberg (1956). The numbers indicate fields, according to O. Vogt (1910a; Fig. 3.9). The significance of the capitals A-E is explained in the text

(to which Braitenberg did not belong), the existence of gradual architectonic changes was categorically denied. Braitenberg made a somewhat infelicitous attempt, to indicate the gradual changes discussed, by including in his maps some borders according to fibre content ('Grenzen nach Faserreichtum'), separating cortical compartments of 'equal darkness' (Fig. 3.17a–e).

In summary, Braitenberg, studying series of Weigert sections through the human frontal lobe with a simple magnifier, divided the cortex of this lobe into six large regions, *r. F1 dors.*, *r. F1 recta*, *r. F2 dors.*, *r. F2 vent.*, *r.* F3, and *r. cingul. ant.*, and two independent smaller entities, one corresponding to area 4 of Brodmann, the other to area 2 of O. Vogt. Within *r. F3*, eleven different areas could be delineated, almost all of which appeared to be directly comparable to myeloarchitectonic areas distinguished by O. Vogt. Passing from the central sulcus to the rostral pole of the frontal lobe, Braitenberg observed several correlated *gradual* structural changes in *r. F1* and *r. F2*.

We now turn to the studies of Adolf Hopf, a collaborator of the Vogts in the Neustadt institute. Hopf's (1956, 1968a) contributions to the myeloarchitecture of the frontal cortex are threefold. Firstly, he performed a new myeloarchitectonic parcellation of this structure; secondly, he prepared maps, showing the main myeloarchitectonic features of the frontal cortex, and thirdly, he made an attempt at the objective registration of the myeloarchitecture of the frontal lobe, using the photometric technique already discussed in a previous section.

Hopf's (1956) renewed myeloarchitectonic parcellation of the human frontal cortex was based on serial sections of two brains, A 18, which had been previously studied by O. Vogt (1910a), and A 39, previously studied by Strasburger (1937a, b). Hopf remained, just like Strasburger, unable to distinguish areas 7 and 40, and to delimit area 45 from 46. Moreover, he regarded Strasburger's areas 43a, 48a, 54b, 56b and 63° as inconspicuous and inconstant variants. The resultant myeloarchitectonic map, encompassing 69 areas, is shown in Fig. 3.18.

In order to visualize the distribution of the various myeloarchitectonic features over the frontal cortex, Hopf (1956) prepared separate maps, showing the overall density of fibres (Fig. 3.19), the density and size of 'Einzelfasern', the length of the radii, and the disposition of the stripes of Baillarger. This 'feature-mapping' showed, *inter alia*, that the precentral cortex possesses a high content of coarse fibres, and that the fibre content decreases in a step-like fashion with increasing distance from the central sulcus (Fig. 3.19).

Hopf's (1968a) publication on the objective registration of myeloarchitectonic features in the human frontal cortex opens with a discussion of the six basic qualitative myeloarchitectonic types, occurring in this cortex (Fig. 3.20). He then presents the results of his registrations of the differential density of fibres, in stretches of cortex involving two different myeloarchitectonic areas. Two of these registrations are shown in Fig. 3.19. Hopf draws the following general conclusions from these registrations: (1) The relative fibre density of the two stripes of Baillarger in relation to each other and to the neighboring sublayers, as well as the general content in myelinated fibres, play a dominant role in these registrations. (2) The existence and reliability of some myeloarchitectonic features can be objectively demonstrated with this new technique. (3) The existence of all of the six basic, qualitatively determined types of frontal cortex (Figs. 3.20 and 3.21) could be confirmed. It is important to note that Hopf used his new myeloarchitectonic registration technique only to substantiate his qualitatively obtained maps (Fig. 3.18), *and not* for the creation of new, objective 'supermaps'.

The last study on the human frontal lobe, to be discussed here, is that of Sanides (1962, 1964), another pupil of the Vogts. This study was principally based on transversely cut serial sections through the left hemisphere of one brain, A43. The sections were stained alternately according to Nissl, for cell bodies, and

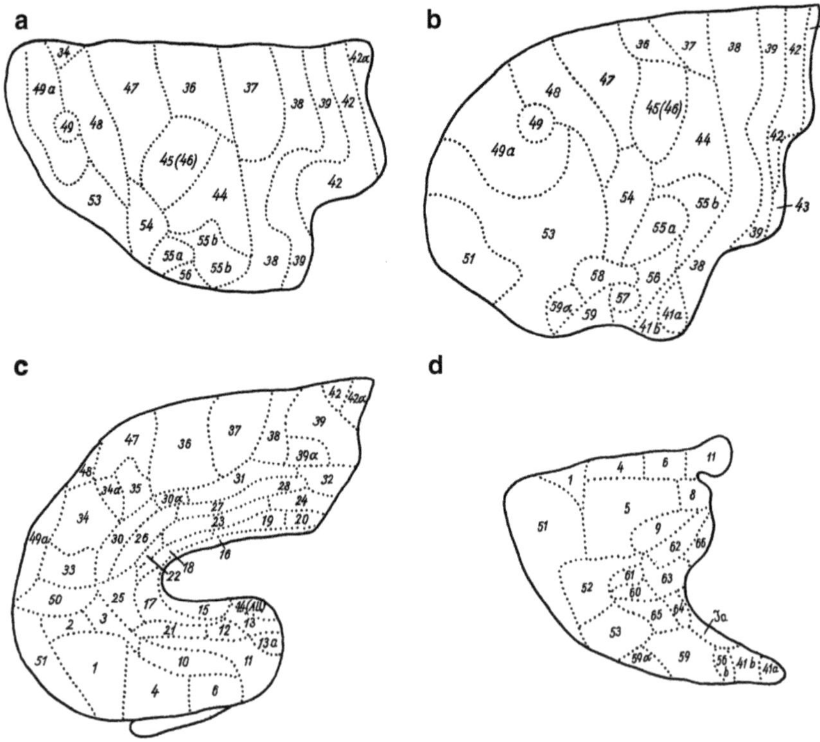

Fig. 3.18 Dorsal (**a**), lateral (**b**), medial (**c**), and basal views (**d**) of the human frontal lobe, showing the myeloarchitectonic parcellation of the frontal cortex, according to Hopf (1956). The numbers correspond to those of O. Vogt (1910a; Fig. 3.9)

according to Heidenhain-Woelcke, for myelinated fibres. Sections of other brains, including A43 and A63, were used for comparison. Sanides' purpose was twofold: (1) To carry out a combined cytoarchitectectonic and myeloarchitectonic analysis of the frontal cortex, hence his choice of material, and (2) to work out a concept, briefly mentioned by C. and O. Vogt (1919, p. 396), according to which the architecture of the cortex shows 'gradations', i. e. discontinuous, stepwise changes of architectonic features.

The results of Sanides' architectonic analysis are shown in Fig. 3.22. He distinguished eight different zones, which were designated as the frontomotor (FmZ), frontopercular (FoZ), frontopolar (FpZ), orbitomesial (OmZ), dorsal paralimbic (PlZd), ventral paralimbic (PlZv), paramotor (PmZ), and paropercular zones (PoZ). These zones are outlined in red in Fig. 3.22. It is important to note that Sanides excluded a considerable portion of the medial surface of the frontal lobe from his analysis (Fig. 3.22d: Pro=proisocortex). This portion corresponds to region II of O. Vogt (1910a; Fig. 3.9c), which encompasses the areas 14 to 33 of that author, and the rostral part of O. Vogt's region III, containing areas 34 and 35. Thus, the portion of the frontal lobe, investigated by Sanides, is in O. Vogt's

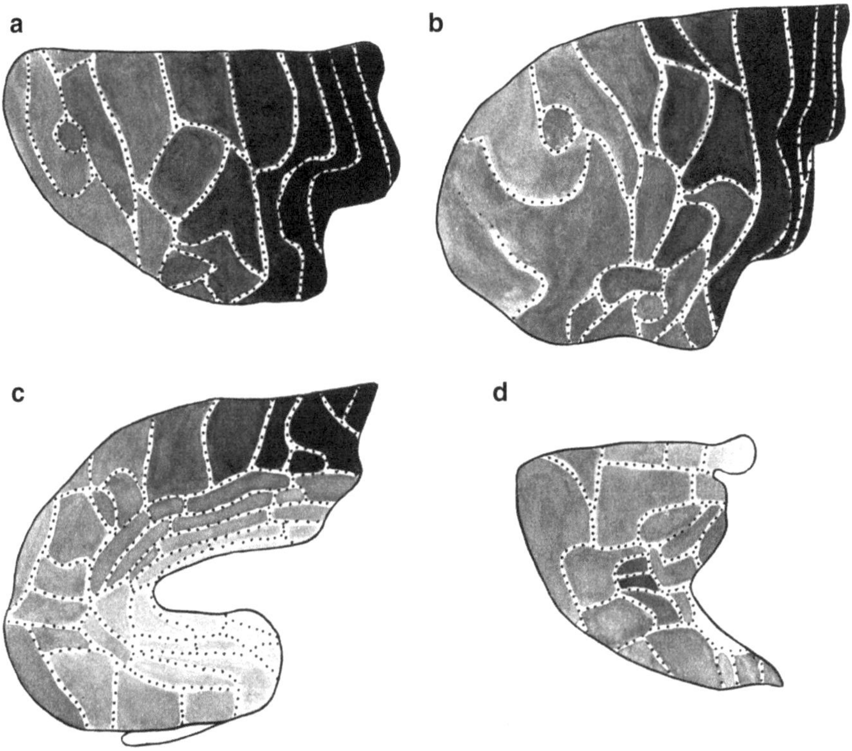

Fig. 3.19 Hopf's parcellation of the human frontal cortex (see Fig. 3.18), used as a matrix for indicating differences in the overall fibre density in the different regions of that cortex (Reproduced from Hopf 1956)

parcellation occupied by 44 (66 minus 22) areas. Sanides delineated a considerable number of separate areas within each of his eight zones, their total number amounting to 62. He established that 35 of these areas are directly comparable to one of the 44 Vogt-areas, and hence, designated them with the same numbers. Four of the Vogt-areas, viz. 51, 52, 53, and 59, were divided into separate dorsal and ventral parts (Fig. 3.22b). Sanides also observed that in many places separate fields are located between adjacent Vogt-areas. These intercalated areas, eight in number, were indicated with the numbers of the two areas involved, separated by a slash (see, *e. g.* 39/40 and 40/47 in Fig. 3.22a). Moreover, he delineated numerous new areas, which he specified by adding letters to the number of an adjacent area (see, *e. g.* 38l, 51p and 39z in Fig. 3.22b). Sanides failed, just like Strasburger and Hopf, to identify Vogt's area 7, and finally, he disagreed with Vogt's delineation of areas 3, 10 and 11, replacing them collectively by two concentric paralimbic areas, Pvl and Pvz (Fig. 3.22b). Sanides emphasized that all of the areas distinguished by him have an equal status, and that the boundaries of all of these areas coincide with changes in both cytoarchitecture and myeloarchitecture. He also emphasized that in

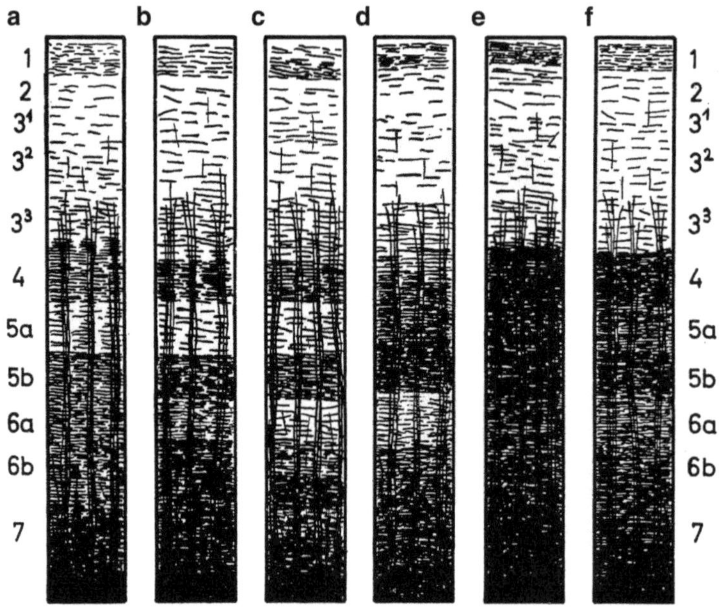

Fig. 3.20 Semidiagrammatic representation of the myeloarchitectonic types of cortex, found in the human frontal lobe. (**a**) Unistrate type, in which only the outer stripe of Baillarger (in layer 4) is clearly visible. (**b**) Propeunisstriate (= nearly unistriate) type, in which the inner stripe of B. (in layer 5b) is somewhat denser than layer 6a. (**c**) Bistriate type, in which both stripes of B. are well demarcated. (**d**) Unitostriate type, in which the two stripes of B. are united by a dense fibre plexus in layer 5a. (**e**) Astriate type, showing a homogeneous fibre density throughout layers 4–7. (**f**) Propeastriate (= nearly astriate) type, in which the stripes of B. are inconspicuous (Reproduced from Hopf 1968a)

the frontal lobe there is a close correlation between structural differentiation and gyrification. According to his observations, most zonal and areal boundaries are located in the depths of the intergyral sulci.

The presence of 'gradations' in the frontal lobes (once again: streams or chains of structurally related, but discrete areas) could be confirmed (see the arrows in Fig. 3.22). Sanides distinguished three of such gradations in the frontal cortex and surmised that they reflect the directions of the evolutionary expansion of that cortex. It is noteworthy that the existence of such gradations has been recently reconfirmed in a study of Broca's region (Amunts and Zilles 2012).

From the foregoing it appears that four authors, O. Vogt, Strasburger, Hopf and Sanides have carried out detailed analyses of the myeloarchitecture of the human frontal cortex. The differences between their results appear to be limited. If we introduce a provisional rule, saying that the presence of a particular area in the frontal cortex is sufficiently secured, if it has been identified by at least three of the four authors involved, it appears that the presence of no less than 61 of the 66 areas, originally delineated by O. Vogt, has been confirmed by later investigations. There

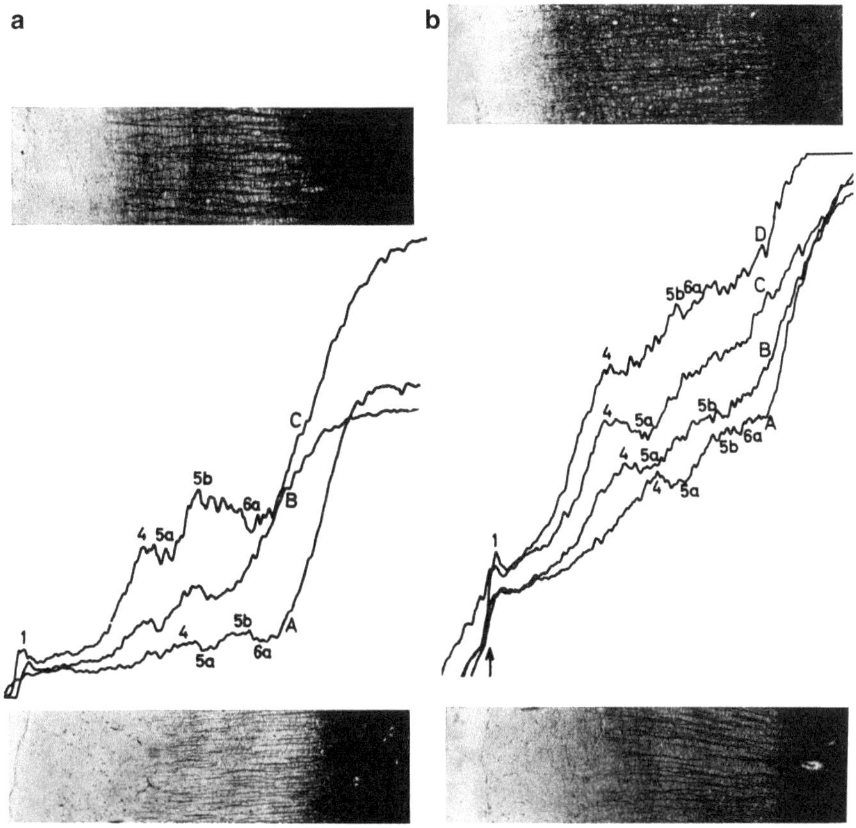

Fig. 3.21 Photometric recordings of the fibre density in bistriate (**a**), and unistriate areas (**b**) of the human frontal cortex. The recordings in (**a**) are taken from area 59, shown in the lower photograph (*curve A*), and from area 56b, shown in the upper photograph (*curve B*). The recordings in (**b**) are taken from area 35, shown in the lower photograph (*curve A*), area 37 (*curve B*), area 38 (*curve C*), and area 39, shown in the upper photograph (*curve D*). Note that there is a clear increase in fibre content from *A* to *D* (Reproduced from Hopf 1968a)

appeared to be no support for the presence of Vogt's areas 7, 29 and 40, and for his delineation of area 45 from area 46. Accepting the suggestion of Hopf, that several of the areas distinguished by Strasburger represent inconspicuous and inconstant variants, the total number of myeloarchitectonic areas, present in the frontal lobe may be estimated at about 70. It is important to note that this outcome differs considerably from those of the myeloarchitectonic pioneers Campbell (1905) and Elliot Smith (1907). The former distinguished only five, the latter 17 areas in the human frontal cortex.

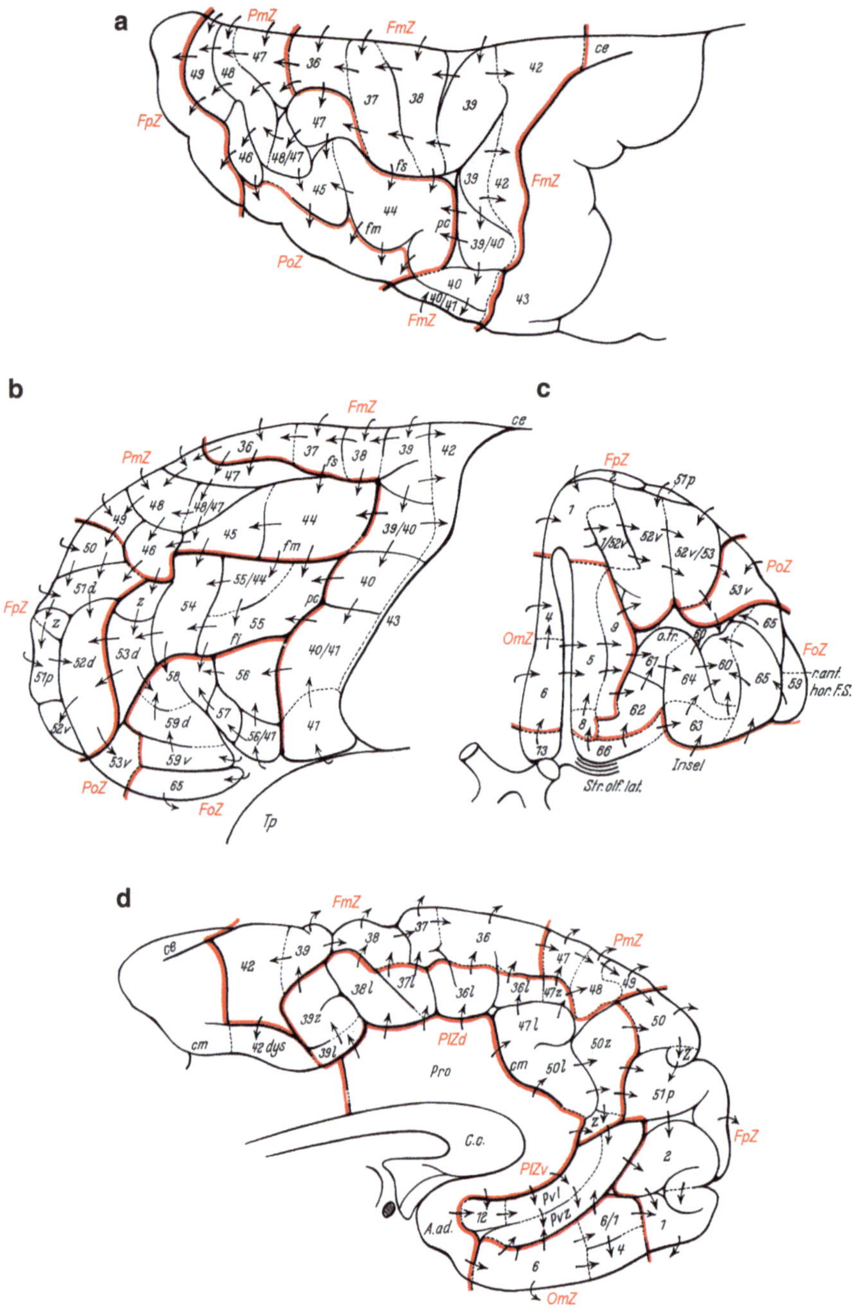

Fig. 3.22 Dorsal (**a**), lateral (**b**), basal (**c**), and medial views (**d**) of the human frontal lobe, showing the cyto-myeloarchitectonic parcellation of the frontal cortex, according to Sanides (1962). The areal boundaries, which do not coincide with sulci, are indicated with dotted lines.

3.4.2 The Insula

The human insular cortex forms a distinct, but entirely hidden lobe, situated in the depth of the Sylvian fissure. The insula is shaped like a triangle, the apex of which is directed basally. A distinct sulcus centralis insulae divides the insula into a larger lobulus anterior and a smaller lobulus posterior. The lobulus anterior is commonly composed of three short gyri, the gyrus brevis primus, -secundus, and –tertius (or -centalis anterior), which converge toward the apex. The lobulus posterior is generally incompletely divided into two gyri, known as the gyrus longus primus (or –centralis posterior) and the gyrus longus secundus. These gross anatomical relations are clearly visible in Fig. 3.23, although not all of the structures mentioned are labeled in this figure.

The myeloarchitecture of the human insular cortex has been studied by C and O. Vogt (1911) and by Brockhaus (1940). The preliminary study of the Vogts was based on a Weigert-Pal series of a single hemisphere, A 18l. They divided the insular cortex into a ventral allocortical zone, and a dorsal isocortical zone. Within the latter, they distinguished six, rostrocaudally arranged areas, which they designated as i1-i6. The bounderies of most of these areas appeared to coincide exactly with the sulci separating the various insular gyri (Fig. 3.23a). This is remarkable, because O. Vogt (1903, 1910a, 1923) has repeatedly emphasized that the relations between the cerebral sulci and the areal boundaries, are variable and inconstant.

The very detailed, combined cytoarchitectonic and myeloarchitectonic study of Brockhaus (1940), another collaborator of the Vogts in the Neustadt institute, was based on six different brains, A 18, A 39, A 40, A 61, A 65, and A66. Brockhaus regarded the presence of the claustrum, a thin sheet of grey matter, situated between the insula and the putamen, as the defining structural feature of the insula, hence he designated the cortex covering this region as claustrocortex. He distinguished three ventrodorsally arranged principal regions within the insula, which he designated as allocortex claustralis (Acl), mesocortex claustralis (Mcl), and isocortex claustralis (Icl), thus intercalating a transitonal zone between the two zones of the Vogts (Fig. 3.23b). So far as the myeloarchitecture of the neocortical (or isocortical) zone of the insula is concerned, Brockhaus' observations tallied with those of the

Fig. 3.22 (continued) The boundaries of zones are in red, just as the abbreviations of the names of these zones. The numbers indicating the areas correspond to those of O. Vogt (1910a). The *arrows* indicate the direction of the 'gradations', discussed in the text. *ce*, central sulcus; *cm*, callosomarginal sulcus; *fs*, superior frontal sulcus; *fm*, middle frontal sulcus; *fi*, inferior frontal sulcus; *o. tr*, transverse orbital sulcus; *pc*, precentral sulcus; *r. ant. hor. F. S.*, anterior horizontal ramus of fissura Sylvii; *Tp*, temporal pole

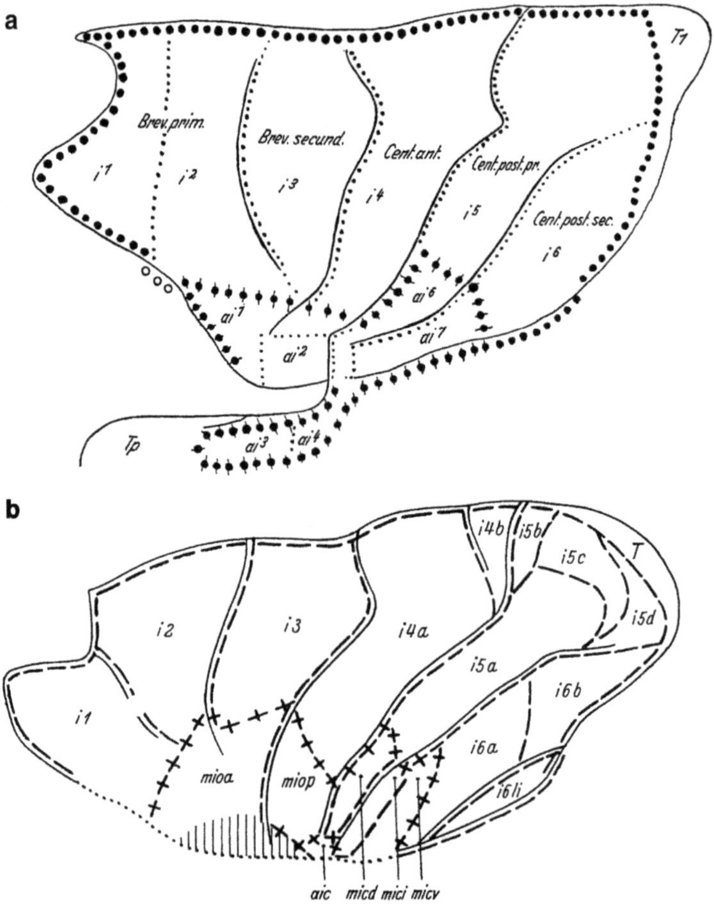

Fig. 3.23 Lateral views of the left human insula. (**a**) Myeloarchitectonic parcellaton according to C. and O. Vogt (1911). (**b**) Cyto- and myelorchitectonic parcellation according to Brockhaus (1940). For explanation, see text. *Brev. prim.*, gyrus brevis primus; *Brev. sec.*, gyrus brevis secundus; *Cent. ant.*, gyrus centralis anterior (=gyrus brevis tertius); *Cent. post. prim.*, gyrus centralis posterior primus (=gyrus longus primus); *Cent. post. sec.*, gyrus centralis posterior secundus (=gyrus longus secundus)

Vogts, with the reservation that he felt justified to subdivide areas i4a, i5a and i6a, into several smaller entities (Fig. 3.23b). All in all, he distinguished 12 neocortical insular areas, within the boundaries of which, there was a complete match of cytoarchitecture and myeloachitecture. Given the fact that the surface of the insular lobe takes up less than 2 % of the total cortical surface area, we consider it likely that the subdivisions of i4a, i5a and i6a, introduced by Brockhaus, represent subareas, rather than areas.

3.4.3 The Parietal Lobe

Contributions to our knowledge of the myeloarchitecture of the human parietal cortex, were made by O. Vogt (1911), Gerhardt (1940), Batsch (1956), Hopf and Vitzthum (1957), and Hopf (1969, 1970b). O. Vogt's study on the myeloarchitecture of the parietal cortex is again preliminary in character, and ends, just as the one on the frontal cortex, abruptly after the description of the last area (cf. Fig. 3.10). It is, however, contrary to that paper, well illustrated with beautiful drawings (Fig. 3.25). The myeloarchitecture of the parietal cortex presented O. Vogt with several problems, which he solved by extending his terminology. Thus, he introduced the terms *eucingulate* and *dyscingulate*. In *eucingulate* areas, layer 2 is, by a paucity of its constituent fibres, sharply demarcated from layer 3 (cf. Fig. 3.3, right panel); in *dyscingulate* areas, this sharp boundary does not exist. O. Vogt's subdivision of the parietal cortex, which was based on the study of three hemispheres, A 18r, A 20l, and A 20r, is even more complex than that of the frontal cortex. Suffice it to mentioning that he distinguished two principal regions within this cortex, which he designated as the euradiate region and the supraradiate region, indicated as VIII and VII, respectively, in Fig. 3.24. The euradiate region was subdivided into eucingulate (VIIIα) and dyscingulate subregions (VIIIβ). Reference to Fig. 3.24 shows that all of these (sub)regions were subdivided into several areas, indicated with Arabic numbers. The numbering of these areas links up with that, used in the frontal cortex (Fig. 3.9). It should be emphasized once again that these numbers have nothing to do with those, used by Brodmann for his cytoarchitectonic areas. All in all, O. Vogt delineated 30 myeloarchitectonic areas in the parietal cortex, numbered 67–96, nine of which (67–75) were situated in subregion VIIIα, fifteen (76–90) in subregion VIIIβ, and six (91–96) in region VII (Fig. 3.24). The myeloarchitecture of some of these areas is shown in Fig. 3.25.

The analysis of the human parietal cortex by Edith Gerhardt (1940), who worked at the KWIH in Berlin-Buch, is primarily cytoarchitectonic in character. Gerhardt indicates, however to have studied several series of which the sections were stained alternately for cell bodies (Nissl) and myelinated fibres (Heidenhain), and emphasizes on this account that the resultant map (Fig. 3.26) is cytoarchitectonic as well as myeloarchitectonic. This map was based on the analysis of a single hemisphere, A 61l. She took O. Vogt's (1911) subdivision as point of departure, but had to deviate at several points from it, for the simple reason that her parcellation was, as she put it, more thorough ('eingehender') than that of Vogt. Thus, she subdivided many Vogt-areas into two or more subareas ('Unterfelder'). This holds in particular for the large areas 83, 85, 89 and 90. Within the area last mentioned, Gerhardt delineated no less than seven subareas, designated as 90ai, 90aip, 90am, 90p, 90t1, 90t1' and 90t1'o (Fig. 3.26).

It is noteworthy, that Gerhardt (1938) also studied the architecture of the parietal cortex in the chimpanzee. She mentioned that this cortex shows a striking resemblance to that in the human, and that only a few of the areas, delineated in the human parietal cortex by Vogt (1911), could not be identified with certainty in the chimpanzee.

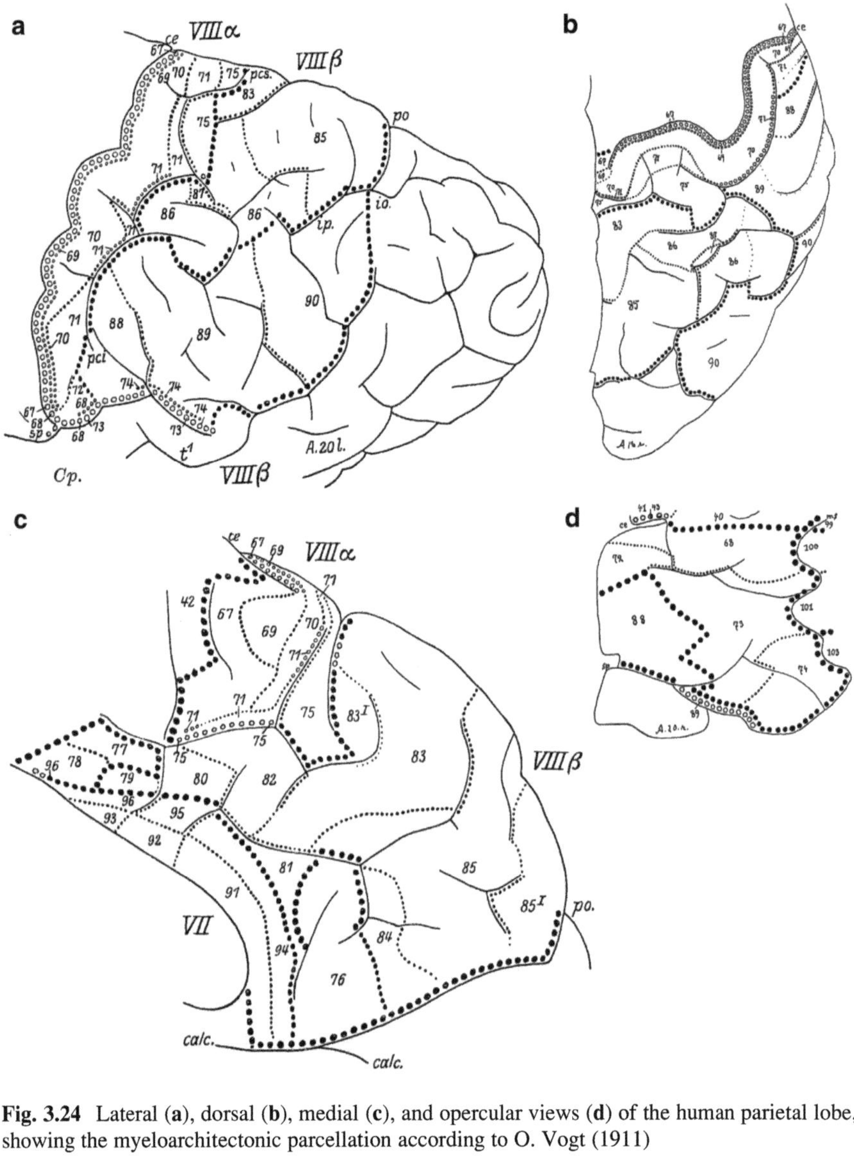

Fig. 3.24 Lateral (**a**), dorsal (**b**), medial (**c**), and opercular views (**d**) of the human parietal lobe, showing the myeloarchitectonic parcellation according to O. Vogt (1911)

The aim of the study of Batsch (1956) was to investigate whether the very detailed parcellation of the parietal cortex, presented by Gerhardt, which was based on the study of a single hemisphere, is applicable to other brains as well. To this end, Batsch, who worked at the Neustadt Institute, studied, apart from hemispheres A 37l and A 61l, 'mehrere Hemisphären', without further specification. Batsch subdivided the parietal cortex into the following eight subregions: (a) subregio postcentralis, (b) s. opercularis, (c) s. parietalis inferior, (d) s. parietalis

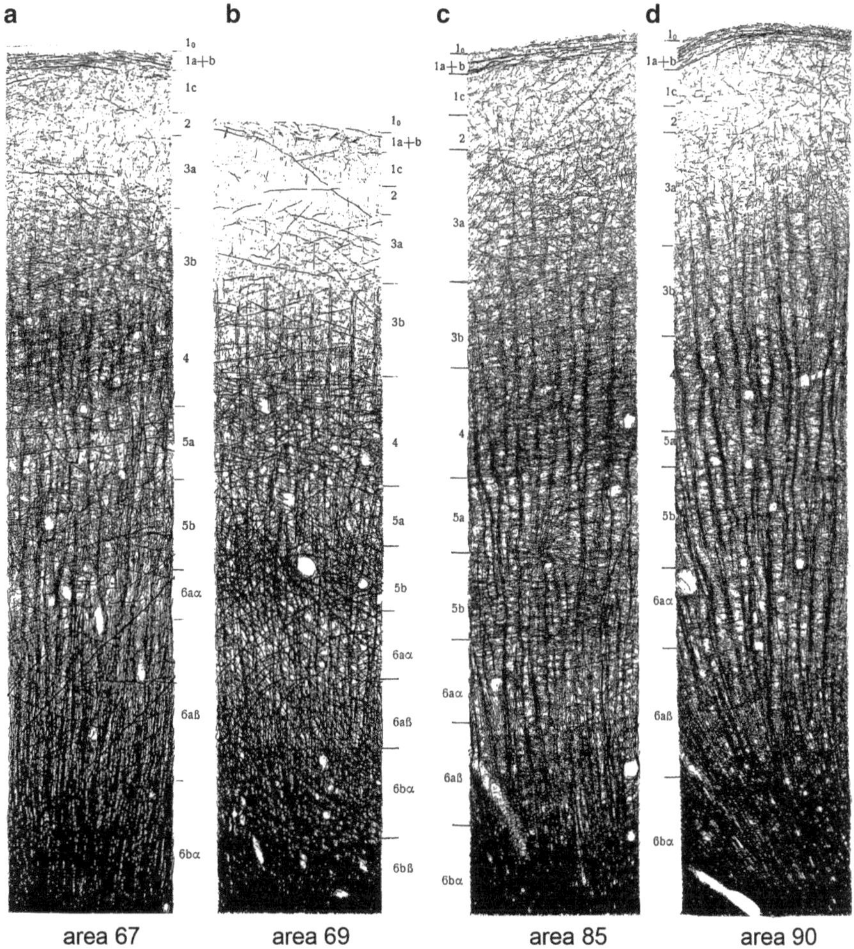

Fig. 3.25 Drawings showing the myeloarchitecture of some parietal cortical areas (Reproduced from O. Vogt 1911)

intermedia, (e) s. parietalis superior-medialis, (f) s. parietalis paracingularis oralis, (g) s. parietalis cingularis caudalis, and (h) s. parietalis cingularis. The basic myeloarchitecture of these subregions is shown in Fig. 3.27. Batsch subdivided all of these eight subregions into a (varying) number of areas. All of these areas corresponded in a one-to-one fashion to the areas 67–96 of O. Vogt, and were numbered accordingly. Sixteen of the 'Vogt-areas' remained undivided. Most of these are located in the parietal paracingular oral (77–82), and parietal cingular subregions (91–96). The remaining 14 'Vogt-areas' were further subdivided into subareas. Thus, area 67 was subdivided into subareas I-IV, area 75 into subareas if and sup, and area 89 into subareas a, ip, m, p and t. In total, Batsch delineated within the human parietal cortex, 45 subareas. The study of Batsch is well documented with

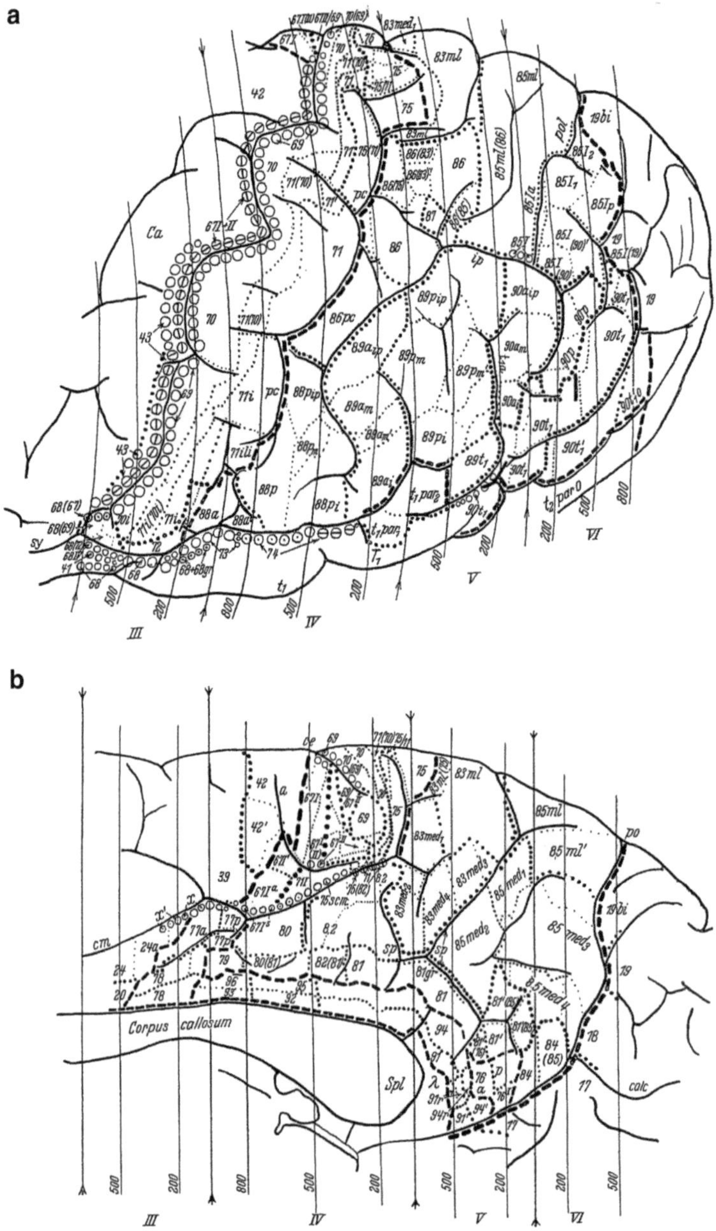

Fig. 3.26 Lateral (**a**), and medial views (**b**), of the human parietal lobe, showing the cyto-myeloarchitectonic parcellation according to Gerhardt (1940)

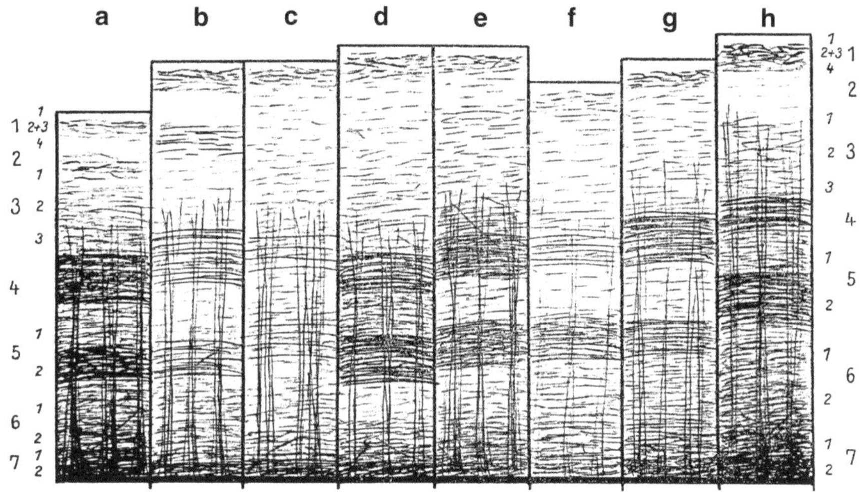

Fig. 3.27 Semidiagrammatic representation of the myeloarchitecture in the various subregions (**a–h**) of the parietal cortex. The names of these subregions are mentioned in the text (Reproduced from Batsch 1956)

photomicrographs, showing the myeloarchitecture of all of the areas and subareas distinguished.

Batsch indicates that his division of the parietal lobe into subregions differs from that of Gerhardt, and that his findings concerning the size and extent of some of the 'Vogt areas', including 71, 73 and 74, also deviate from those of Gerhardt. In general, it may be said that the parcellations of Batsch and Gerhardt correspond, at the area level, closely to each other, and to that of O. Vogt as well. However, although the total number of subareas distinguished by the two authors first mentioned is about the same, there appeared to be no close one-to-one correspondence at this level.

The last publications on the myeloarchitecture of the parietal cortex, to be discussed here, are those of Hopf and Vitzthum (1957) and Hopf (1969, 1970). Hopf and Vitzthum (1957) visualized the distribution of the various myeloarchitectonic characteristics over the parietal lobe in a series of 'feature maps', whereas Hopf (1969, 1970) registered the relative content of myelinated fibres in the various cortical layers, in each of the postcentral and parietal areas. The 'feature maps' were based on slightly modified versions of the (very complex) cortical maps of Batsch (1956). These 'modified-Batsch-maps' are reproduced in Fig. 3.28. The only differences with the originals are that the dorsal extent of subareas 70^{I} and 7^{II} is reduced, and that the anterior surface of the postcentral gyrus is folded anteriorly. By this transformation, areas 67 and 69, which are normally hidden in the central sulcus, are exposed. The studies of Hopf and Vitzthum (1957) and Hopf (1969, 1970) showed that the most characteristic myeloarchitectonic featues in the parietal lobe include the overall fibre density,

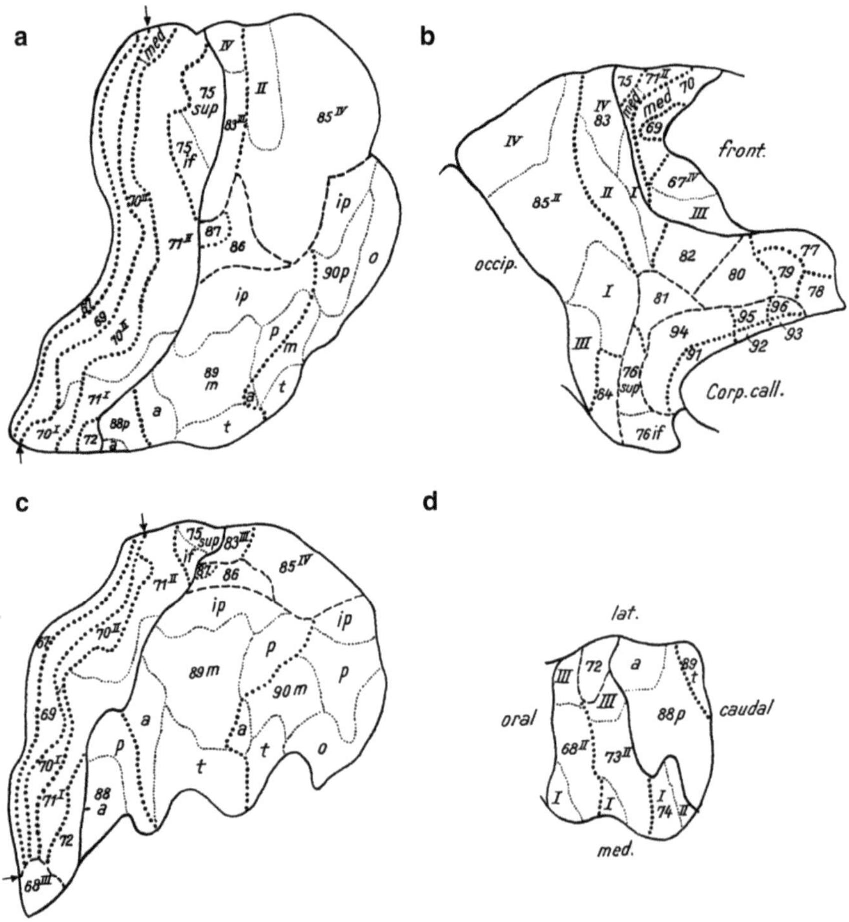

Fig. 3.28 Dorsolateral (**a**), medial (**b**), lateral (**c**), and opercular views (**d**), of the human parietal lobes, showing the myeloarchitectonic parcellation, according to Batsch (1956), slightly modified by Hopf and Vitzthum (1957). The *arrows* in (**a**) and (**c**) indicate the locations where areas 67 and 69 have been folded anteriorly in order to expose them

the strength of the radial bundles, and the relation of the stripes of Baillarger to each other and to their neighbouring layers. The highest overall fibre content was found in the postcentral region, the lowest in the inferior parietal lobule. Hopf (1969, 1970), who included eight different hemispheres in his registrations of the relative laminar fibre content in the parietal lobe, found only minor interindividual variations. It is of note that the paper of Hopf and Vitzthum (1957, pp. 98–102) contains a very useful comparison of the various subdivisions of the parietal cortex in tabular form.

3.4.4 The Occipital Lobe

As is well known, Brodmann (1909) divided the human occipital lobe into three, concentrically arranged cytoarchitectonic areas, the areae striata (17), – occipitalis (18), and –praeoccipitalis (19). These three areas have also been recognized by C. and O. Vogt (1919), in their preliminary reconnaissance of the myeloarchitecture of the human cortex. The boundary between the striate area and the occipital area is, cytoarchitectonically, as well as myeloarchitectonically, by far the most distinct one in the entire neocortex. Reference to Fig. 3.29a, b shows that this boundary is marked by the sudden transition of cytoarchitectonic sublayers IVa-c in area 17, into the undivided layer IV in area 18, as well as by the equally sudden ending of the outer stripe of Baillarger (=line of Gennari; = line of Vicq d'Azyr). This striking myeloarchitectonic difference between the striate and occipital areas is also clearly seen in the more detailed pictures shown in Fig. 3.29c, d.

The only detailed myeloarchitectonic study on the human occipital cortex is that of Lungwitz (1937), who worked at the KWIH in Berlin-Buch. His study, which was based on sections through both hemispheres of brain A 37, stained according to Weigert-Kulschitzky, was confined to the preoccipital area, which he designated, following O. Vogt, as PrO. Lungwitz characterized his study as a "myeloarchitectonische Unterfelderung" of that area. He delineated 17 subareas within PrO, which he designated with combinations of two-to-four letters (Fig. 3.30). Although it is indicated in the text that these letter combinations refer to conceptualizations of O. Vogt, their significance is not fully clear (at least not to the present reviewer). All of the subfields are described in a peculiar sort of 'myelo-shorthand', and documented with detailed drawings (The pictures of three subareas, scet, sct and scd, are shown in Fig. 3.29e, f). In order to give the reader some idea of the style of the paper under discussion, the description of one subarea, scd, is shown in full in Fig. 3.31.

At the end of his paper, Lungwitz puts forward the question whether PrO has to be considered a Regio, that is, a complex of areas, or rather as an area with numerous subareas. He then emphasizes that this question can only be adequately answered in light of knowledge concerning the physiological differentiation of the entity in question, and he adds that this knowledge is currently lacking: "Darüber wissen wir jedoch zur Zeit nichts sicheres" (Lungwitz 1937, p. 638). In this respect, the situation has changed considerabbly in the mean time. Physiological studies have shown that the preoccipital region harbours a large number of functional areas (see: Wandell et al. 2007). Taking these new data into consideration, it seems reasonable to assume that the myeloarchitectonic entities delineated by Lungwitz, represent areas, rather than subareas.

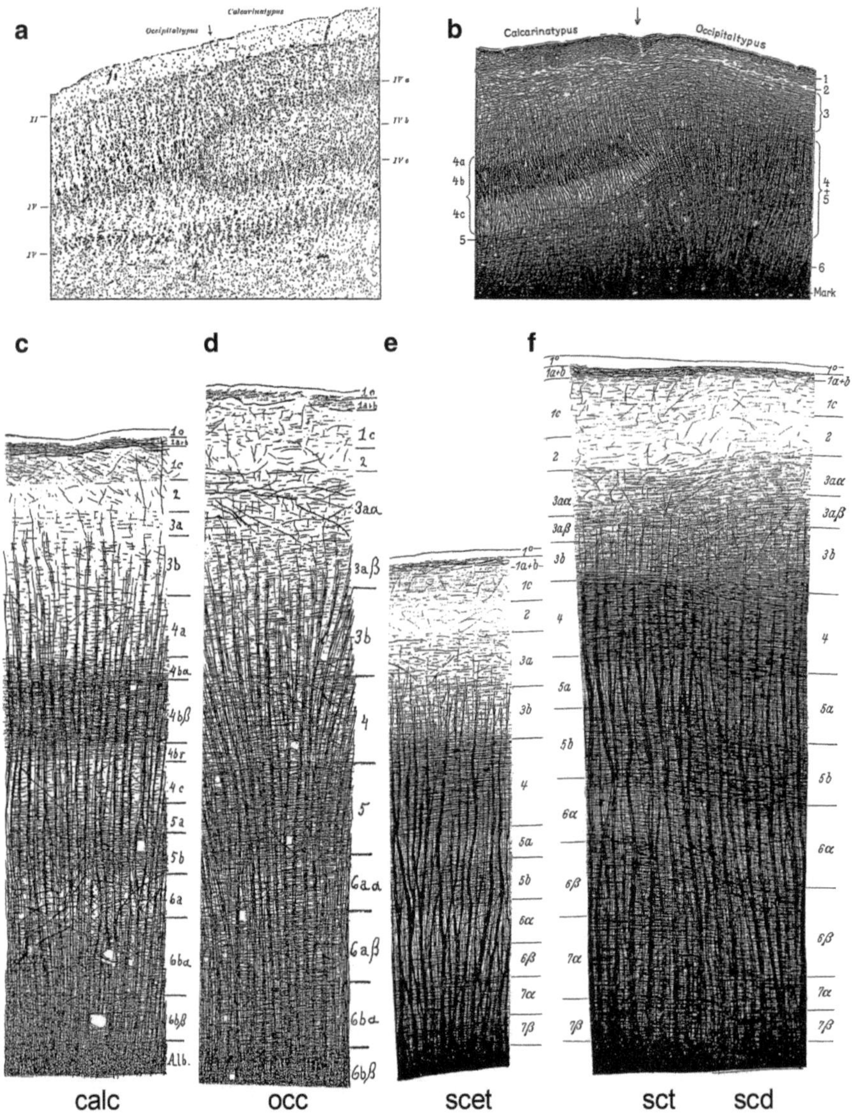

Fig. 3.29 The architecture of the human occipital cortex. Cytoarchitecture (**a**), and myeloarchitecture (**b**), of the transition of the calcarine area into the occipital area. Note that (**a**) and (**b**) are each other's mirror image (Reproduced from Brodmann (1914)). (**c**, **d**) Detailed pictures of the myeloarchitecture of the calcarine and occipital areas, reproduced from C. and O. Vogt (1919). (**e**, **f**). Detailed pictures of the myeloarchitecture of subareas scet, sct and scd, situated within the preoccipital area (Reproduced from Lungwitz 1937)

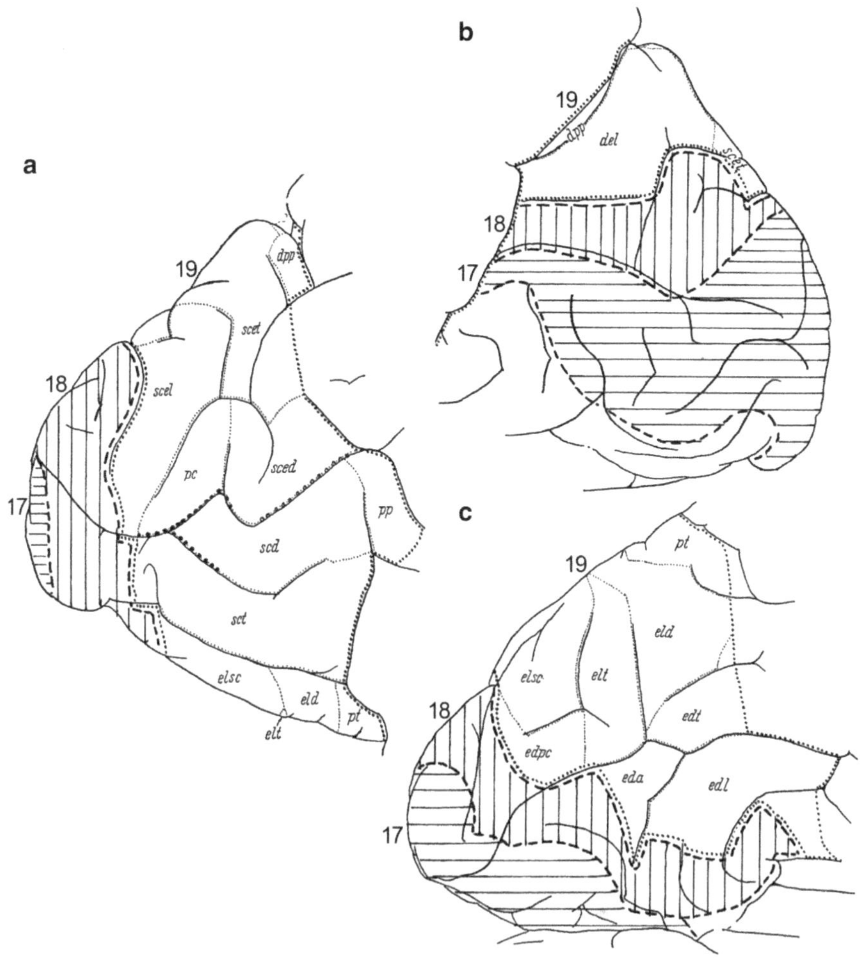

Fig. 3.30 Lateral (**a**), medial (**b**), and basal views (**c**), of the human occipital lobe, showing the myeloarchitectonic parcellation of the preoccipital area (=area 19 of Brodmann), according to Lungwitz (1937). *17*, *18*, *19*, areas of Brodmann

3.4.5 The Temporal Lobe

The myeloarchitecture of the human temporal lobe, has been studied by two authors, Beck (1925, 1928, 1930), and Hopf (1954a, 1955, 1968b).

The very detailed studies of Beck are confined to the dorsal surface of the temporal lobe. His first paper (Beck 1925) is entitled (in translation): "On the exactness of the myeloarchitectonic parcellation of the cerebral cortex". In this paper he describes the results of an analysis of the dorsal temporal cortex, based on sections of a single hemisphere, A 27l, stained according to Weigert-Kultschitzky.

scd

1^0 faserlos.

1a + b schmal. Heller als in **sct.** Feiner Ff. Sehr feine Tf. Gut gegen *1c* abgegrenzt.

1c gering breiter als *1a + b*. Heller als in **sct.** Keine Hf.

2 sehr hell. Faserärmer als in **sct.**

3a α feinste Ef. Ff gleich dicht *1c*. Gleich breit *1c*.

3a β kaum dunkler als *3a α*. Annähernd gleich breit. Pinselförmige Aufsplitterung der Radii.

3b dunklerer Ff.

4 annähernd gleich breit *5a + b*. Dichter Ff. Gut abgegrenzt. Vereinzelt Sf. Keine Hf. Sf von wenig starkem Kaliber.

5a breiter als in **sct.** Hebt sich besser ab als dort.

5b nur gering heller als *4*. Annähernd gleich breit *5a*. Einige Hf stärkeren Kalibers.

6 α und *6 β* heller als *5a*. Breiter als **sct.**

7 α dunkler. Gut abgegrenzt.

Radii gebündelt bis in *3a β*. Ziemlich dicht stehend. Bündel von wechselnder Breite. Radiäre Ef.

Trizonal — eufasciär — stark tenuifasciär — schwach multo-, gering crassofibrös. Eutangential — schwach eucingulär. Schwach trisingulofibrös — ausgesprochen separatus. Schwach grossofibrös — externodensior — externolatior. Subconjunctostriär — tenuiintrastriär. Intimolatior — tenuilimitatus. Eu-, medio-, ziemlich denso-, mixtoradiär.

Fig. 3.31 Description of subarea scd in Lungwitz (1937, p. 624)

This choice was intentional, because he knew that O. Vogt had previously studied the same region of the same brain, but had not published the results as yet. Beck was able to distinguish 28 sharply defined myeloarchitectonic areas in the region investigated (Fig. 3.32a). He also shows the results of O. Vogt's unpublished parcellation (Fig. 3.32b), emphasizing to have worked completely independently from the latter. Beck observed that there is a striking resemblance between the results of the two studies. He points out that his map does not only contain all of the areas distinguished by O. Vogt, but that these areas also correspond in position and size, to such an extent (Beck 1925, p. 285): "daß man fast von einer mathematischen Exaktheit sprechen könnte." And he continued: "Bei solchen Resultaten muß jeder Zweifel (on the reliability of the myeloarchitectonic method, R. N.) verstummen." The fact that his map contained more areas than that of O. Vogt (28 versus 20) is, according to Beck, due to the fact that he divided some of Vogt's areas into subareas.

In two subsequent papers, published in1928 and 1930, Beck presents a detailed myeloachitectonic analysis of the same region, reportedly based on no less than 24 (non-specified) hemispheres, stained according to Kultschitzky-Wolters. The study is well documented with numerous photomicrographs, a part of one of which is reproduced in Fig. 3.34b. Beck found that the medial part of the dorsal temporal lobe contains several allocortical regions, including the Regio temporalis insularis (ti), R. praepiriformis (prpy), R. entorhinalis (e), and R. periamygdalaris (Pam). The remaining, neocortical superior temporal cortex appeared to be divisible into six

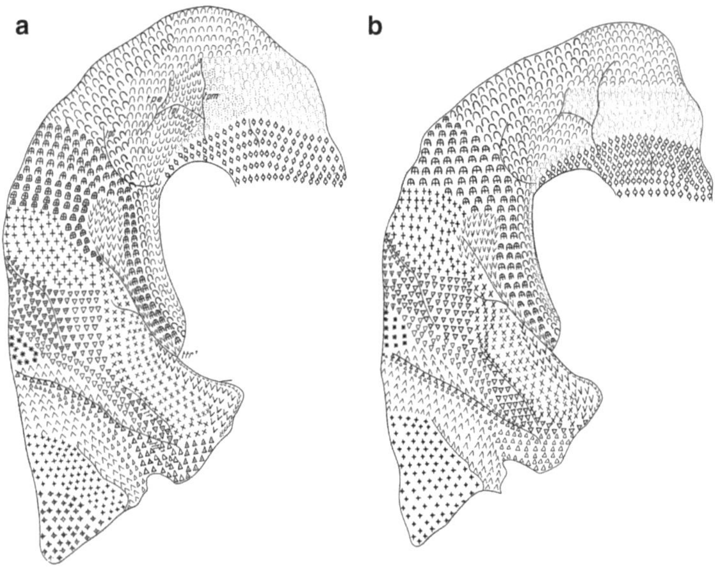

Fig. 3.32 Myeloarchitectonic parcellation of the cortex covering the dorsal surface of the human temporal lobe. Both maps are based on serial sections of one and the same hemisphere (A 271) and prepared independently by Beck and O. Vogt (see text) (Reproduced from Beck 1925)

subregions (Fig. 3.33a), each of which containing a number of parts, which in their turn could be further divided into a number of areas (Fig. 3.34b), as follows:

Subregio temporopolaris (tp), two parts (tpm, tpl), 17 areas
S. temporalis superior (ts), two parts (tsm, tsl), 12 areas
S. parainsularis (tpar), 3 areas
S. temporalis transversa prima (ttrI), five parts (ttrIin, ttrIi, ttrIe, ttrIex, ttrIl), 24 areas
S. temporalis transversa secunda (ttrII), two parts (ttrIIm, ttrIIl), 11 areas
S temporalis transversa tertia (ttrIII), two parts (ttrIIIm, ttrIIIl), 7 areas

So, Beck distinguished within the superior temporal neocortex, six subregions, with 13 parts, and 74 areas. If we add the 15 areas, delineated by him within the four allocortical regions analyzed (Fig. 3.34b), the total number of superior temporal areas distinguished by that author appears to amount to 89. This number is much higher than that found in his previous study (see above). Beck explains this difference from the fact that many of the 28 areas of his old scheme (Fig. 3.32a), appeared to be divisible on closer scrutiny into two or more subareas (Fig. 3.34a). He emphasized that the boundaries between all of the areas distinguished, were sharp and omnilaminar (*i. e.*, involved all of the seven layers of the cortex), and that all of the areas could be recognized in all of the hemispheres investigated.

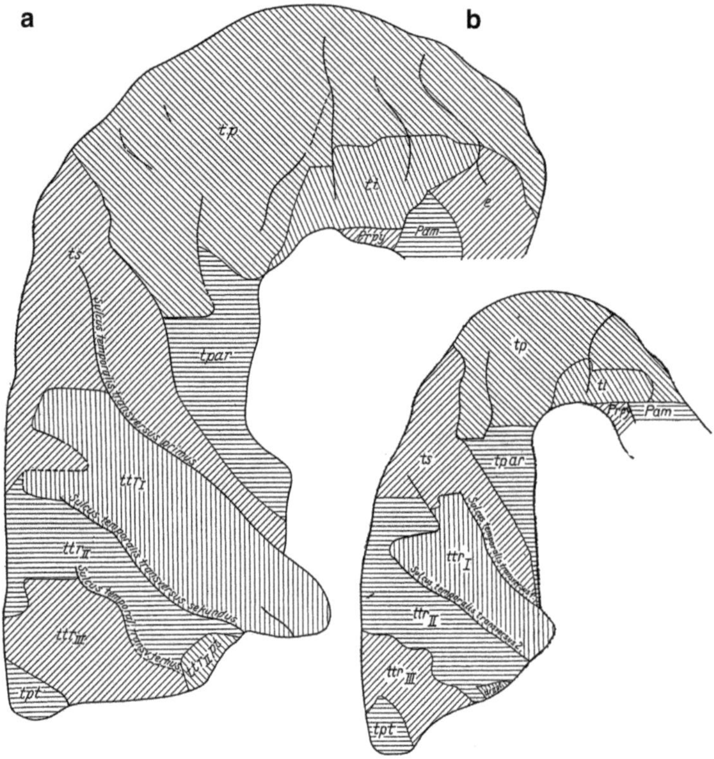

Fig. 3.33 Myeloarchitectonic parcellation of the cortex covering the dorsal surface of the temporal lobe into subregions, (**a**) in the human, according to Beck (1928), and (**b**) in the chimpanzee, according to Beck (1929)

At the end of his last paper on the subject, Beck (1930, p. 258) addresses the question, whether it makes sense to delimit such a large number of areas in a relatively small cortical region. He answers this question in the positive, for the simple reason that all of these areas are true anatomical entities. He concludes the paper by stating (p. 259): "Alles zusammengenommen glauben wir, daß wir eher zu wenig als zu viel Felder abgegrenzt haben".

It is noteworthy that Beck (1929) has devoted a separate study to the myeloarchitecture of the superior temporal cortex of the chimpanzee. A detailed discussion of this paper is beyond the scope of the present review. Suffice it to mention that Beck (1929, p. 406) found that the graphical reconstruction of this region of the chimpanzee and that of the human, show an astounding similarity ("eine verblüffende Ähnlichkeit"), both at the subregional (Fig. 3.33a, b), and at the areal levels (Fig. 3.34a, c). The only notable difference between these two species was that in

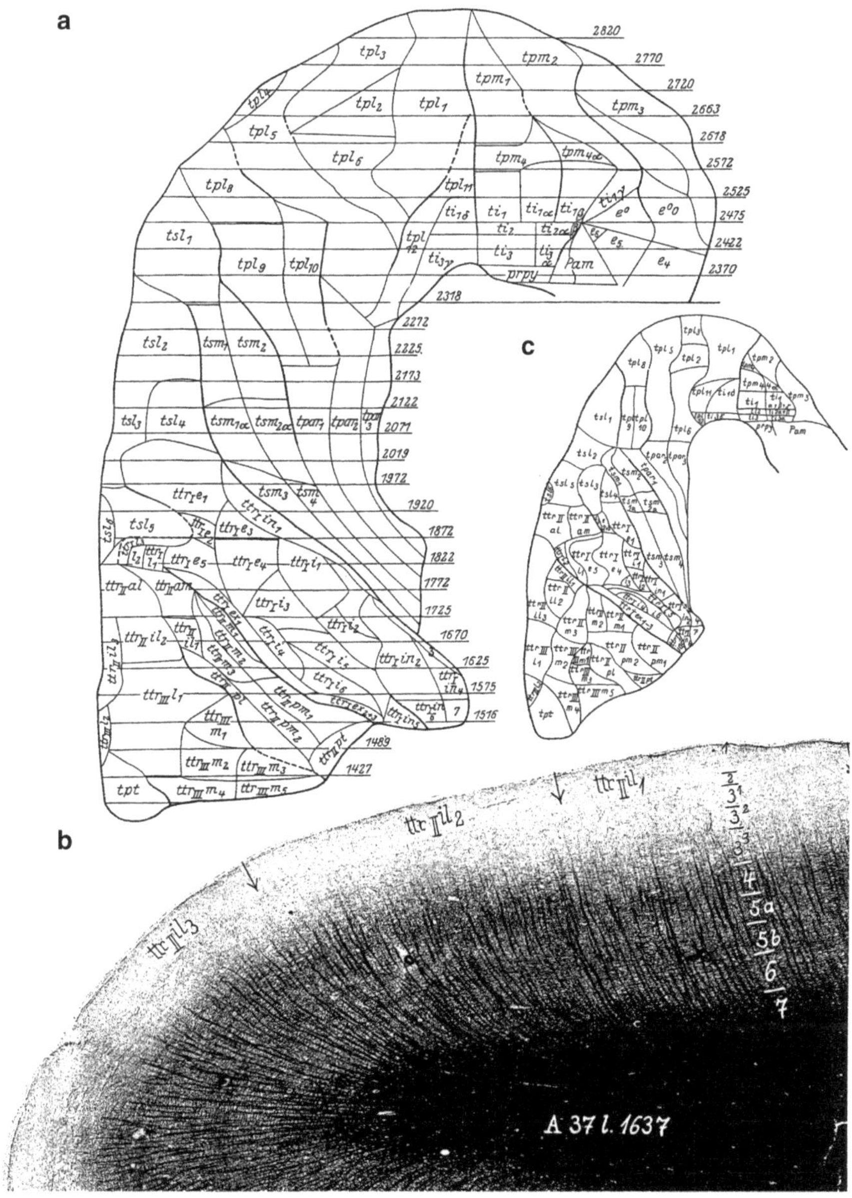

Fig. 3.34 The myeloarchitecture of the cortex covering the dorsal surface of the temporal lobe. (**a**) Parcellation in the human, according to Beck (1928), based on hemisphere A 37l. (**b**) Transverse section through the caudal part of the temporal lobe of the same human hemisphere, reproduced from Beck (1930). (**c**) Parcellation in the chimpanzee, according to Beck (1929)

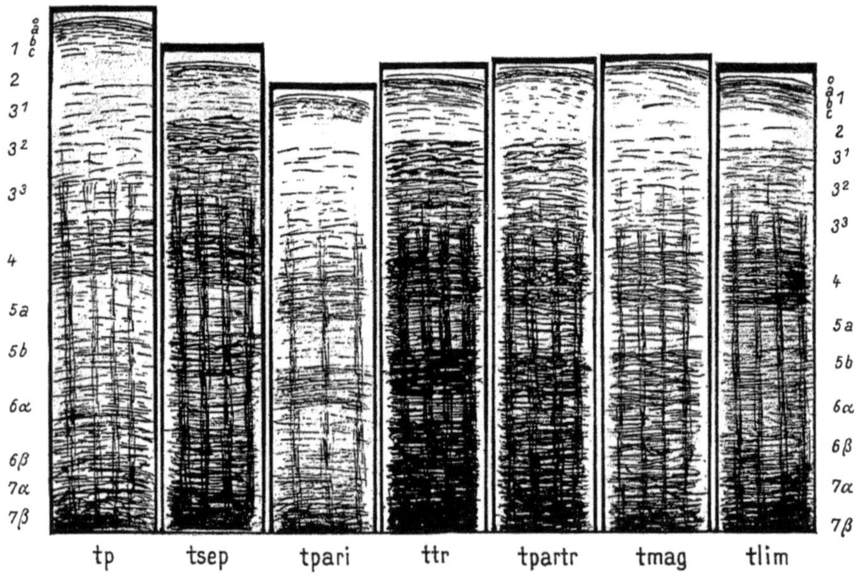

Fig. 3.35 Semidiagrammatic representation of the myeloarchitecture of the seven regions of the human temporal lobe. The abbreviations are explained in the text (Reproduced from Hopf 1954b)

the human temporal cortex the total content of myelinated fibres ("Markfasergehalt") was much higher than in the chimpanzee.

Hopf (1954a) has been the first, and so far only investigator, who subjected the entire human temporal cortex to a detailed myeloarchitectonic analysis. His study is reportedly based on serial sections of several brains, stained according to Weigert-Wolters. The graphical reconstructions, illustrating his results, were all based on a single brain, specified as MB 59.

Hopf divided the temporal cortex into seven regions, each of which showing a typical and characteristic myeloarchitecture (Fig. 3.35). Most of these regions appeared to be subdivisible into two or more subregions, and within each of these subregions a number of areas could be delineated. All of the entities distinguished received full Latin names. The names of the seven regions are presented below. The names of the subregions belonging to the first region will also be mentioned. As regards the areas, we confine ourselves to mentioning the names of those lying within the first subregion.

1. Regio temporopolaris (tp), four subregions, 13 areas

   Subregio medialis (tp.m)

   Area medialis interna (tp.m.i)
   Area medialis externa (tp.m.e)
   Area medialis posterior (tp.m.p)
   Area medialis postica (tp.m.pt)
   Area medialis inferior (tp.m.if)

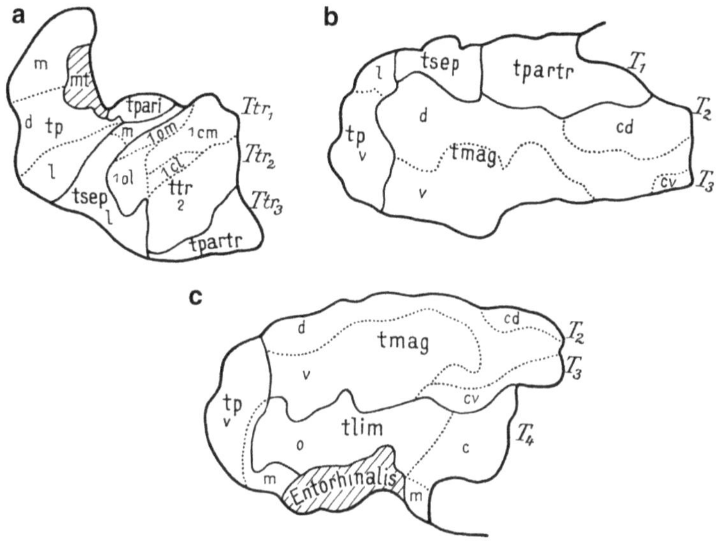

Fig. 3.36 Dorsal (**a**), lateral (**b**) and basal views (**c**) of the human temporal lobe, showing the location of the various myeloarchitectonic regions and subregions (Reproduced from Hopf 1954a). Abbreviations of regions: *tp*, regio temporopolaris; *tsep*, regio temporalis separans; *tpari*, regio temporalis parainsularis; *ttr*, regio temporalis transversa; *tpartr*, regio temporalis paratransversa; *tmag*, regio temporalis magna; *tlim*, regio temporalis limitans. Abbreviations of subregions: *c*, caudal; *d*, dorsal; *l*, lateral; *m*, medial; *o*, oral; *v*, ventral. Abbreviations of gyri: *T1, T2, T3*, superior, middle and inferior temporal gyri; *T4*, lateral occipitotemporal (fusiform) gyrus; *Ttr1, Ttr2, Ttr3*, first, second and third transverse gyri of Heschl

Übergangsfeld (tp/mti)
Übergangsfeld (tp/mtm)

Subregio ventralis (tp.v)
Subregio lateralis (tp.l)
Subregio dorsalis (tp.d)

2. Regio temporalis separans (tsep), two subregions, six areas
3. Regio temporalis parainsularis (tpari), one subregion, three areas
4. Regio temporalis transversa (ttr), five subregions, 13 areas
5. Regio temporalis paratransversa (tpartr), one subregion, four areas
6. Regio temporalis magna (tmag), four subregions, 13 areas
7. Regio temporalis limitans (tlim), three subregions, eight areas

Thus, according to Hopf (1954a), the cortex of the temporal lobe can be myeloarchitectonically subdivided into seven regions, 20 subregions, and 60 areas (Figs. 3.36 and 3.37). Hopf pointed out that the various regions could be macroscopically distinguished in his preparations, and that the subregions and areas delimited were detectable in all of the series studied.

a

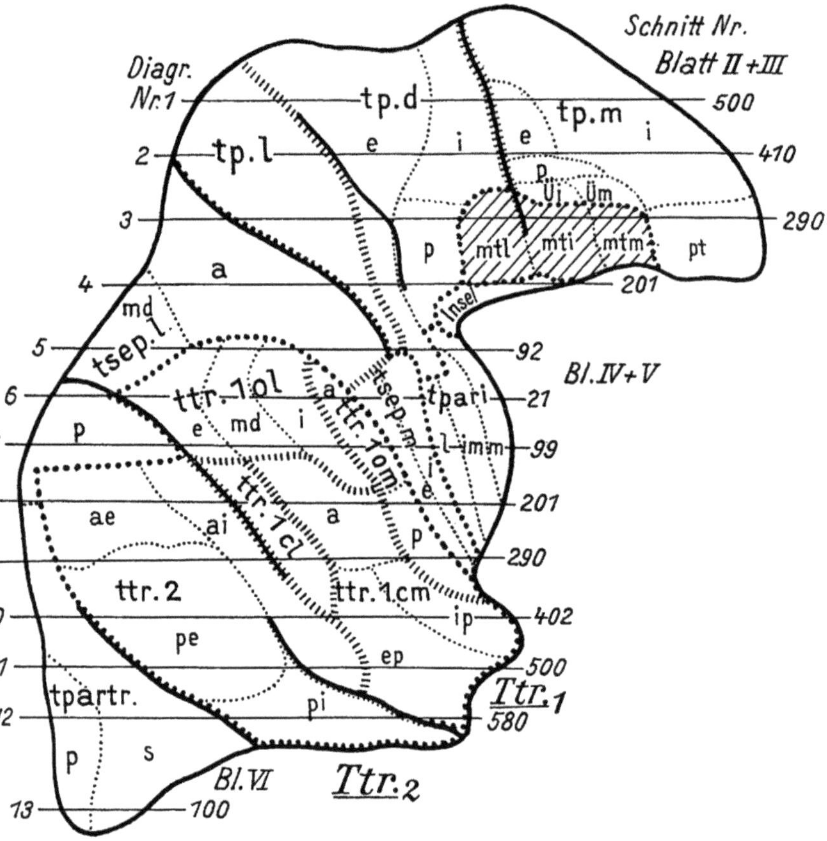

• • • • = border of a region
•••••••• = border of a subregion
.... = border of an area
—— = sulcus
large letters = symbols for subregions
small letters = symbols for areas

Fig. 3.37 (continued)

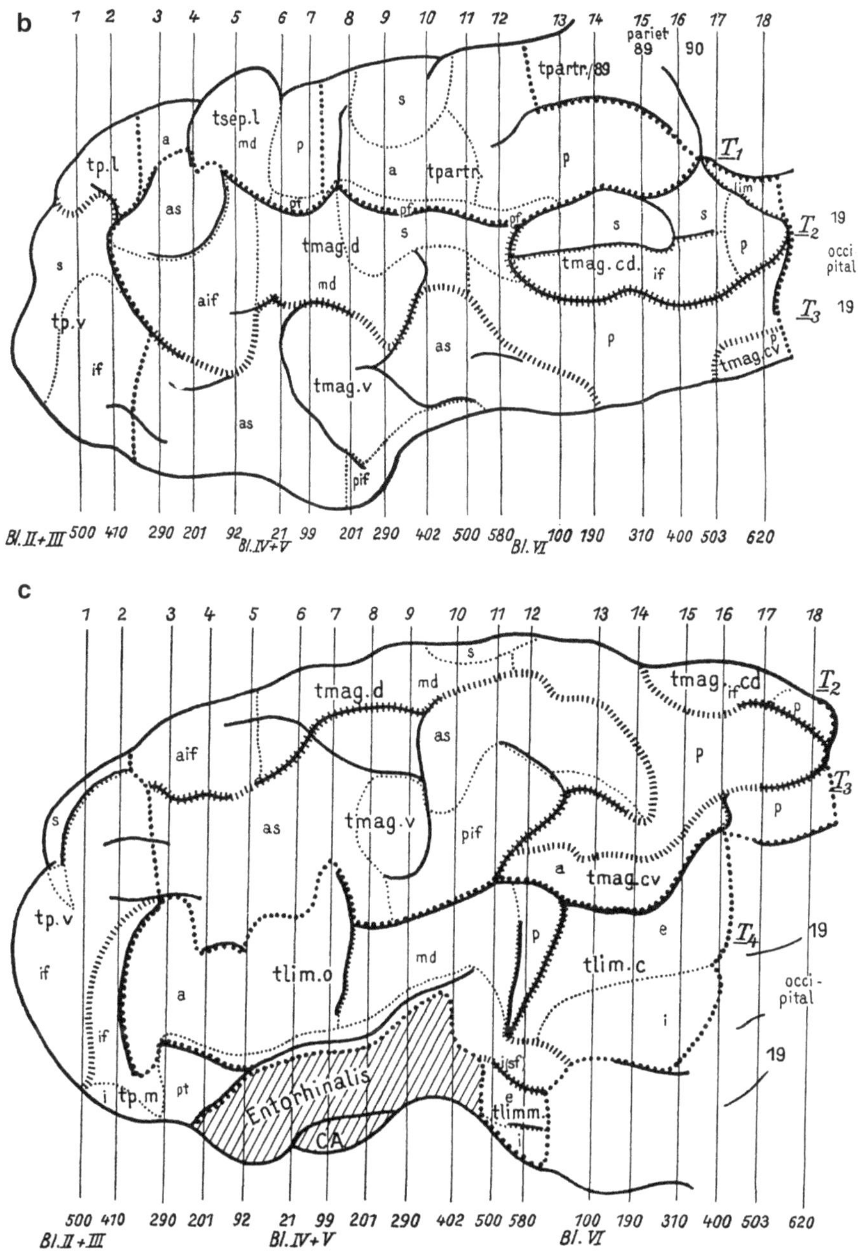

Fig. 3.37 Dorsal (**a**), lateral (**b**) and basal views (**c**) of the temporal lobe, showing the myeloarchitectonic parcellation, according to Hopf (1954a)

The number of temporal areas distinguished by Hopf is much higher than that recognized in the pioneering myeloarchitectonic studies of Campbell (1905; Fig. 3.6a, b) and Elliot Smith (1907; Fig. 3.7a, b). The former delineated only two, the latter seven temporal areas. The areas delineated by Elliot Smith show, so far as the lateral surface of the temporal lobe is concerned, a certain resemblance with the subregionalization of Hopf (cf. Fig. 3.7a with Fig. 3.36b). As regards the dorsal surface of the temporal lobe, the five regions distinguished here by Hopf correspond in a one-to-one fashion with the subregions delimited by Beck. The only difference at this level is that the Regio temporalis transversa (ttr) of Hopf is subdivided by Beck into three subregions, the subregio temporalis prima (ttrI), secunda (ttrII), and tertia (ttrIII) (cf. Fig. 3.33a with Fig. 3.36a). The 12 subregions which Hopf distinguished in the superior temporal cortex, correspond fairly well to the 13 parts of Beck. It is only at the level of the areas that the studies of Hopf and Beck differ considerably: 29 versus 73! Not surprisingly, Hopf ascribes this difference to the fact that Beck in some regions, particularly the regio temporalis polaris (tp) and the regio temporalis transversa (ttr) has descended with his parcellation to the subareal level (Figs. 3.34a and 3.37a).

Finally, it should be mentioned that Hopf, following the myeloarchitectonic mapping study, discussed above (Hopf 1954a), has devoted two additional publications to the the temporal cortex, one to the distribution of the principal myeloarchitectonic features over this cortex (Hopf 1955), the other to a photometric analysis of the myeloarchitecture of the same cortex (Hopf 1968b). Special attention is paid in these studies to the total content of myelinated fibres in the various temporal regions, as well as to the density of individual fibres ('Einzelfasern'), the presence of a stripe of Kaes-Bechterew, the disposition of the stripes of Baillarger and the differentiation of the bundles of radial fibres. It was found that the regio temporalis transversa (ttr), which covers the transverse gyri of Heschl (Fig. 3.37a: Ttr1, Ttr2), and which forms the end station of the auditory projection, shows by far the highest content of myelinated fibres of all temporal regions. There appeared to be a step-like decrease in the content of myelinated fibres and radial bundles, with increasing distance from the primary auditory cortex. The existence of the seven myeloarchitectonically different regions in the temporal cortex, delineated by microscopic inspection (Fig. 3.35), could be photometrically confirmed.

3.4.6 *Summary and Conclusions*

From the foregoing, it appears that all parts of the neocortex have been the subject of one or more myeloarchitectonic studies. Most of these studies were based on an analysis of one or two hemispheres, but in some, notably those of Strasburger (1938), on the frontal lobe, and Beck (1928, 1930) on the temporal lobe, a much larger material was used.

The cortices of all five telencephalic lobes, frontal, insular, parietal, occipital and temporal, have been parcellated into number of separate myeloarchitectonic entities. In the frontal, parietal, and temporal lobes, hierarchical subdivisions into regions, subregions, (divisions, subdivisions), and areas were used.

The data available do not allow for a definitive assessment of the exact number of myeloarchitectonic areas in the neocortex, because of the following uncertainties: (1) The findings of the various authors concerning the number of areas, present in a given lobe, may show considerable differences. (2) Most authors label the smallest units in their parcellations areas, but some designate them as subareas. (3) Criteria for distinguishing areas from subareas are lacking. (4) The data available for some lobes are scant. Nevertheless, in light of the data and considerations presented above, we venture to estimate the total number of myeloarchitectonic areas in the human neocortex to be about 185: 70 frontal, 6 insular, 30 parietal, 19 occipital, and 60 temporal.

It should be emphasized that the data reviewed, are adequate and sufficient for the composition of a myeloarchitectonic map of the human neocortex, which would be at least as reliable as any of the classic architectonic maps, i.e. the maps of Campbell (1905), Elliot Smith (1907), Brodmann (1909, 1914), von Economo and Koskinas (1925), Bailey and von Bonin (1951) and Sarkissov et al. (1955).

3.5 A General Concept of the Architecture of the Cerebral Cortex

3.5.1 Retrospect and Prospect

In the preceding pages, we reviewed the entire literature on the myeloarchitectonic parcellation of the human neocortex, in total 31 publications, no less than 27 of which stemming from the Vogts themselves or from their direct collaborators. Almost all of these publications, many of them of extraordinary length, appeared in the 'home journals' of the successive 'Vogt-Vogt-Institutes', *i. e.* the "Journal für Psychologie und Neurologie", and the "Journal für Hirnforschung". Surveying this body of literature, one cannot help but admire the patience and perseverance with which the authors concerned, and their technical assistants, have accomplished this enormous amount of work. These publications, and the companion cytoarchitectonic studies were, as we have seen, aimed at the division of the cerebral cortex into units, designated as fields or areas. A special noun, 'Felderung', and a special verb, 'feldern, felderte, gefeldert' (or even 'durchfeldern') were created to designate these special activities. It should be remembered that the results of these, extremely time-consuming, activities were not merely 'contributions to the architectonics of the cerebral cortex', but represented at the same time the realization of the research program formulated by O. Vogt (1903), outlined in the introductory part of the present review. It is important to note that the Vogts did not only participate in the realization of this program, but also developed, in light of the results obtained, a general concept of the structural and functional architecture of the cerebral cortex, and of the central nervous system (CNS) in general. The present chapter provides the opportunity to articulate for the first time, this general concept, the elements of which are dispersed over numerous publications of the pertinent authors. In what follows, an annotated synopsis of the 'Vogt-Vogt concept' will be presented. This synopsis is preceded however by a brief note on techniques.

3.5.2 Note on Techniques

The research program of the Vogts and their co-workers, and the resultant concept to be discussed below, were exclusively based on observations derived from two histological staining procedures, the Nissl- and Weigert techniques. The question arises whether these techniques are appropriate for an analysis of the architecture of the cortex. The answer to this question is: yes. The essence of the Nissl and Weigert techniques is that they reduce the baffling intricacy of the cortex (and of the CNS in general) to manageable, i. e., to analyzable proportions. The Nissl technique 'strips' the neurons from their processes, reducing them to spots, differing in size, shape and arrangement (Fig. 3.3, left panel), whereas the Weigert technique stains only the myelin sheaths of the axons, which are concentrated in radial bundles and tangential plexuses (Fig. 3.3, right panel). All of these are "leicht erkennbare Merkmale" (Vogt 1903, p. 161). A difficulty, inherent to the Weigert method is that slight variations in the time of differentiation may lead to considerable differences in the pattern of staining (Fig. 3.11c, d). Several other 'architectonics', based on other staining procedures, have been attempted, such as fibrillo-architectonics, based on silver-reduction techniques, glia-architectonics, based on glia stains, and angio-architectonics, based on the injection of dyes in blood vessels, but none of these procedures has been really successful. This does not hold true, however, for the more recently developed receptor-architectonics, *i. e.*, the study of the regional density and laminar distribution of different transmitter receptor types, as visualized by autoradiography or immunohistochemistry. The borders of cortical areas defined by cytoarchitecture, are generally perfectly matched by those derived from receptor mapping (Zilles and Amunts 2009, 2010; Zilles and Palomero-Gallagher 2001). Other complementary techniques, such as the histochemical staining for acetylcholinesterase, and the immunohistochemical stainings for calcium-binding proteins, have also been shown to be useful in architectonic studies (Carmichael and Price 1994; Öngür et al. 2003).

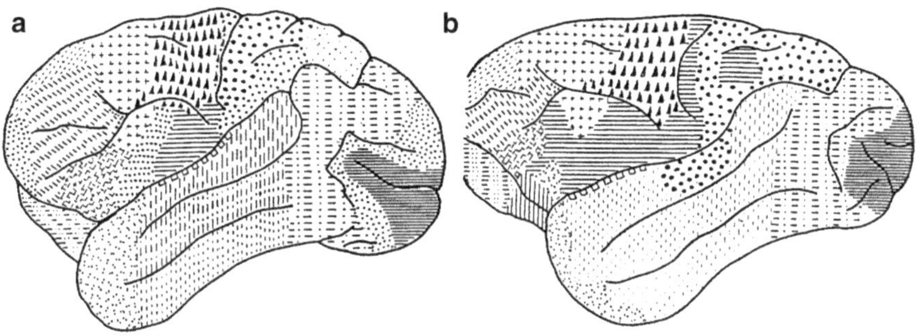

Fig. 3.38 Parcellation of the cortex of two different specimens of the spider monkey *Ateles geoffroyi*, made independently by K. S. Lashley (**a**) and G. Clark (**b**) (Reproduced from Lashley and Clark 1946)

3.5.3 An Annotated Synopsis of the Vogt-Vogt Concept of the Organization of the Cerebral Cortex

In what follows, the views of the Vogts are summarized in ten theses. These theses, including the sources on which they are based, are printed *in italics*.

1. *The entire cerebral cortex is divisible into a number of juxtaposed cytoarchitectonic areas* (Vogt 1906, p. 74, 1927, p. 251).

It is clear that O. Vogt relied here on the results of his collaborator Brodmann, who had successfully parcellated the human cerebral cortex, and that of a considerable number of other mammals as well. The reliability of the results of Brodmann, and those of classic cytoarchitectonics in general, has been challenged repeatedly, most ardently by Lashley and Clark (1946). These authors independently studied serial Nissl sections through the brains of two spider monkeys, both producing a map of the cerebral cortex. The two maps showed little agreement (Fig. 3.38). They then compared their two specimens and found that there was considerable variation in size and appearance of corresponding areas from brain to brain, and that some areas could be seen in the one brain, but not in the other. Moreover, they remained unable to locate most of the neocortical areas recognized by other investigators in this monkey and in the rhesus macaque. Lashley and Clark (1946, p. 231) concluded that standard cytoarchitectonic maps are unreliable and "represent little more than the whim of the individual student". However, numerous recent studies, including Amunts et al. (1999, 2003), Caspers et al. (2006, 2008), Scheperjans et al. (2008a, b), using quantitative and observer-independent procedures, have confirmed the presence of many of the areas delineated by Brodmann, and detected additional, previously not described areas.

2. *The entire cerebral cortex is divisible into a number of juxtaposed myeloarchitectonic areas.* (Vogt 1910b, p. 418).

The evidence for this thesis is presented in the first part of this review. The following points may be recalled:

(i) Elliot Smith (1907; Fig. 3.7a, b), who practised a sort of myeloarchitectonics 'avant la lettre', was able to distinguish no less than 50 different, sharply delimited areas in the human cortex.
(ii) Many experts maintain that most areal borders can be observed with the naked eye in their myelin-stained preparations. This is line with the finding of C. and O. Vogt, (1928, p. 467): "Die myeloarchitektonische Gliederung der Grisea ist viel augenfälliger als die cytoarchitektonische."
(iii) The fact that the results of Beck and O. Vogt, who mapped the human dorsal temporal cortex independently from each other, show a striking similarity (Fig. 3.32), pleads strongly for the reliability of the myeloarchitectonic approach. Their findings are in marked contrast with those of Lashley and Clark (1946; Fig. 3.38) on cytoarchitectonics.
(iv) The presence of most of the cortical areas, delineated by O. Vogt (1910a, 1911) in his early myeloarchitectonic studies, has been confirmed by several later investigators.
(v) The thorough myeloarchitectonic explorations of the human cerebral cortex by Adolf Hopf (1968a, b, 1969, 1970b), were not only based on visual inspection, but also on objective registrations of myeloarchitectonic features with a photometric technique.
(vi) It remains enigmatic that the results of myeloarchitectonic parcellations of particular cortices, or parts thereof, may show considerable differences. The rather global subdivision of the cortex of the orang-utan by Mauss (1911; Fig. 3.13), and the much more detailed parcellations of the frontal, parietal, and temporal cortices of the chimpanzee by Strasburger (1937b; Fig. 3.16), Gerhardt (1938), and Beck (1929; Fig. 3.34c), respectively, are striking cases in point. It is also highly remarkable that Beck (1925; Fig. 3.32a), who initially found 28 areas in the human dorsal temporal cortex, subdivided the same cortical region in a later study (Beck 1928; Fig. 3.34a) in no less than 89 areas!

3. *Because the boundaries of the cytoarchitectonic and the myeloarchitectonic cortical areas* ***coincide completely****, it is correct to designate these entities simply as* ***architectonic areas*** (Vogt and Vogt 1919, p. 361, 1954, p. 16, 1956, p. 409).

The Vogts were deeply convinced of the correctness of this thesis. The discrepancy between the relatively low number of cytoarchitectonic areas in the human cortex, distinguished by Brodmann (1909), and the much higher number, resulting from their own myeloarchitectonic studies, was explained by claiming that Brodmann had missed numerous boundaries (Vogt 1918, p. VI). C. and O. Vogt (1919, p. 365) emphasized to have found cytoarchitectonic counterparts of all of their myeloarchitectonic cortical areas. It may be recalled in this context that Sanides (1962), Brockhaus (1940), and Gerhardt (1940), who carried out combined cytoarchitectonic and myeloarchitectonic analyses of the human frontal, insular, and parietal cortices, respectively, all reported complete concordance of the results obtained with the two techniques.

Combination of cytoarchitectonics with other types of 'architectonics' may also lead to concordant results. Thus, it has been shown that in the human cerebral cortex, receptor-architectonic borders quite often precisely match cytoarchitectonic borders. The finding that receptor mapping not only corroborates classical borders of cytoarchitectonic areas, but can also reveal more detailed subdivisions (Zilles and Amunts 2009), may well mark a shift from the 'Brodmann' to the 'Vogt-Vogt-level of parcellation' of the cortex (see below).

4. *The boundaries of the architectonic areas are hair-sharp and extend through all cortical layers* (Vogt and Vogt 1919, p. 365; Vogt 1951, p. 117).

This is one of the most controversial points of the Vogt-Vogt concept. As for cytoarchitectonics, there can be no doubt that some boundaries, such as that between the primary motor and the somatosensory areas, and that between the calcarine and occipital areas (Fig. 3.29a), are very distinct, but this does not hold true for the borders of many other areas. Brodmann (1909) indicated that several of the boundaries in his map of the human cerebral cortex, particularly those between the various prefrontal and parietal areas, are not sharp. Von Economo and Koskinas (1925) found that in the human neocortex, gradual transitions, rather than distinct areal boundaries, are the rule. At this point, they were fiercely attacked by Marthe Vogt (1928a, b), the eldest daughter of the Vogts, who defended the existence of hair-sharp boundaries in the cortex with numerous examples. Von Economo (1928, p. 322)maintained, however, that the cortical areas and their borders do not constitute a totally rigid ("absolut starres") system, and that the 'limitrophic adaptations', described by the Vogts (see below), point in the same direction.

Even among the direct collaborators of the Vogts, there was no unanimity as to the sharpness of the interareal boundaries, as the following example may show. Mauss (1911) analyzed the myeloarchitecture of the cortex of the orang-utan, This study resulted in a beautiful map (Fig. 13), in which the cortex is parcellated into about 40 areas. Mauss (1911, p. 437) mentioned, that the preparation of this map had compelled him: "...mehr oder weniger scharfe Grenzen zu ziehen, wo es sich um allmähliche Übergänge, oft sogar nur relativ ausgedehnte Mischzonen handelt". In 1938, Edith Gerhardt subjected the parietal cortex of the chimpanzee to a detailed cyto- and myeloarchitectonic analysis. She delineated a larger number of areas in this region, than Mauss had done in the entire cortex of the orang-utan. At the end of her study, Gerhardt (1938, p. 385) compares her results with those of Mauss (1911): "Es ist interessant, daß dort, wo Mauss von unscharfen Grenzen und Übergangsgebieten spricht, sich meist scharf gegeneinander abgesetzte kleinere Struktureinheiten finden ließen, ..."

Omnilaminarity, i. e., the phenomenon that the boundaries between architectonic units extend through all cortical layers, played an important role in the views of the Vogts. They believed that, within a given architectonic unit, all neurons, and hence all neuronal layers, are adapted to the central function of that particular unit (Vogt

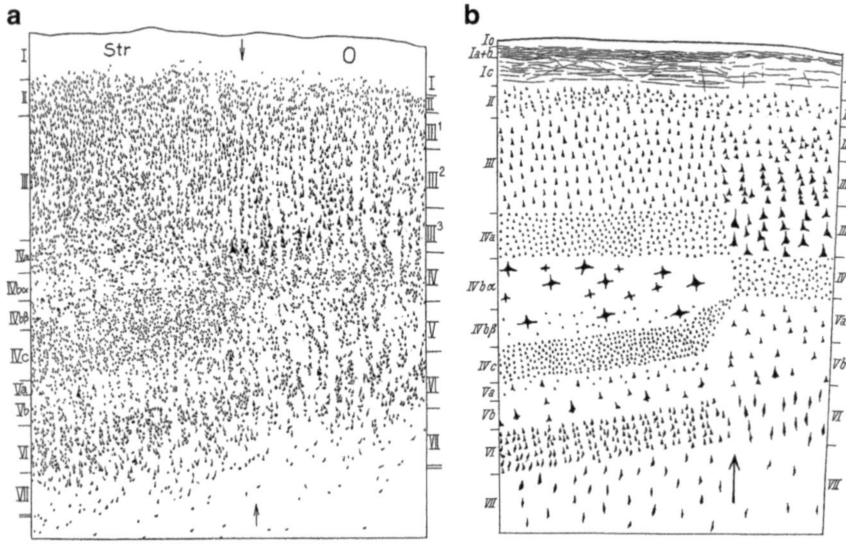

Fig. 3.39 The boundary (*arrows*) between the striate (*Str*) and occipital areas (*O*) in the human cortex. (**a**) Cytoarchitecture, based on a Nissl preparation. (**b**) Scheme, showing that this boundary is marked by cytoarchitectonic changes in all of the cortical layers (Reproduced from C. and O. Vogt 1936)

1927, p. 251). The present author has always thought that this 'omnilaminarity-claim' does no justice to the cytoarchitectonic reality. Thus, I believed that the boundary between the striate and occipital cortical areas is essentially confined to the middle cortical layers, and that there is continuity in the superficial and deep layers (Fig. 3.39a). The analytical scheme of the situation, presented by C. and O. Vogt (1936, p. 266; Fig. 3.39b), indicates, however, that changes in all cortical layers mark the boundary between the two areas.

With regard to myeloarchitectonics, it may be recalled that Elliot Smith (1907) emphasized that the bounderies between his macroscopically observed myeloarchitectonic areas in the human cortex are very distinct. A similar conclusion was reached by the Vogts and their numerous collaborators, who studied the myeloarchitecture of the human cortex at the microscopic level. In order to enable the reader to form an opinion about these boundaries, we reproduced here a number of illustrations from the literature (Figs. 3.15b, c, 3.16h, 3.29b, f, and 3.34b), showing the transition of two or more myeloarchitectonic areas into each other.

5. *Architectonic areas are, structurally, not always homogeneous throughout; their border zones may show* ***'limitrophic adaptations'*** *to adjacent areas* (Vogt and Vogt 1919, p. 369, 1928, p. 468).

In their own words (Vogt and Vogt 1919, p. 369): "Es soll aber nicht verschwiegen werden daß die Architektonik in der ganzen Ausdehnung des Feldes durchaus nicht eine absolut gleiche ist. Speziell im Grenzgebiet des einzelnen

Feldes tritt eine Annäherung an den jedesmaligen Bau des benachbarten Feldes auf".

6. *The cortical architectonic units may be expected to have specific afferent and efferent connections* (Vogt 1906, p. 74; Vogt and Vogt 1956, p. 408).

O. Vogt (1918, p. VI) referred in this context to the fact that focal cortical lesions always lead to sharply delimited thalamic degenerations. However, the hodological substantiation of their architectonic findings, has never formed part of the research program of the Vogts. Several authors, including Lashley and Clark (1946), Bailey and von Bonin (1951), Le Gros Clark (1952), and Jones (1987, 2008), have indicated that in their opinion, the afferent and efferent connections form the most reliable basis for a rational subdivision of the cerebral cortex. Rose and Woolsey (1948), studying the structure and connections of the cingulate cortex in rabbit and cat, found that the cingulate cortex in these animals can be subdivided into three areas, and that each of these areas co-extends with the distribution field of a particular nucleus within the anterior thalamic nuclear group. Such close correlations between cytoarchitecture and the projection fields of individual thalamic nuclei were later also found in many other regions of the cerebral cortex (Rose and Woolsey 1949; Jones and Burton 1976; Seltzer and Pandya 1978). Using data collected in the macaque connectivity database CoCoMac (Stephan et al. 2001), Passingham et al. (2002) demonstrated that in this monkey, each cortical area has a unique pattern of cortico-cortical connections, a defining '*connectional fingerprint*'. Finally, it may be mentioned that recently, several authors, including Behrens and Johansen-Berg (2004), Johansen –Berg et al. (2005), and Anwander et al. (2007), using diffusion imaging, have presented evidence, suggesting that in the human cortex, particular areas have a distinct '*connectional architecture*'.

7. *The fact that all of the cortical areas distinguished display a specific structural organization indicates that all of these areas subserve a specific function* (Vogt 1903, p. 160; Vogt and Vogt 1954, p. 116).

This categorical statement refers to an important motive for the architectural studies of the Vogts, which included, before all things, the performance of physiological preparatory work. "Vor allem wollen wir physiologische Vorarbeit liefern" (Vogt 1911, p. 379). We have seen that Campbell (1905) had precisely the same intention.

In order to substantiate their functional notions, the Vogts carried out extensive electrical stimulation studies in monkeys (Vogt and Vogt 1907, 1907, 1919, 1942). These studies, and parallel experiments of Foerster (1936), in patients who underwent brain operations, showed that areas from which particular motor responses can be elicited, generally match with particular architectonic areas. C. and O. Vogt (1928, p. 468) emphasized that only one single architectonic parcellation can be the physiologically correct one. They indicated that it would be the task of neurophysiologists to determine the precise functions of all of the architectonic units distinguished, and never speculated on the outcome of that inquiry.

Other authors have also advanced the hypothesis that the cerebral cortex is composed of units, which represent structural as well as functional units. Thus, Carmichael and Price (1994) carried out a detailed multiarchitectonic study of the orbital and medial prefrontal cortex (OMPFC) of the rhesus monkey. This study resulted in the identification of no less than 22 discrete areas in the OMPFC of this species. Subsequent experimental hodological studies (Carmichael and Price 1995a, b, 1996) revealed that all of the areas distinguished have specific connections. The authors mentioned concluded that each of the cortical areas delineated, represents a module with specific input–output relations, and a unique role in information processing. They considered it likely that much of the cortex consists of such *discrete structural and functional modules.* Roland and Zilles (1998) expressed the expectation that combining the results of quantitative architectonic studies of the cortex that apply objective, observer-independent procedures, with functional neuroimaging data, will lead to the detection of *functional cortical fields.* They advanced the hypothesis that the organization of the cortex is based on such functional fields, and speculated that all neurons and synapses within these fields perform a co-operative computation. Recently, Zilles and Amunts (2009) have pointed out that receptor mapping, particularly quantitative multireceptor mapping, may provide important clues to both the structural and the functional organization of the cerebral cortex.

8. *The fact that each of the cortical architectonic units subserves a specific function indicates that these units are to be considered as separate organs, and that the cortex as a whole represents a complex of organs ("Organkomplex")* (Vogt and Vogt 1922, p. 163).

This view, which was strongly supported by Brodmann (1909, p. 237), goes back, as we shall see, to the phrenologists Gall and Spurzheim.

9. *Cortical architectonic units may show a specific vulnerability for particular pathological processes. These are manifestations of a general phenomenon to be designated as* ***topical pathoclisis*** (Vogt and Vogt 1922, p. 163. 1936, p. 452, 1942, p. 368, 1956, p, 405).

This component of the topistic approach to the cortex of the Vogts has met with little response. Von Economo (2009, p. 178) noted that amyotrophic lateral sclerosis, at least initially, specifically affects the primary motor area, but that he was not acquainted with any other disease of the central nervous system showing a similar areal circumscription.

10. *The total number of cortical fields ("Rindenfelder") or topistic units ("topistische Einheiten") in the human cerebral cortex amounts to about 200* (Vogt and Vogt 1919, p. 364; Vogt 1951, p. 117).

This large number of areas, has been experienced by numerous previous authors, including Bailey and von Bonin (1951, pp. VII, 231), Le Gros Clark (1952, p. 104), and Jones (2003, p. 19), as being out of all proportions. It was felt that Brodmann's subdivision into some 40 areas, comes somewhere near what seems reasonable. It is felt appropriate to sum up here the evidence indicating that, according to the present state of knowledge, the estimation of the Vogts was absolutely realistic.

(i) C. and O. Vogt (1919, p. 365) maintained that Brodmann has missed numerous cytoarchitectonic borders, and that they had found cytoarchitectonic counterparts of all of their numerous myeloarchitectonic areas.
(ii) The analyses of the human cerebral cortex of Von Economo and Koskinas (1925) and Sarkissov et al. (1955) resulted in a much larger number of cytoarchitectonic areas than that of Brodmann.
(iii) The review of the myeloarchitectonic literature, presented in the first section of this paper, led us to the conclusion that the human neocortex contains about 185 myeloarchitectonic areas. If we take into consideration that the Vogts included, apart from neocortical areas, also allocortical areas in their estimation, their number appears to be practically identical to ours.
(iv) Nieuwenhuys et al. (2008, p. 535) plotted the results of a detailed multiarchitectonic study of the orbital and medial prefrontal cortex (Öngür at al. 2003), and of a number of functional imaging studies on the localization of visuotopic areas (Hadjikani et al. 1998; Press et al. 2001; Van Essen, 2005), on a flattened version of Brodmann's map (Van Essen 2006). It appeared that the 30 areas delineated and plotted, together occupy 20 % of the neocortical surface. Extrapolating from these data, it was estimated that that the neocortex is composed of some 150 areas or units.
(v) During the last decades, the C. & O. Vogt Institute of Brain Research in Düsseldorf and the Institute of Neuroscience and Medicine, Research Centre Jülich, have produced a large number of detailed studies on the architecture of the human cerebral cortex. These studies, in which the results of quantitative cytoarchitectonic analyses are generally combined with receptor-architectonic data, have shown that many of the cytarchitectonic areas, distinguished by Brodmann (1909, 1914), are divisible into two or more smaller units. In what follows, the results of a number of these studies is recorded in a very condensed form, as for instance: prefrontal cortex, 3 |BAs 9, 10, 11| 1, 7, which means in full: The pertinent authors investigated a sector of the prefrontal cortex, occupied by three Brodmann areas (BAs), namely: BAs 9, 10 and 11; they confirmed the presence of one BA, and subdivided the remaining BAs in this sector into seven smaller units.

Geyer et al. (1996, primary motor cortex: 1 |BA 4| 0, 2)
Amunts et al. (2010, Broca's region: 2 |BAs 44, 45| 0, 4)
Uylings et al. (2010, orbitofrontal cortex: 2 |BAs 11,46| 1, 5)
Palomero-Gallagher et al. (2008, anterior cingulate cortex: 3 |BAs 24, 25, 32| 0, 10)
Kurth et al. (2010, posterior insular cortex: 1 |BA J.post| 0, 5)
Scheperjans et al. (2008a, superior parietal cortex: 2 |BAs 5, 7| 0, 12)
Caspers et al. (2006, inferior parietal cortex: 2 |BAs 39, 40| 0, 7)
Eickhoff et al. (2008, visual cortex: 3 |BAs 17,18, 19| 0, 8)

These data suggest that there may well be almost four times as many architectonic areas in the human neocortex as Brodmann indicated. Jones (2008, p. 2231) noted about the recent efforts of the C.& O. Vogt Institute that their numbers, though not yet finished, "seem well on the way to approximating those of the Vogts".

(vi) Physiological studies, using single cell recordings, have shown that monkeys have more than 30 cortical areas for processing visual information, at least 15 for somatosensory information, and some 20 for auditory information (Kaas 2002). Extrapolation of these numbers to the human, point to the presence of a much larger number of cortical areas than the 44 of Brodmann's parcellation scheme.

(vii) Most recently, Glasser and van Essen (2011) estimated the total number of architectonic areas per hemisphere in humans at 150–200, on the basis of as yet unpublished observations of D. C. Van Essen, M. F. Glasser, D. L. Dierker, J. Harwell, and T. Coalson.

Before closing this section, some additional remarks on the Vogt-Vogt concept should be made.

1. The Vogts believed that their concept does not hold only for the cortex, but is also applicable to other parts of the brain, such as the striatum and the thalamus (Vogt and Vogt 1922, p. 23).
2. One of the plans of the Vogts that never has been realized was the implementation of an "Individualanatomie" of the cortex, based on a large collection of brains of geniuses ('Elitegehirne') and of mentally retarded people. It was expected that in the brains of highly gifted individuals, such as great musicians or great mathematicians, particular cortical areas would be strongly developed, whereas in feeble-minded people, particular areas would be underdeveloped or even rudimentary (Vogt 1910b, p. 419). It is noteworthy that C. and O. Vogt (1956, p. 426) emphasized on the last page of their last publication that, within the context of such an 'Individualanatomie', the brains of asocials would also deserve examination. We have seen that Theodor Kaes (1907) pursued a similar goal.
3. C. and O. Vogt (1922, p. 12, 1929, p. 154) expected that their topistic analyses of the cortex could be conducive to a pathological-anatomical classification of mental diseases, whereas Brodmann (1913) considered knowledge on the segregation of the cortex into architectonic units of great potenial importance for the tackling of anthropological problems: "Gibt es im besonderen am Gehirn verschiedener Menschenrassen lokalisatorische Tatsachen, die sich als Zeichen eines primitiveren Zustandes deuten und somit vielleicht auch für das Rassenproblem fruchtbar machen lassen?" (Brodmann 1913, p. 202)
4. Finally, it may be mentioned that not all of the collaborators of the Vogts agreed with all of the details of their concept. Thus, Brodmann (1909, p. 306) emphasized that some of the boundaries of some neocortical areas are not sharp. He also believed that higher cortical functions result from the conjoint activities of a large number of areas distributed more or less widely over the cortical surface. Hopf (1954b, p. 464) warned in a general paper on the

architectonics of the cortex, for making the Vogt-Vogt concept absolute: "In jeder Area ein ganz isoliertes Organ mit einer einzigen bestimmten Funktion zu sehen, ist natürlich völlig unsinnig".

3.5.4 Summary

1. During their long scientific career, the Vogts gradually developed a general concept of the organization of the human cerebral cortex.
2. This concept may be epitomized as follows:

   The human cerebral cortex is segregated into around 200 discrete, juxtaposed structural and functional modules or units.

3. There is converging quantitative cytoarchitectonic, receptorarchitectonic, myeloarchitectonic, hodological, as well as functional evidence, indicating that this concept is essentially correct.

3.6 The myeloarchitectonic studies of the Vogt-Vogt school, and the current explorations of the functional organization of the cerebral cortex with neuroimaging techniques

We have just seen that the extensive architectonic studies of the Vogts and their collaborators, have resulted in the concept that the human cerebral cortex is composed of about 200 structural and functional units. The Vogts and their associates knew, of course that some of these units are involved in particular sensory or motor functions, but in general they refrained from speculating on the specific functions of the remaining units. Other authors have been less reluctant in this respect. Thus, Franz Joseph Gall (1758–1828) and Johann Spurzheim (1758–1832) maintained, as early as the beginning of the nineteenth century, that the cerebral cortex is composed of discrete organs or regions that represent different mental faculties, and that there are as many such organs as there are mental faculties. They distinguished some 35 of these localized faculties, including arithmetic, hope, conscientiousness, language, constructiveness, destructiveness and parental love. They based their theory on the examination of the skulls of a great variety of people, from eminent and highly gifted to lunatics and criminals, assuming: (i) that the intellectual and moral abilities are differentially developed in each individual; (ii) that these differences are reflected in the size of the cortical organs related to these abilities, and (iii) that the size of the pertinent cortical organs is in their turn reflected in prominences of the overlying skull, i. e., in cranial

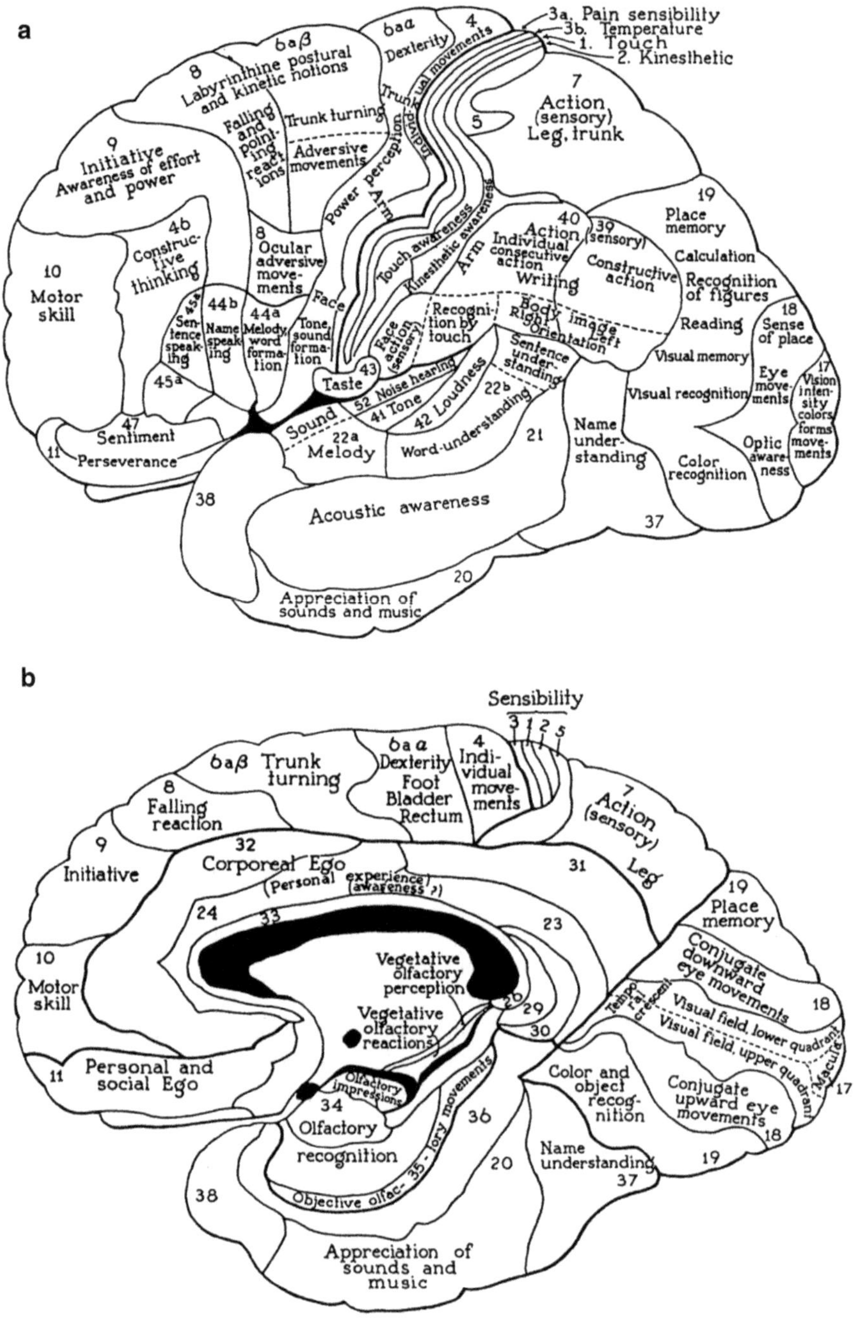

Fig. 3.40 Lateral (**a**), and medial views (**b**) of the human cerebral hemispheres, showing the localization of functions in the cerebral cortex, according to Kleist (1934). The *numbers* indicate Brodmann's (1909) cytoarchitectonic areas

bumps. The highly speculative and dubious branch of science founded by Gall and Spurzheim. is now generally known as phrenology (from Gr. phrèn=mind).

About a hundred years later, the neurologist Karl Kleist (1879–1960) made another attempt at a far going localization of functions in the human cerebral cortex. He had a vast amount of clinical material at his disposal, including, apart from numerous regular clinical cases, about 300 persons who had sustained local brain injuries during World War I (Kleist 1934). He summarized his findings in a map, shown here in Fig. 3.40. It will be seen that Kleist subdivided the cortex according to Brodmann (1909), and that he provided almost all of the 44 cytoarchitectonic areas distinguished by the latter with a functional label. The overall functions of the primary sensory and motor areas are correctly indicated. However, Kleist went far beyond that by attributing all sorts of higher cognitive and mental functions and faculties to many other areas. Thus, he associated temporal area 21 with auditory awareness ("akustische Aufmerksamkeit"), prefrontal area 10 with motor skill ("motorische Handlungsfolgen"), and orbitofrontal area 11 with personal and social ego ("Selbst- und Gemeinschafts-Ich"). Because of this detailed localization of psychic functions, many of his colleagues disposed of Kleist's map as 'brain mythology' (cf. Creutzfeldt 1983). Uttal (2001, p. 109) qualified this map as "a modern manifestation of phrenological thinking." It is ironic that Brodman, *n. b.* the creator of the map used by Kleist, as we mentioned already, did not believe that higher cognitive functions can be related to individual cortical areas.

From Kleist to the present! It is no exaggeration to say that the modern imaging techniques, such as positron emission tomography (PET) and functional magnetic resonance imaging (fMRI), have revolutionized our capacity to localize functions and functional complexes in the cerebral cortex. As eloquently put into words by Mesulam (2011, p. 2): "We are now in the midst of yet another revolution, a revolution powered by spectacularly successful methods for the non-invasive functional and structural imaging of the human brain. Impressive advances in signal acquisition, data analysis, and task design have collectively empowered a multidisciplinary army of investigators to map the cerebral cartography of vision, language, love, lust, greed, altruism, empathy, conflict, and virtually any other mental faculty that can be delineated." Mesulam used this passage to introduce a pressing call to renewed and refined studies of human cortical connectivity. The present author is of the opinion that this activity, however significant, should be preceded by renewed and refined studies on the problem as to how to establish the morphological identity of the entities, which form the edges, nodes and hubs in this connectivity, i.e., the foci of cortical activity, observed in functional neuroimaging studies.

At present, modified versions of Brodmann's map are commonly used for the structural interpretation of neuroimaging data. However, it has become increasingly clear that these "Brodmann" maps do not provide the neuroanatomical precision and accuracy for an adequate mapping of fMRI data (Geyer et al. 2011). Brodmann's map is a pioneering work, but necessarily contains false delineations (e.g., his area 19 does not match any of the extrastriate areas shown by retinotopic mapping), missing delineations (most of the intrasulcal cortex was never mapped by him) and various other problems (for review see Zilles and Amunts 2010).

The data, reviewed in the present paper, have shown that there is converging evidence, indicating that the human cerebral cortex is divisible into some 200 structural and functional units, and that the maps, resulting from the meticulous myeloarchitectonic parcellations of the Vogts and their associates, have yielded similar results. It is for these reasons that we strongly recommend an attempt at combining and synthesizing the results of cytoarchitectonic mapping studies of Brodmann on the human cerebral cortex, with those of the Vogts. The resultant ‘supermap’ would not only mark the consummation of the enterprise on which these eminent scientists embarked, more than a century ago, but would probably also yield an optimal frame of reference for the localization of functions, as revealed by neuroimaging studies.

It stands to reason that, because of the considerable interindividual variability of the areal borders in the human cortex, interpretations based on the futuristic ‘supermap’ just referred to, could only be performed probabilistically.

The problems of functional localization in the human cortex would be greatly reduced if it were possible to map the specific structural correlates of functional activity, in a non-invasive way, directly in each individual living brain under study. Recent studies, using high-resolution MRI, have shown that such direct correlations between cortical architecture and function in living brains are now within our range. Such in-vivo explorations of the histology of the cortex have revealed that local differences in the total fibre-content (Glasser and Van Essen 2011), and in the laminar patterns of the myelinated fibres (Geyer et al. 2011) give excellent MRI contrast. This being so, it may be expected that the meticulous myeloarchitectonic studies of the Vogt-Vogt school, will play a prominent role in the interpretation of the results of these new in-vivo mappings

Acknowledgments The author thanks Drs. Bob Turner and Karl Zilles for critically reading an earlier version of this paper, Mr. Ton Put for help with the illustrations, and Suzanne Bakker M.Sc. for moral support and reference management. Finally, the author wants to acknowledge especially the invaluable and continuous assistance of Dr. Jenneke Kruisbrink, the librarian of our Institute. Without her help, this article would not have been possible.

References

Amunts K, Zilles K (2012) Architecture and organizational principles of Broca’s region. Trends Cogn Sci 16(8):418–426. doi:org/10.1016/j.tics.2012.06.005

Amunts K, Schleicher A, Bürgel U, Mohlberg H, Uylings HB, Zilles K (1999) Broca’s region revisited: cytoarchitecture and intersubject variability. J Comp Neurol 412(2):319–341. doi:10.1002/(SICI)1096-9861(19990920)412:2<319::AID-CNE10>3.0.CO;2-7

Amunts K, Schleicher A, Ditterich A, Zilles K (2003) Broca’s region: cytoarchitectonic asymmetry and developmental changes. J Comp Neurol 465(1):72–89. doi:10.1002/cne.10829

Amunts K, Lenzen M, Friederici AD, Schleicher A, Morosan P, Palomero-Gallagher N, Zilles K (2010) Broca’s region: novel organizational principles and multiple receptor mapping. PLoS Biol 8(9):e1000489. doi:10.1371/journal.pbio.1000489

Anwander A, Tittgemeyer M, von Cramon DY, Friederici AD, Knosche TR (2007) Connectivity-based parcellation of Broca’s area. Cereb Cortex 17(4):816–825. doi:10.1093/cercor/bhk034

Bailey P, Bonin G (1951) The isocortex of man. University of Illinois Press, Urbana
Batsch EG (1956) Die myeloarchitektonische Untergliederung des Isocortex parietalis beim Menschen. J Hirnforsch 2:225–258
Beck E (1925) Zur Exaktheit der myeloarchitektonischen Felderung des Cortex cerebri. J Psychol Neurol 31:281–288
Beck E (1928) Die myeloarchitektonische Felderung des in der Sylvischen Furche gelegenen Teils des menschlichen Schläfenlappens. J Psychol Neurol 36:1–21
Beck E (1929) Der myeloarchitektonische Bau des in der Sylvischen Furche gelegenen Teiles des Schläfenlappens beim Schimpansen (Troglodytes niger). J Psychol Neurol 38:309–420
Beck E (1930) Die Myeloarchotektonik der dorsalen Schläfenlappenrinde beim Menschen. J Psychol Neurol 41:129–262
Behrens TE, Johansen-Berg H (2005) Relating connectional architecture to grey matter function using diffusion imaging. Philos Trans R Soc Lond B Biol Sci 360(1457):903–911. doi:10.1098/rstb.2005.1640
Braak H (1980) Architectonics of the human telencephalic cortex. Springer, Berlin/Heidelberg/New York
Braitenberg V (1956) Die Gliederung der Stirnhirnrinde auf Grund ihres Markfaserbaus (Myeloarchitektonik). In: Rehwald E (ed) Das Hirntrauma. Thieme, Stuttgart, pp 183–203
Braitenberg V (1962) A note on myeloarchitectonics. J Comp Neurol 118:141–156
Brockhaus H (1940) Die Cyto- und Myeloarchitektonik des Cortex claustralis und des Claustrum beim Menschen. J Psychol Neurol 49:249–348
Brodmann K (1903a) Beiträge zur histologischen Lokalisation der Grosshirnrinde: Regio Rolandica. J Psychol Neurol 2:79–107
Brodmann K (1903b) Beiträge zur histologischen Lokalisation der Grosshirnrinde. Zweite Mitteilung: Der Calcarinatypus. J Psychol Neurol 2:133–159
Brodmann K (1905a) Beiträge zur histologischen Lokalisation der Grosshirnrinde. Dritte Mitteilung: Die Rindenfelder der niederen Affen. J Psychol Neurol 4:177–226
Brodmann K (1905b) Beiträge zur histologischen Lokalisation der Grosshirnrinde. IV. Mitteilung: Der Riesenpyramidentypus und sein Verhalten zu den Furchen bei den Karnivoren. J Psychol Neurol 6:108–120
Brodmann K (1906) Beiträge zur histologischen Lokalisation der Grosshirnrinde: Fünfte Mitteilung: Über den allgemeinen Bauplan des Cortex Pallii bei den Mammaliern und zwei homologe Rindenfelder im besonderen. Zugleich ein Beitrag zur Furchenlehre. J Psychol Neurol 6:275–400
Brodmann K (1908a) Beiträge zur histologischen Lokalisation der Grosshirnrinde. Sechste Mitteilung: Die Cortexgliederung des Menschen. J Psychol Neurol 10:231–246
Brodmann K (1908b) Beiträge zur histologischen Lokalisation der Grosshirnrinde. VII. Mitteilung: Die cytoarchitektonische Cortexgliederung der Halbaffen (Lemuriden). J Psychol Neurol 10:287–334
Brodmann K (1909) Vergleichende Lokalisationslehre der Grosshirnrinde in ihren Prinzipien dargestellt auf Grund des Zellenbaues. J.A. Barth, Leipzig
Brodmann K (1913) Neue Forschungsergebnisse der Großhirnrindenanatomie mit besonderer Berücksichtigung anthropologischer Fragen. Gesellschaft deutscher Naturforscher und Ärtze 85:200–240
Brodmann K (1914) Physiologie des Gehirns. In: Von Bruns P (ed) Neue deutsche Chirurgie, vol 11 Pt. 1. Enke, Stuttgart, pp 85–426
Cajal SR (1894) The Croonian Lecture: the fine structure of the nerve centres. Proc R Soc [Lond] 55:444–468
Campbell AW (1905) Histological studies on the localisation of cerebral function. Cambridge University Press, Cambridge
Carmichael ST, Price JL (1994) Architectonic subdivision of the orbital and medial prefrontal cortex in the macaque monkey. J Comp Neurol 346(3):366–402. doi:10.1002/cne.903460305
Carmichael ST, Price JL (1995a) Limbic connections of the orbital and medial prefrontal cortex in macaque monkeys. J Comp Neurol 363(4):615–641. doi:10.1002/cne.903630408

Carmichael ST, Price JL (1995b) Sensory and premotor connections of the orbital and medial prefrontal cortex of macaque monkeys. J Comp Neurol 363(4):642–664. doi:10.1002/cne.903630409
Carmichael ST, Price JL (1996) Connectional networks within the orbital and medial prefrontal cortex of macaque monkeys. J Comp Neurol 371(2):179–207. doi:10.1002/(SICI)1096-9861(19960722)371:2<179::AID-CNE1>3.0.CO;2-#
Caspers S, Geyer S, Schleicher A, Mohlberg H, Amunts K, Zilles K (2006) The human inferior parietal cortex: cytoarchitectonic parcellation and interindividual variability. Neuroimage 33 (2):430–448. doi:10.1016/j.neuroimage.2006.06.054
Caspers S, Eickhoff SB, Geyer S, Scheperjans F, Mohlberg H, Zilles K, Amunts K (2008) The human inferior parietal lobule in stereotaxic space. Brain Struct Funct 212(6):481–495. doi:10.1007/s00429-008-0195-z
Creutzfeldt OD (1983) Cortex cerebri. Leistung, strukturelle und funktionelle Organisation der Hirnrinde. Springer, Berlin/Heidelberg/New York
Eickhoff SB, Rottschy C, Kujovic M, Palomero-Gallagher N, Zilles K (2008) Organizational principles of human visual cortex revealed by receptor mapping. Cereb Cortex 18 (11):2637–2645. doi:10.1093/cercor/bhn024
Elliot Smith G (1907) A New Topographical Survey of the Human Cerebral Cortex, being an Account of the Distribution of the Anatomically Distinct Cortical Areas and their Relationship to the Cerebral Sulci. J Anat Physiol 41(Pt 4):237–254
Flores A (1911) Die Myeloarchitektonik und die Myelogenic des Cortex Cerebri beim Igel. J Psychol Neurol 17:215–247
Foerster O (1936) Motorische Felder und Bahnen. Sensible cortical Felder. In: Bumke O, Foerster O (eds) Handbuch der Neurologie, vol 6. Springer, Berlin, pp 1–448
Gerhardt E (1938) Der lsocortex parietalis beim Schimpanzen. J Psychol Neurol 48:329–386
Gerhardt E (1940) Die Cytoarchitektonik des Isocortex parietalis beim Menschen. J Psychol Neurol 49:367–419
Geyer S, Ledberg A, Schleicher A, Kinomura S, Schormann T, Burgel U, Klingberg T, Larsson J, Zilles K, Roland PE (1996) Two different areas within the primary motor cortex of man. Nature 382(6594):805–807. doi:10.1038/382805a0
Geyer S, Weiss M, Reimann K, Lohmann G, Turner R (2011) Microstructural Parcellation of the Human Cerebral Cortex – From Brodmann's Post-Mortem Map to in vivo Mapping with High-Field Magnetic Resonance Imaging. Front Hum Neurosci 5:19. doi:10.3389/fnhum.2011.00019
Glasser MF, Van Essen DC (2011) Mapping human cortical areas in vivo based on myelin content as revealed by T1- and T2-weighted MRI. J Neurosci 31(32):11597–11616. doi:10.1523/JNEUROSCI.2180-11.2011
Hadjikani N, Liu AK, Dale AM, Cavanagh P, Tootell RB (1998) Retinotopy and color sensitivity in human visual cortical area V8. Nat Neurosci 1:235–241
Hopf A (1954a) Die Myeloarchitektonik des Isocortex temporalis beim Menschen. J Hirnforsch 1:208–279
Hopf A (1954b) Zur architektonischen Gliederung der menschlichen Hirnrinde. J Hirnforsch 1(6):442–496
Hopf A (1955) Über die Verteilung myeloarchitektonischer Merkmale in der isokortikalen Schläfenlappenrinde beim Menschen. J Hirnforsch 2:36–54
Hopf A (1956) Über die Verteilung myeloarchitektonischer Merkmale in der Stirnhirnrinde beim Menschen. J Hirnforsch 2(4):311–333
Hopf A (1966) Über eine Methode zur objektiven Registrierung der Myeloarchitektonik der Hirnrinde. J Hirnforsch 8(4):301–313
Hopf A (1968a) Photometric studies on the myeloarchitecture of the human temporal lobe. J Hirnforsch 10(4):285–297
Hopf A (1968b) Registration of the myeloarchitecture of the human frontal lobe with an extinction method. J Hirnforsch 10(3):259–269

Hopf A (1969) Photometric studies on the myeloarchitecture of the human parietal lobe. I. Parietal region. J Hirnforsch 11(4):253–265
Hopf A (1970a) Oskar Vogt. 100th anniversary of his birthday. J Hirnforsch 12(1):1–10
Hopf A (1970b) Photometric studies on the myeloarchitecture of the human parietal lobe. II. Postcentral region. J Hirnforsch 12(1):135–141
Hopf A, Vitzthum HG (1957) Uber die Verteilung myeloarchitektonischer Merkmale in der Scheitellappenrinde beim Menschen. J Hirnforsch 3(2–3):79–104
Johansen-Berg H, Behrens TE, Robson MD, Drobnjak I, Rushworth MF, Brady JM, Smith SM, Higham DJ, Matthews PM (2004) Changes in connectivity profiles define functionally distinct regions in human medial frontal cortex. Proc Natl Acad Sci USA 101(36):13335–13340. doi:10.1073/pnas.0403743101
Jones EG (1987) Brodmann's Areas. In: Adelman G (ed) Encyclopedia of Neurosciences, vol 1, Birkhäuser. Boston, Basel/Stuttgart, pp 180–181
Jones EG (2003) Two minds. Nature 421(6918):19–20
Jones EG (2008) Cortical maps and modern phrenology. Brain 131(8):2227–2233
Jones EG, Burton H (1976) Areal differences in the laminar distribution of thalamic afferents in cortical fields of the insular, parietal and temporal regions of primates. J Comp Neurol 168 (2):197–247. doi:10.1002/cne.901680203
Kaas JH (2002) Neocortex. In: Ramachandran VS (ed) Encyclopedia of the Human Brain, vol 3. Academic, Amsterdam, pp 291–303
Kaes T (1907) Die grosshirnrinde des menschen in ihren Massen und in ihrem Fasergehalt. Ein gehirnanatomischer Atlas. G. Fischer, Jena
Klatzo I (2002) Cécile and Oskar Vogt: the visionaries of modern neuroscience. Springer, Wien/New York
Kleist K (1934) Gehirnpathologie. J.A. Barth, Leipzig
Kurth F, Eickhoff SB, Schleicher A, Hoemke L, Zilles K, Amunts K (2010) Cytoarchitecture and probabilistic maps of the human posterior insular cortex. Cereb Cortex 20(6):1448–1461. doi:10.1093/cercor/bhp208
Lashley KS, Clark G (1946) The cytoarchitecture of the cerebral cortex of Ateles; a critical examination of architectonic studies. J Comp Neurol 85(2):223–305
Le Gros Clark WE (1952) A note on cortical cyto-architectonics. Brain 75(1):96–104
Lungwitz W (1937) Zur myeloarchitektonischen Untergliederung der menschlichen Area praeoccipitalis (Area 19 Brodmann). J Psychol Neurol 47:607–639
Mauss F (1908) Die faserarchitektonische Gliederung der Grosshirnrinde. J Psychol Neurol 13:263–325
Mauss F (1911) Die faserarchitektonische Gliederung des Cortex cerebri der anthropomorphen Affen. J Psychol Neurol 18:410–467
Mesulam M (2011) The evolving landscape of human cortical connectivity: facts and inferences. Neuroimage: Available online 22 Dec 2011. doi:10.1016/j.neuroimage.2011.12.033
Meynert T (1884) Psychiatrie: Klinik der Erkrankungen des Vorderhirns begründet auf dessen Bau, Leistungen und Ernährung. W. Braumüller, Wien
Nieuwenhuys R, Voogd J, van Huijzen C (2008) The human central nervous system. Springer, Heidelberg
Öngür D, Ferry AT, Price JL (2003) Architectonic subdivision of the human orbital and medial prefrontal cortex. J Comp Neurol 460(3):425–449. doi:10.1002/cne.10609
Palomero-Gallagher N, Mohlberg H, Zilles K, Vogt B (2008) Cytology and receptor architecture of human anterior cingulate cortex. J Comp Neurol 508(6):906–926. doi:10.1002/cne.21684
Passingham RE, Stephan KE, Kotter R (2002) The anatomical basis of functional localization in the cortex. Nat Rev Neurosci 3(8):606–616. doi:10.1038/nrn893
Press WA, Brewer AA, Dougherty RF, Wade AR, Wandell BA (2001) Visual areas and spatial summation in human visual cortex. Vision Res 41(10–11):1321–1332
Roland PE, Zilles K (1998) Structural divisions and functional fields in the human cerebral cortex. Brain Res Brain Res Rev 26(2–3):87–105

Rose JE, Woolsey CN (1948) Structure and relations of limbic cortex and anterior thalamic nuclei in rabbit and cat. J Comp Neurol 89(3):279–347. doi:10.1002/cne.900890307

Rose JE, Woolsey CN (1949) The relations of thalamic connections, cellular structure and evocable electrical activity in the auditory region of the cat. J Comp Neurol 91(3):441–466

Sanides F (1962) Die Architektonik des menschlichen Stirnhirns. In: Müller M, Spatz H, Vogel P (eds) Monographien aus dem Gesamtgebiete der Neurologie und Psychiatrie, vol 98. Springer, Berlin/Göttingen/Heidelberg

Sanides F (1964) The cyto-myeloarchitecture of the human frontal lobe and its relation to phylogenetic differentiation of the cerebral cortex. J Hirnforsch 47:269–282

Sarkissov S, Filimonoff I, Kononowa E, Preobraschenskaja I, Kukuew L (1955) Atlas of the cytoarchitectonics of the human cerebral cortex. Medgiz 20: Moscow

Scheperjans F, Eickhoff SB, Homke L, Mohlberg H, Hermann K, Amunts K, Zilles K (2008a) Probabilistic maps, morphometry, and variability of cytoarchitectonic areas in the human superior parietal cortex. Cereb Cortex 18(9):2141–2157. doi:10.1093/cercor/bhm241

Scheperjans F, Hermann K, Eickhoff SB, Amunts K, Schleicher A, Zilles K (2008b) Observer-independent cytoarchitectonic mapping of the human superior parietal cortex. Cereb Cortex 18 (4):846–867. doi:10.1093/cercor/bhm116

Seltzer B, Pandya DN (1978) Afferent cortical connections and architectonics of the superior temporal sulcus and surrounding cortex in the rhesus monkey. Brain Res 149(1):1–24

Stephan KE, Kamper L, Bozkurt A, Burns GA, Young MP, Kötter R (2001) Advanced database methodology for the Collation of Connectivity data on the Macaque brain (CoCoMac). Philos Trans R Soc Lond B Biol Sci 356(1412):1159–1186. doi:10.1098/rstb.2001.0908

Strasburger EH (1937a) Die myeloarchitektonische Gliederung des Stirnhirns beim Menschen und Schimpansen – I. J Psychol Neurol 47:460–491

Strasburger EH (1937b) Die myeloarchitektonische Gliederung des Stirnhirns beim Menschen und Schimpansen – II. J Psychol Neurol 47:565–606

Strasburger EH (1938) Vergleichende myeloarchitektonische Studien an der erweiterten Brocaschen Region des Menschen. J Psychol Neurol 48:477–511

Uttal WR (2001) The new phrenology: The limits of localizing cognitive processes in the brain. MIT Press, Cambridge, MA

Uylings HB, Sanz-Arigita EJ, de Vos K, Pool CW, Evers P, Rajkowska G (2010) 3-D cytoarchitectonic parcellation of human orbitofrontal cortex correlation with postmortem MRI. Psychiatry Res 183(1):1–20. doi:10.1016/j.pscychresns.2010.04.012

Van Essen DC (2005) A Population-Average, Landmark- and Surface-based (PALS) atlas of human cerebral cortex. Neuroimage 28:635–662

Van Essen DC (2006) SumsDB (2006) http://sumsdb.wustl.edu:8081/sums/index.jsp

Vogt O (1903) Zur anatomischen Gliederung des Cortex cerebri. J Psychol Neurol 2:160–180

Vogt O (1906) Über strukturelle Hirnzentra mit besonderer Berücksichtingung der strukturellen Felder des Cortex pallii. Anat Anz 29:74–114

Vogt O (1910a) Die myeloarchitektonische Felderung des Menschlichen Stirnhirns. J Psychol Neurol 15:221–232

Vogt O (1910b) Considerations generales sur la myelo-architecture du lobe frontal. Rev Neurol 19:405–420

Vogt O (1911) Die myeloarchitektonik des isocortex parietalis. J Psychol Neurol 18:379–390

Vogt O (1918) Korbinian Brodmann. J Psychol Neurol 24:I–X

Vogt O (1923) Furchenbildung und Architektonische Rindenfelderung. J Psychol Neurol 29:438–439

Vogt O (1927) Architektonik der menschlichen Hirnrinde. Jahresversammlung d. deutschen Verein fuer Psychiatrie Düsseldorf, 23./24.9.1926. Allg Z Psychiat 86:247–266

Vogt M (1928a) Über omnilaminaire Strukturdifferenzen und lineare Grenzen der architektonischen Felder der hinteren Zentralwindung des Menschen. J Psychol Neurol 35:177–193

Vogt M (1928b) Erwiderung zu dem vorstehenden Aufsatz von Economos. J Psychol Neurol 36:320–322

Vogt O (1943) Der heutigen Stand der cerebralen Organologie und die zukünftige Hirnforschung. Anat Anz 94:49–73

Vogt O (1951) Die anatomische Vertiefung der menschlichen Hirnlokalisation. Klin Wochenschr 29(7–8):111–125

Vogt C, Vogt O (1907) Zur Kenntnis der elektrisch erregbaren Hirnrindengebiete bei den Säugetieren. J Psychol Neurol 8:277–456

Vogt C, Vogt O (1919) Allgemeinere Ergebnisse unserer Hirnforschung. J Psychol Neurol 25:279–468

Vogt O, Vogt C (1922) Erkrankungen der Grosshirnrinde im Lichte der Topistik, Pathoklise und Pathoarchitektonik. J Psychol Neurol 28:8–171

Vogt C, Vogt O (1928) Die Grundlagen und die Teildisziplinen der mikroskopischen Anatomie des Zentralnervensystems. In: Handbuch des mikroskopischen Anatomie des Menschen, vol 4 Teil 1. Springer, Berlin, pp 448–477

Vogt C, Vogt O (1929) Über die Neuheit und den Wert des Pathoklisenbegriffes. J Psychol Neurol 38:147–154

Vogt C, Vogt O (1936) Sitz und Wesen der Krankheiten im Lichte der topistischen Hirnforschung und des Variierens der Tiere. J Psychol Neurol 47:237–457

Vogt C, Vogt O (1942) Morphologische Gestaltungen unter normalen und pathogenen Bedingüngen. J Psychol Neurol 50:161–524

Vogt C, Vogt O (1954) Gestaltung der topistischen Hirnforschung und ihre Forderung durch den Hirnbau und seine Anomalien. J Hirnforsch 1:1–46

Vogt C, Vogt O (1956) Weitere Ausführungen zum Arbeitsprogramm des Hirnforschungsinstitutes in Neustadt/Schwarzwald. J Hirnforsch 2:403–427

Von Economo C (1928) Bemerkungen zu dem Aufsatz von Marthe Vogt. J Psychol Neurol 36:320–322

Von Economo C (2009) Cellular structure of the human cerebral cortex. (trans:Triarhou LC). Karger, Basel

Von Economo C, Koskinas GN (1925) Die Cytoarchitektonik der Hirnrinde des erwachsenen Menschen. Springer, Wien/Berlin

Wandell BA, Dumoulin SO, Brewer AA (2007) Visual field maps in human cortex. Neuron 56 (2):366–383. doi:10.1016/j.neuron.2007.10.012

Zilles K, Amunts K (2009) Receptor mapping: architecture of the human cerebral cortex. Curr Opin Neurol 22(4):331–339. doi:10.1097/WCO.0b013e32832d95db

Zilles K, Amunts K (2010) Centenary of Brodmann's map conception and fate. Nat Rev Neurosci 11:139–145. doi:10.1038/nrn2776

Zilles K, Palomero-Gallagher N (2001) Cyto-, myelo- and receptor architectonics of the human parietal cortex. Neuroimage 14:8–20

Part II
The Challenge of Mapping Cortical Areas Noninvasively in Living Brains

Chapter 4
Estimating the Location of Brodmann Areas from Cortical Folding Patterns Using Histology and Ex Vivo MRI

Bruce Fischl

Abstract The human cerebral cortex can be parcellated into a mosaic of microscopically (i.e., architectonically) definable areas based on localizable and more or less pronounced changes in the laminar distribution of neuronal cell bodies (cytoarchitecture) and/or intracortical myelinated fibers (myeloarchitecture) (Brodmann K (1909) Vergleichende Lokalisationslehre der Großhirnrinde in ihren Prinzipien dargestellt auf Grund des Zellenbaues. Verlag von Johann Ambrosius Barth, Leipzig; Vogt (J Psychol Neurol 18:107–118, 1911); von Economo C (1929) The cytoarchitectonics of the human cerebral cortex. Oxford University Press, London; Sarkissov S, Filimonoff I, Kononova IP, Preobrazenskaja NS, Kukueva L (1955) Atlas of the cytoarchitectonics of the human cerebral cortex. Medgiz, Moscov). The most famous of these parcellations is the one proposed by Korbinian Brodmann (1909) Vergleichende Lokalisationslehre der Großhirnrinde in ihren Prinzipien dargestellt auf Grund des Zellenbaues. Verlag von Johann Ambrosius Barth, Leipzig) a century ago. Most current imaging studies of the human cortex report the location of effects as a "Brodmann area" (BA). Although these attributions are common, they are not typically accompanied by any rigorous statistical analysis of the uncertainty associated with the localization. More commonly researchers identify Brodmann areas based on an ad hoc assessment of the location of interest relative to surrounding folding patterns. This approach is problematic as (1) there is no means to rigorously test the uncertainty of the localizations, and (2) until recently, little has been known about the relationship between the Brodmann areas and the cortical folds. In this chapter, I discuss methods for using imaging and computational tools to more accurately localize Brodmann areas, as well as to quantify the uncertainty associated with the localization. This includes the use of ultra-high-resolution ex vivo MRI, the geometry of cortical folding patterns, and the minimum

B. Fischl (✉)
Athinoula A. Martinos Center, Massachusetts General Hospital/Harvard Medical School, Bldg. 149, 13th St., Charlestown, MA 02129, USA
e-mail: fischl@nmr.mgh.harvard.edu

S. Geyer and R. Turner (eds.), *Microstructural Parcellation of the Human Cerebral Cortex*, DOI 10.1007/978-3-642-37824-9_4, © Springer-Verlag Berlin Heidelberg 2013

interior distance that axonal fibers would need to traverse in order to connect two cortical areas. The results both improve state-of-the-art accuracy in the in vivo estimation of architectonic boundaries, but also provide some insight into the relationship between the variability of areal boundaries with respect to cortical folding patterns and an area's place in a putative cortical processing hierarchy.

4.1 Introduction

The human cerebral cortex can be parcellated into a mosaic of microscopically (i.e., architectonically) definable areas based on localizable and more or less pronounced changes in the laminar distribution of neuronal cell bodies (cytoarchitecture) and/or intracortical myelinated fibers (myeloarchitecture) (Brodmann 1909; Vogt 1911; von Economo 1929; Sarkissov et al. 1955). The most famous of these parcellations is the one proposed by Korbinian Brodmann (Brodmann 1909) a century ago. Most current imaging studies of the human cortex report the location of effects as a "Brodmann area" (BA). Although these attributions are common, they are not typically accompanied by any rigorous statistical analysis of the uncertainty associated with the localization. More commonly researchers identify Brodmann areas based on an ad hoc assessment of the location of interest relative to surrounding folding patterns. This approach is problematic as (1) there is no means to rigorously test the uncertainty of the localizations, and (2) until recently, little has been known about the relationship between the Brodmann areas and the cortical folds. While computational analysis of neuroimaging data has advanced remarkably in the last decade (Gerig and Kikinis 1990; Gerig et al. 1991; Gerig et al. 1992; Miller et al. 1993; Dale 1994; Davatzikos and Prince 1995; Prince et al. 1995; Davatzikos and Bryan 1996; Davatzikos et al. 1996; Kikinis et al. 1996; Thompson et al. 1996; Wells et al. 1996; Ashburner et al. 1997; Dale and Buckner 1997; Davatzikos 1997; Thompson et al. 1997; Ashburner et al. 1998; Leventon and Grimson 1998; MacDonald 1998; Fischl et al. 1999a; Zeng et al. 1999; Dale et al. 2000; Thompson et al. 2000; Fischl et al. 2001; Zhang et al. 2001; Fischl et al. 2002; Fischl et al. 2004; Han et al. 2004; Miller 2004; Smith et al. 2004; Styner et al. 2004; Tosun et al. 2004), none of this research has focused explicitly on relating the cytoarchitecture to the observable geometry.

The myelo- and cytoarchitectural differences between adjacent areas that are the basis of the definition of the Brodmann areas vary considerably in terms of their spatial scale and subtlety. For example, probably the most salient histological feature of the cortex was first observed by Francisco Gennari in 1782 and is named after him: the stria of Gennari, a highly myelinated stripe in layer IV of primary visual cortex (Brodmann area 17). The stria of Gennari is one of the few histological features of the cortex that has been detected in vivo, at least for part of its extent (Clark et al. 1992), and is easily resolved using ex vivo MRI (Hinds et al. 2008). Another prominent cytoarchitectural feature of the cortex is the layer II islands in entorhinal cortex (EC, Brodmann area 28) that give rise to the perforant pathway through which most neocortical information travels to the hippocampus.

More recently, there have been some successes in using ex vivo MRI at ultra-high field to robustly visualize these cell dense regions throughout the extent of EC (Augustinack et al. 2005; Fischl et al. 2009), as well as the perforant path itself (Augustinack et al. 2010).

The most widely used coordinate system in neuroimaging is the one developed by Talairach and Tournoux (Talairach et al. 1967; Talairach and Tournoux 1988), which provides stereotaxic maps for inferring the architectonic localization of cortical effects (e.g. functional or structural differences between populations or conditions). Unfortunately, while popular tools exist for estimating BAs from Talairach coordinate (Lancaster et al. 1997; Lancaster et al. 2000), this coordinate system has been shown to be a poor predictor of the locations of both primary sensory (Rademacher et al. 1992; Rademacher et al. 1993; Amunts et al. 2000; Geyer et al. 2000; Morosan et al. 2001; Rademacher et al. 2001) and higher-order cortical areas (Amunts et al. 1999; Amunts et al. 2005). An alternative and even more widespread approach is to make an ad hoc estimation of the BA containing a given cortical effect by visually comparing individual folding patterns with those in Brodmann's drawings, a determination that comes with no defined estimate of precision or uncertainty. Further, Brodmann's drawings give no means of assessing the variability of the relationship between the folds and the cytoarchitectonic boundaries.

The variability of the architectonics has been characterized in several studies, particularly the landmark work of Rajkowska and Goldman-Rakic, in which 7 human left hemispheres were analyzed to characterize the variability in areas 9 and 46 (Rajkowska and Goldman-Rakic 1995a, b), with reconstructions of the lateral portion of the hemispheres carried out in five cases. In this study, considerable variability was found in the morphology of frontal sulcal patterns. Further, by overlaying their architectonic maps on the Talairach atlas, Rajkowska and Goldman-Rakic were able to point out the ambiguity in other published results that reported findings in a particular BA (e.g. an effect reported in area 9 could have been 45 or 46 instead). It has not been clear whether the well-documented inaccuracy of the use of the Talairach coordinate system for localizing BAs reflects the true variability of the underlying architectonic areas, or if higher dimensional nonlinear coordinate systems based on other types of macroscopically observable features could be used in order to increase the accuracy of the localization of the underlying cyto- and myeloarchitecture.

Cortical folding patterns are intriguing in terms of their spatial complexity and their cross-subject variability. The deepest cortical folds, such as the sylvian fissure, the calcarine sulcus, etc...., are the phylogenetically oldest, the first to form in human development and the most stable across subjects. In human beings, cortical development begins prenatally with the majority of neurons being generated before birth. The development of the folds starts at about 9 weeks in gestation, proceeds rapidly until birth, and continues with subtle effects into late adolescence (Chi et al. 1977; Lewis 1997). The mechanisms involved in the formation of folding patterns, and their relationship to architectonic boundaries, remains unclear, although some combination of neural differentiation, migration and axon formation must play a role. One theory suggests that the differential growth of the outer layers of the forming cortex relative to the inner layers results in cortical buckling (Caviness

1975) and fold formation. Another hypothesizes that the mechanical tension generated by the creation of axonal connections leads to folding (Van Essen 1997). In terms of temporal sequencing, there is the notion that large scale cortical folds develop earlier, and the secondary and tertiary folds develop later (Chi et al. 1977), a theory that we have experimentally confirmed using spherical wavelets and Gompertz functions to model the development of cortical folding patterns (Yu et al. 2007) in a set of neonates.

Recent research has aimed to develop algorithms that allow the explicit assessment of the uncertainty associated with the in vivo localization of a set of Brodmann areas with respect to any particular coordinate system (Schleicher and Zilles 1990; Geyer et al. 1997; Amunts et al. 1999; Schleicher et al. 1999; Amunts et al. 2000; Geyer et al. 2000; Morosan et al. 2001; Rademacher et al. 2001; Rademacher et al. 2002; Amunts et al. 2005; Malikovic et al. 2007; Fischl et al. 2008; Hinds et al. 2008; Fischl et al. 2009; Yeo et al. 2010). In this chapter we summarize the current state of the art regarding estimating architectonic boundaries using cortical folding patterns as well as other features that are accessible in vivo. The estimation is based on information gleaned either from standard histological analysis, or from using ultra-high field MRI of fixed brains. The remainder of this chapter is organized as follows. Section 4.2 presents techniques for estimating architectonic boundaries using cortical folding patterns, and a quantification of the accuracy of volumetric and surface-based coordinate systems, as well as a comparison of how well predicted areas are by folding patterns. Section 4.3 introduces ex vivo magnetic resonance imaging (MRI) and its advantage and challenges, and shows how data acquired in this way can be used in place of more typical histological processing in order to delineate architectonic boundaries. In Sect. 4.4 we introduce other measures that can help constrain the location of architectonic boundaries, and in Sect. 4.5 we discuss the possibility of in vivo localization. Finally in Sect. 4.6 we finish with a discussion of future directions and concluding remarks.

4.2 Localizing Brodmann Areas Based on Cortical Folding Patterns

4.2.1 Registration Based on Cortical Folding Patterns

Despite the widespread use of cortical folding patterns to perform ad hoc estimations of the locations of the Brodmann areas in individuals, little is understood regarding the relationship of the folds to the Brodmann areas, or whether there is a hierarchy in the predictability of the Brodmann areas. The architectonics are of course important as the mosaic of functionally defined regions that are arrayed across the cortical sheet, e.g. (Allman and Kaas 1971; Tootell et al. 1983; Felleman and Van Essen 1991; Sereno and Allman 1991), are strongly linked to the underlying anatomy. In collaboration with Drs. Amunts and Zilles we used whole-brain histology for the automated identification of cortical areas (Schormann and

Zilles 1998; Amunts et al. 1999; Zilles et al. 2002)) and surface-based analysis (Dale et al. 1999; Fischl et al. 1999a; Fischl et al. 1999b) to explicitly test how well the folds predict the locations of the areas.

More specifically, we have developed a technique that uses cortical folding patterns to drive inter-subject registration as opposed to the more commonly used image intensities. The advantages of this technique are: (1) folding patterns are better predictors of functional and architectonic properties than image intensities, and (2) it reduces the problem from a three dimensional one of finding a set of correspondences in the volume, to a two dimensional one of finding correspondences on the surface. One can think of this as moving the folds around on the surface of the sphere, but not allowing them to move inwards or outwards. An immediate question that arises in surface-based registration is what geometric features to use in generating the registration? One school of thought (Thompson and Toga 1996; Drury and Van Essen 1997) is that the folding patterns are so variable that manual labeling is required to identify correspondences across subject, with the remainder of the cortex driven into register by interpolating the registration of the landmarks. Another approach, which we take, is to use the entire pattern of cortical geometry, but measure the variance of the folds across subjects and weight the registration by the inverse of the variance. From a probabilistic perspective this amounts to a maximum likelihood solution in which the noise in the folds are modeled with a Gaussian distribution with mean and variance computed from a sample population. It is worth pointing out that from this probabilistic perspective, the landmark-based approach amounts to assuming the variance is identity at the landmarks and infinite away from them. Instead of making this unrealistic assumption, we instead measure the variance and use it as a natural weighting to diminish or remove the effects of folds that are too variable to base cross-subject registration upon.

A related, but equally important question is whether all folds have the same variance across space. As noted in the introduction, there is the notion that deeper folds are more stable, which implies that the noise in the folding patterns is *heteroscedastic* – that is, it is different for different points in the cortex. In order to quantify the stability of the folds and the relationship between depth and cross-subject consistency we processed 435 subjects in the publically available OASIS dataset (Marcus et al. 2007). For each subject, we binarized the average convexity or sulcal depth to remove any bias involving the absolute value of the depth, then performed rigid spherical registration to our standard atlas. We then computed the cross-subject variance of the sulcal depth for each point in the cortex. If the variance of the noise in the folds is the same across space, we would expect to see a linear increase in the standard deviation of the sulcal depth with the mean, as scaling the mean of a Gaussian results in a linear scaling of the standard deviation. However, as shown by Fig. 4.1, this linear scaling appears to hold for the shallower folds (those with small mean average convexity), however, the deepest folds are significantly less variable than would be predicted by identically distributed Gaussian noise, implying that the postulated relationship between depth and stability holds, and that

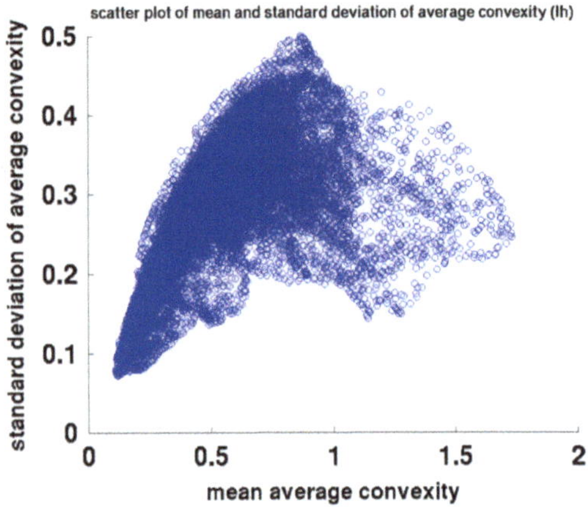

Fig. 4.1 Plot of the mean of the absolute value sulcal depth or average convexity versus the standard deviation across the 435 subjects in the OASIS dataset (Marcus et al. 2007)

measuring the variance at each point in space will weight deeper folds more strongly, leading to a more stable cross-subject registration procedure.

In order to investigate the relationship between folding patterns and architectonic location, we used the reconstructed histological volumes provided by Drs. Amunts and Zilles to generate surface models of the gray/white interface for each of the subjects that they had obtained histologically-defined architectonic labels (Dale et al. 1999; Fischl et al. 1999a; Fischl et al. 2001). The 8 labeled Brodmann area maps (areas 2, 4a, 4p, 6, 44, 45, 17 and 18) were sampled onto the surface models for each hemisphere, and errors in this sampling were manually corrected (e.g. when a label was erroneously assigned to both banks of a sulcus). The 10 left and 10 right hemispheres were morphed into register using the high-dimensional nonlinear morphing technique that aligns cortical folding patterns (Fischl et al. 1999b) as described above, with heteroscedastic variance estimates. Note that no specific optimization was performed for aligning the Brodmann areas presented in this report. Rather, a set of parameters that had been determined to be optimal for aligning V1 in a separate ex vivo data set (Hinds et al. 2008) were used with no modification.

Using the surface-based registration, we constructed spatial probability maps for the 8 Brodmann areas. The results of this analysis are shown in Fig. 4.2, which displays the average convexity of the in vivo atlas that is used as a common space in dark (sulci) and light (gyri) gray, with the probability maps overlaid using a heat scale. These include primary and secondary visual areas BA 17 and BA 18, respectively, BA 44 and BA 45 (subdivisions of Broca's area), somatosensory area BA 2, primary motor areas 4a and area 4p, and finally the premotor area BA 6. Frequency estimates of the probability that each point was part of each BA were constructed in a surface-based coordinate system by tabulating the number of times that a label occurred at a given point and dividing by the total number of subjects for each label. Each point in the surface-based coordinate system can then be probed to determine the probability that it is part of any of the set of labeled BAs. As can be seen, the folding-based surface alignment also results in extremely accurate alignment of the underlying BAs, particularly primary visual, somatosensory and motor cortex, thus highlighting the utility of the surface-based coordinate system as the space in which to develop techniques designed for the automated detection of architectonic boundaries in the human cerebral cortex.

A perfect coordinate system with respect to the architectonics would yield probability maps that are step functions – perfect certainty that a given coordinate is contained within an area transitioning to perfect certainty that one is outside an area. Unfortunately, current volumetric coordinate systems in use in neuroimaging are far from this ideal, yielding spatial probability maps with few if any coordinates that achieve perfect agreement in the training data (e.g. (Geyer et al. 1997; Amunts et al. 1999; Amunts et al. 2000; Geyer et al. 2000; Rademacher et al. 2002)). In contrast, the maps shown in Fig. 4.2 contain significant regions of 100 % probability, validating and quantifying the use of folding patterns to estimate the location of architectonic boundaries, a procedure that has been in use for decades.

In order to quantify the distance error in the folding-pattern-based registration, we used the registration to map the Brodmann Area labels across subjects and computed the symmetric mean Hausdorff distance between the mapped label and the true one, as shown in Fig. 4.3. The Hausdorff distance is a set-theoretic metric that measures the minimum distance from each point in one set to any point in the other, then typically is assigned the maximum over all points. Symmetrizing it means returning the average in each direction (i.e. the min using set 1 as the source and set 2 as the target, then the min using set 2 as the source and set 1 as the target). The mean Hausdorff distance uses the mean over all minimum distances as opposed to the max, and gives a better intuitive estimate of how close the majority of one boundary is to the other. Examining Fig. 4.3 we can see that in general the folding patterns are excellent estimators of the areal boundaries for primary motor and somatosensory areas, with V1 exhibiting an average boundary error of approximately 2.3 mm, or less than the size of a typical functional MRI (fMRI) voxel, with the estimation error increasing as one transitions up the putative cortical hierarchy to over 6 mm for BA44 and 45 in the left hemispheres, and even more than that in the right.

4.2.2 Using Optimal Weighting of Cortical Folding Patterns

In the previous section we showed that the alignment of cortical folding patterns produced significantly better registration of the underlying architectonics than standard volumetric techniques. The underlying algorithm used to generate this result computes summary statistics from a population of subjects, then models the folding patterns using a Gaussian distribution, with the mean and variance of the Gaussians estimated from the training data at each point on the surface. The maximum *a posteriori* estimate of the registration function uses the inverse of the

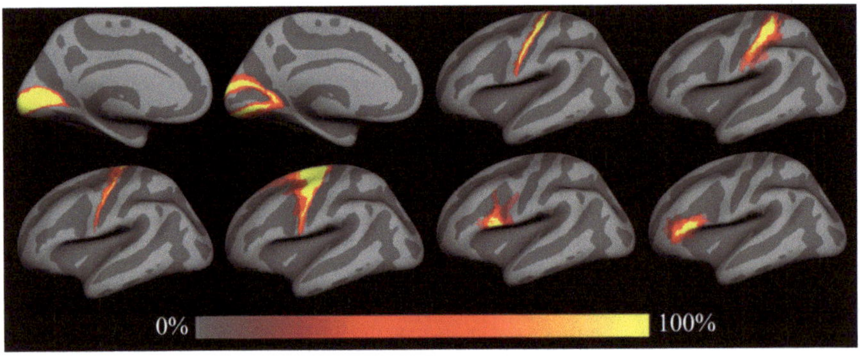

Fig. 4.2 Spatial probability maps of different Brodmann areas. *Top row*: left hemisphere areas 17, 18, 4p and 2 (from left to right). *Second row*: areas 4a, 6, 44 and 45. *Third row*: color scale used for spatial probability maps

variance of the folds as a natural weighing. While the inverse variance is optimal for aligning folds given this specific probabilistic model (Gaussian distributed mean folding patterns), there is no reason to expect it to optimally align the underlying architectonics. In this section we discuss a recently developed computational technique to compute a weighting that is optimal for architectonic alignment.

Specifically, in the registration one uses the cross-subject mean image as a target, weighted by the cross-subject variance. In (Yeo et al. 2010) we generalized this method to learn both a target to replace the mean, and an optimal weighting to replace the inverse variance. The learning of such a set of patterns is extremely computationally intensive as for each step in the learning process all the training data must be registered with the current target and weighting, then the target and weighting are adjusted to more optimally align the architectonics, then the registration is run again and the process is iterated until it converges. However, while the training is slow, the use of the generalized target and vertex-wise weighting to register a new subject is exactly the same computational cost as the inverse variance weighting presented in the previous section.

The results of learning such a target and weighting and a comparison with the inverse variance weighting is given in Fig. 4.4. The red outlines in this figure represent the true outline of the underlying architectonic region as provided by histological analysis ((Amunts et al. 2000) for V1 and V2, (Amunts et al. 1999) for BA44 and 45, (Geyer et al. 1997; Geyer and Ledberg 1996) for area 2, and (Malikovic et al. 2007) for MT/V5), while the green outline shows the subject with the median accuracy for inverse variance weighting with a mean target (top row) and for optimal target/weighting in the bottom row. (Fig. 4.4)

Directly comparing the methods in terms of their mean symmetric Hausdorff distance in Fig. 4.5 reveals the increased accuracy of the computed optimal weighting with respect to the geometric weighting by inverse variance.

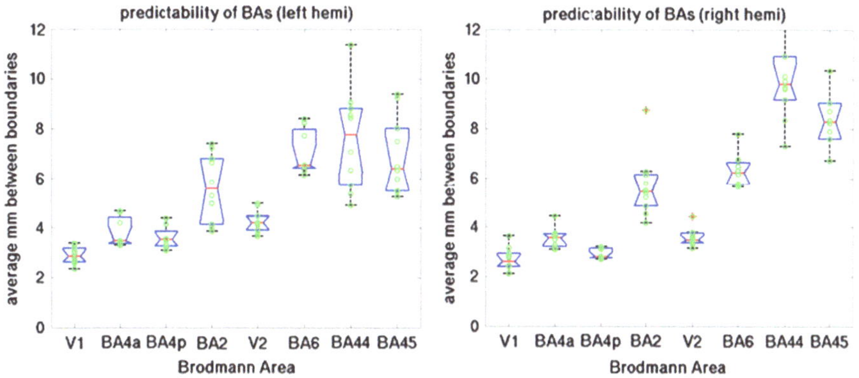

Fig. 4.3 Hausdorff distance from predicted boundary to true boundary for the left hemisphere (*on left*) and right hemisphere (*on right*) of ten subjects for eight histologically defined Brodmann Areas

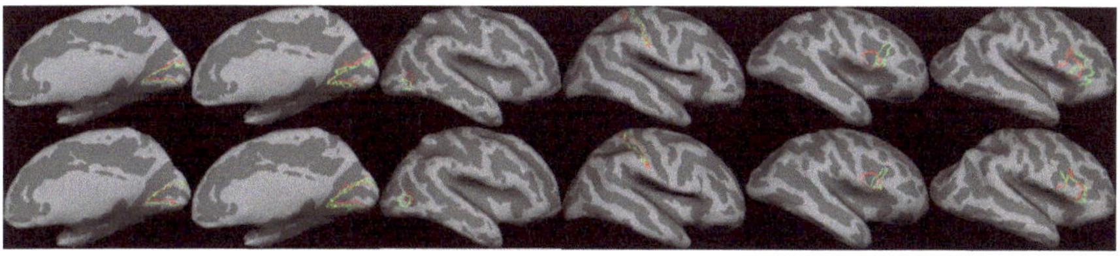

Fig. 4.4 Localization provided by inverse variance weighting (*top row*) and optimal weighting (*bottom row*). *Red* outline is the true (histologically-based) Brodmann Area, and *green* is the one provided by the registration. The subject was chosen to be the one with the median accuracy of each technique. From *left* to *right*: V1, V2, MT/V5, BA2, BA44, BA45

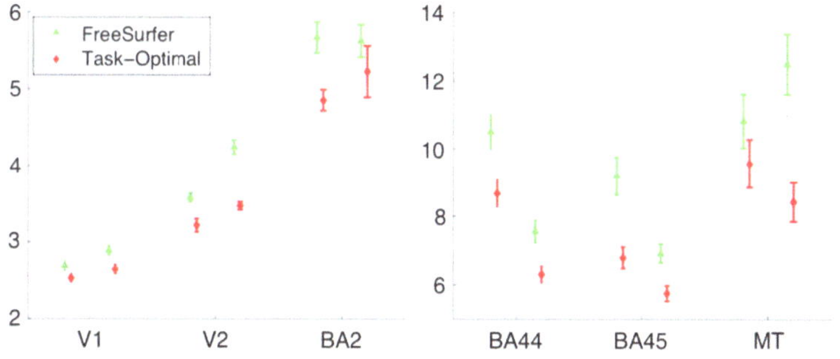

Fig. 4.5 Mean symmetric Hausdorff distance for inverse variance weighting ("FreeSurfer", *green*) and optimal weighting (*red*). Note the different scales on the two plots, reflecting the accuracy of the alignment of areas closer to the sensory periphery (*left*) than those further in way in terms of number of synapses

4.3 Using Ex Vivo MRI to Visualize Architectonic Boundaries

In MRI one pays a steep price for increased resolution, a fact of imaging physics that has made the quest for high-resolution images a difficult one. The signal in a voxel is proportional to its volume, and therefore goes down with the third power of the length of the sides of an isotropic voxel. Isotropic voxel resolution is of course important so that we do not preference an arbitrary direction in the brain. In addition, signal-to-noise ratio (SNR) goes up with the square root of time, assuming temporally uncorrelated noise in subsequent acquisitions. This implies, for example, that to obtain the same SNR at ½ mm isotropic as at 1 mm isotropic, one must scan $(2^3)^2 = 64$ times as long! It is therefore easy to see that simply scanning longer to recover the SNR required to get to high resolutions is not a tenable strategy. What resolution do we require to directly visualize histological properties suitable for detecting and labeling architectonic boundaries? If the cortex averages 2.5 mm in thickness, then each of the six layers is approximately 400 μms thick (although the total thickness is by no means equally distributed among the layers), and Shannon's sampling theorem would suggest that 200 μm would be the coarsest resolution at which resolving the layers is possible. In practice, given the unequal laminar and spatial distributions of thickness we estimate that 100 μm resolution is required over much of the cortex. Given the basic physics relating SNR and resolution it is apparent that obtaining the resolution required to begin to visualize architectonic properties is exceedingly challenging.

It is worth pointing out the importance of developing techniques for delineation of areas with ex vivo MRI. The statistical modeling of relationships between macroscopic features and microscopic boundaries would benefit greatly from the availability of large numbers of examples. The handful that are available currently are simply insufficient, for example, to examine whether second order properties of the folding patterns are predictive of underlying cytoarchitectural boundaries. As one example: Do pairs of folds predict the location of an area, or, if tension based theories of morphogenesis (Van Essen 1997) have explanative power, then are the location of an area such as MT constrained by the locations of V1, to which it is directly connected, as well as the posterior callosum, through which fibers connecting MT to the contralateral MT flow? These and other questions can only be addressed if we have sufficient sample sizes of detailed structural/cytoarchitectural data, something that will be far easier to accomplish if we can directly image the necessary architectonic details without requiring mounting and staining of tissue sections.

While histology remains the gold standard of neuroanatomy, high resolution ex vivo imaging is gaining increasing importance (Johnson et al. 1986; Johnson et al. 2002; Augustinack et al. 2004; Hinds et al. 2008; Fischl et al. 2009; Yushkevich et al. 2009; Augustinack et al. 2010; Augustinack et al. 2012). While it is unlikely that ex vivo MRI can approach the types of resolution or specificity one obtains under a microscope in the near future (e.g. $< = 1$ μm), a series of factors make it feasible to increase SNR by up to three orders of magnitude relative to in vivo scans, an increase that can be directly traded for an order of magnitude in resolution. These factors include the lack of blurring induced by cardiac and respiratory cycles, the fact that small coils can be placed within a cm or so of the sample with no intervening skull to attenuate the signal and load the coil, and of course the possibility of scanning for many days as a means of recovering SNR is perfectly feasible (as opposed to the 90 min or so that is the maximum length of an in vivo scan session). Relative to histology, ex vivo MRI has some notable advantages:

1. It is possible to obtain multiple contrasts of the same exact tissue (e.g. T1, T2, proton density, diffusion, magnetization transfer, etc.. . ..).
2. Obtaining whole-brain or whole-hemisphere MRI data requires no more effort than a small sample, and importantly, orders of magnitude less effort than whole brain histology.
3. MRI is an intrinsically three dimensional technique that yields isotropic voxels.
4. MRI does not incur irreversible distortions such as the tearing and warping induced by cutting, mounting and staining in histology.

These last two points are of particular importance, as they facilitate the accurate processing of large samples, including whole hemisphere and whole brains, something that is difficult or impossible to do with histological samples in with each section has been distorted differently, and the two in-plane directions are fundamentally different than the through-plane one. This is of particular importance when looking for discontinuities in the through-plane direction, or for example when trying to track axons or fascicles through-plane. An example of the types of data that one can acquire with ex vivo MRI together with the anatomical structures that are visible in it is given in Fig. 4.6.

In order to assess our ability to automatically detect architectonic boundaries with ex vivo MRI, we sampled the MRI data along line segments connecting the gray/white boundary and the pial surface throughout the extent of the sample. Independently, a neuroanatomist with great expertise in the medial temporal lobe (Dr. Jean Augustinack) manually delineated the borders of a set of medial temporal lobe regions including entorhinal cortex, parasubiculum, presubiculum and the subiculum proper (see Fig. 4.7). We then computed the distance between the laminar distribution of intensities along the line segments, ordering them from lateral to medial, and compared them with the manually defined borders, as is done in the histological processing shown previously (Schleicher et al. 1999). The results of this study are given in Fig. 4.8. As can be seen, each of the manually defined borders coincides with peaks in the distance function, indicating that these borders can be detected using the high resolution ex vivo MR data.

Thus, these types of data can be used to build models of anatomical structures and infer the locations of architectonic boundaries in much the same way as the histological data has been. An example of this type of procedure applied to the boundaries of entorhinal cortex is given in Fig. 4.11. In the next section we discuss the challenges of acquiring MRI data ex vivo, and the types of image acquisitions that are optimal for fixed tissue.

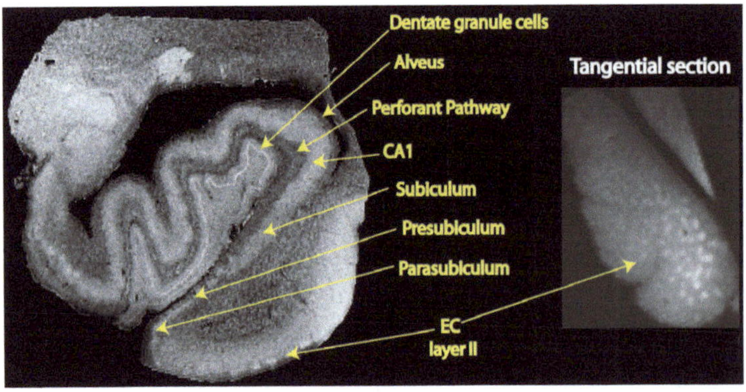

Fig. 4.6 Image of the human medial temporal lobe (*left*) acquired at 7 T with 100 μm isotropic voxels (TR = 20, TE = 5, α = variable). *Right*: oblique slice of same specimen through layer II of EC showing the layer II islands as brighter regions. Layers in the CA fields of the hippocampus and the dentate gyrus are clearly visible

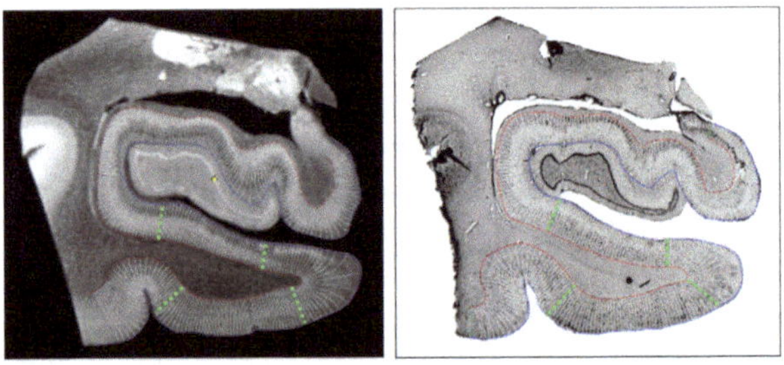

Fig. 4.7 Line profiles shown over a high-resolution 7 T ex vivo MRI (*left*) and the Nissl stain of the corresponding slice (*right*). The *green lines* show locations of local maxima in the properties of adjacent blocks of line profiles (*red line* = gray/white boundary, *blue line* = pial surface, *Abbreviations: EC*, entorhinal cortex*; PC*, perirhinal cortex*; para-sub*, parasubiculum*; pre-sub*, presubiculum)

4.3.1 Optimizing Image Acquisition and Reducing Distortions

The imaging of ex vivo tissue samples at high field involves different challenges than those posed by more standard in vivo structural imaging. The samples are almost always imaged after fixation, a process that dramatically changes their MR properties. Specifically, fixation reduces the T1 relation time of both gray and white matter, and brings them closer to each other. The result of this reduction is to make standard T1-weighting an extremely inefficient way to image fixed tissue samples. Here, we build on techniques we have developed for high SNR, low distortion imaging (Fischl et al. 2004) and adapt them for use in ex vivo samples with the goal

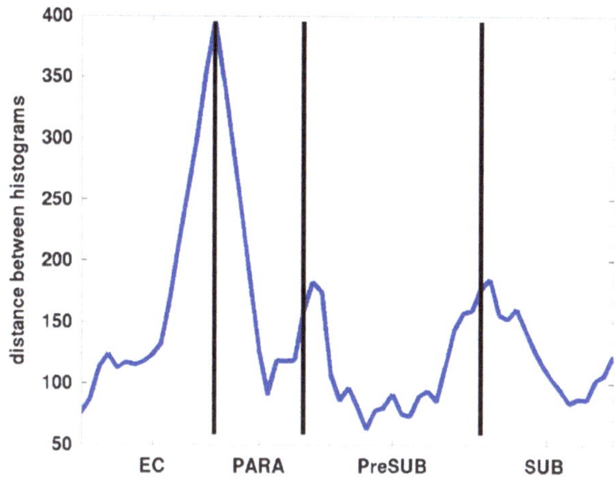

Fig. 4.8 Distance between adjacent histograms of laminar intensity profiles from lateral (*left*) to medial (*right*). Note the peaks in profile distances indicating dissimilarity between the laminar distribution of intensities at manually defined border of EC and the para subiculum (PARA), the PARA and the pre subiculum (PreSUB), as well as the PreSUB and the subiculum (SUB) proper

of generating sufficient CNR to extend the number of cytoarchitectonic borders visible in MR images.

4.3.1.1 Optimizing Ex Vivo MRI Acquisition

Much effort has been devoted in the MRI community towards finding pulse parameters that are optimal for various tasks (Grief et al. 1985; Edelstein et al. 1986; Baker 1991; Constable and Henkelman 1991; Epstein et al. 1994; Constable et al. 1995; Venkatesan and Haacke 1997), including optimization for segmentation (Prince et al. 1995). In recent work, we phrased the segmentation problem in terms of models of image acquisition (Fischl et al. 2004). An advantage this type of approach is that it provides a natural framework for formulating an energy functional for sequence optimization specifically for segmentation. In particular, we can compute the probability of mistakenly labeling a voxel as class c_1 when the true class is c_2 as a function of the MR parameters **m**, using information from a probabilistic atlas:

1.
$$p(c_1(\mathbf{r})|c_2(\mathbf{r}),\mathbf{m}) = \int p(c_i(\mathbf{r})|\mathbf{I}(\mathbf{m}))p(\mathbf{I}(\mathbf{m})|c_2(\mathbf{r}))d\mathbf{I} = p(c_1(\mathbf{r})) \times \int p(\mathbf{I}(\mathbf{m})|c_i(\mathbf{r}))p(\mathbf{I}(\mathbf{m})|c_2(\mathbf{r}))d\mathbf{I}$$

We then seek the combination of MR parameters **m** that minimize the probability of misclassification over all pairs of tissue classes that occur together anywhere in the atlas:

2.
$$\hat{\mathbf{m}} = \arg\min_{\mathbf{m}} A(\mathbf{m}) = \iiint_{\mathbf{r}} \sum_{c_2 \neq c_1} \sum_{c_1} p(c_1|c_2, \mathbf{m}, \mathbf{r}) d\mathbf{r}$$

The likelihood terms $p(\boldsymbol{I}/c_i(\boldsymbol{r}))$ are assumed to be normally distributed with means and covariances given by:

3.
$$\hat{\mu}_c(\mathbf{m}, \mathbf{r}) = S(\mathbf{m}_{\text{predicted}}, \beta(\mathbf{m}_{\text{training}}, \mu_c(\mathbf{r})), \ \hat{\Sigma}_c(\mathbf{m}) = J_{\text{predicted}}(J^{+}_{\text{training}} \Sigma_c J^{+T}_{\text{training}}) J^{\text{T}}_{\text{predicted}} + \lambda Id$$

where $\mathbf{m}_{\text{training}}$ are the MR parameters used in the construction of the atlas, and $\mathbf{m}_{\text{predicted}}$ are the MR parameters of the synthesized image being assessed, and the function S is the solution to the steady state Bloch equations (Bloch et al. 1946). The covariance structure is predicted by decomposing the noise into two parts. The first is anatomical variability in the intrinsic tissue properties of the various brain structures. The second is white noise inherent in the imaging process. In this formulation, J is the jacobian matrix of S, J^{+} denotes the pseudo-inverse of matrix J, Id is the identity matrix, and λ is a constant that reflects the component of the noise that is scan-dependent, encapsulating factors such as averaging multiple acquisitions and the bandwidth of the scan. Numerically, equation (1) is integrated over the region of intensity that is nonzero for both classes c_1 and c_2, typically $+ -5$ standard deviations from the mean. Minimizing equation (2) thus amounts to reducing the amount of overlap of the distributions for tissue classes that are likely to occur at the same location.

The ambiguity measure captures the difficulty of segmentation in several crucial ways. First, it allows the intrinsic properties of the tissue classes to vary over space, as the ambiguity is computed separately for each atlas location. This is important, as the tissue characteristics show considerable spatial variability (Ogg and Steen 1998; Steen et al. 2000; Fischl et al. 2004). Second, only tissue classes that co-occur in a given atlas voxel contribute to the ambiguity. Thus, the difficulty of segmenting for example cortical gray matter from dura would affect the sequence optimization, but not cortical gray matter from the caudate, as these structures never occur in the same region of atlas space. Finally, acquiring datasets for each of the possible parameters would not be tenable, something that is obviated by the use of the forward model of image formation S.

The parameter estimation procedure is based on a standard gradient echo saturation recovery acquisition protocol such as FLASH/SPGR, which is available on the vast majority of clinical scanners. As such, this sequence is limited in that it doesn't take advantage of recent advances in imaging hardware and acquisition techniques. In particular, it is typically relatively low bandwidth, implying that distortions due to magnetic field inhomogeneities and susceptibility artifacts can be substantial. The low bandwidth is of course used to increase SNR. In general, there is a trade-off between high-bandwidth, low distortion, low SNR images, and low-bandwidth, high-distortion, high-SNR images. That is, SNR and distortion both go down with bandwidth.

In order to avoid this trade-off, we have developed a high-bandwidth multi-echo flash (MEF) sequence that minimizes distortion while maximizing SNR (Fischl et al. 2004). In a single 6.5 min scan, the same amount of time required for a $1 \times 1 \times 1$ mm single-echo FLASH scan, this sequence provides 8 high-bandwidth images at different echo times. While the individual scans can be noisy, the information in the ensemble is significantly greater than the low bandwidth FLASH scans. In addition, the higher bandwidth of the MEF sequence, coupled with the fact that alternating echoes are collected with opposite read-out directions, results in less distortion in the images due to B0 effects (chemical shift and susceptibility distortion) (Fischl et al. 2004). This is particularly important for longitudinal studies in which different shim settings can result in substantial differential distortions between scan sessions for low bandwidth sequences. Physiologic and bulk motion during the readout also result in fewer artifacts due both to the shorter readouts of the multi-echo sequence, and to the averaging of the readouts with alternating directions. Finally, image reconstruction techniques can exploit the alternating readout direction to recover parts of the image previously lost due to susceptibility artifacts (Kadah and Hu 1998; Chen and Wyrwicz 1999; Schmithorst et al. 2001), an important consideration given the difficulty of removing all air bubbles when preparing an ex vivo sample of imaging.

We have recently developed a set of techniques for using MEF scans acquired at varying flip angles and/or repetition times (TR) to estimate the underlying tissue parameters that are the source of image contrast in standard gradient echo sequences (i.e. T1, proton density PD and T2*) (Fischl et al. 2004). Once these parameters have been estimated, they can then be used as input to simulations designed to maximize CNR noise per unit time. We applied these techniques to a set of ex vivo tissue samples, imaged in a 3 turn solenoid coil (28.5 mm i.d. $\times$ 44 mm in length) on a 7 T human scanner (Siemens Medical Systems, Erlangen, Germany). The data was acquired using a MEF sequence, with TR $= 40$ ms, $\alpha = 15°{\cdot}20°{\cdot}25°$, 4 echoes, TE $= 8$ ms, 16 ms, 24 ms, 32 ms) at 100 μm isotropic resolution. The multiple flip angles were used to estimate the T1 and proton density of the underlying tissue using the techniques described in (Fischl et al. 2004). In addition, the multiple echoes were used to estimate $T2^*$ using a log-linear fit to the data. ROIS were then manually drawn in the gray and white matter, and the mean tissue parameters were computed to be: T1 $=$ (770 ms, 1078 ms), PD $=$ (18,024,19,996), $T2^* =$ (11 ms, 26 ms) for (wm,gm)). An example of the data (first 4 images) and the parameter maps (last 3 images) is given in Fig. 4.10. Interestingly, the dominant source of contrast in these images is given by T2* differences in the tissue.

Using these parameters, the steady state Bloch equations were applied (Bloch et al. 1946) and the CNR was computed over TR $=$ [5,80], and flip angle $\alpha =$ [3,40]. For each TR/α pair, the CNR was computed by averaging the synthesized echoes, using an echo spacing of 3 ms, and assuming that 2 ms are required for spoiling at the end of each TR. The CNR was assumed to increase with

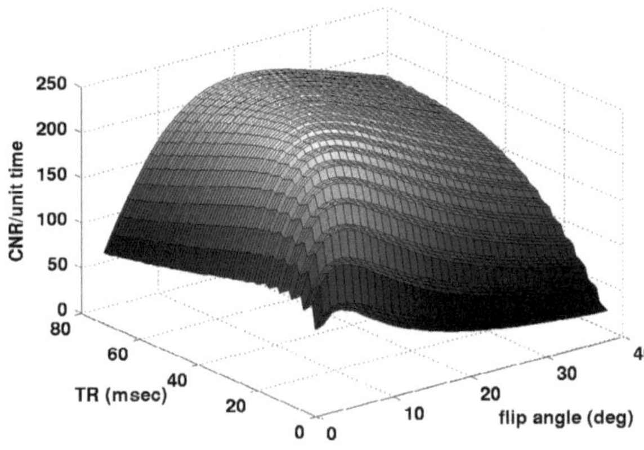

Fig. 4.9 Plot of CNR/unit time versus TR and α for ex vivo MEF

the ½ power of the # of echoes that can be acquired for a given TR and echo spacing, and to decrease with sqrt(TR), corresponding to the assumption that the noise is time-independent (that is, a longer TR implies that fewer scans can be collected and averaged to increase SNR). Note that one advantage of the multi-echo FLASH is that bandwidth effects are no longer important, as the bandwidth is fixed at the maximum value in order to be continually reading data during the sequence. The optimum acquisition parameters from this simulation were computed to be $TR = 48$ ms, $\alpha = 16$, which we will use in future acquisitions. One point to note is that this analysis does not take advantage of the change in contrast properties with echo time. That is, equal weighting of the echoes is clearly not optimal, as the contrast evolves in time due to the differing T2* properties of gray and white matter, as can be seen in Fig. 4.9.

Middle Temporal Area MT/V5

The middle temporal area MT (also known as V5), is a motion sensitive area in the primate brain. There is disagreement on where precisely MT is located in both the anatomical and functional MRI literature and even within modality (Dumoulin et al. 2000; Annese et al. 2005). Nonetheless, we can roughly localize MT as partially in Brodmann area 19 and partially in Brodmann area 37 (Zilles and Clarke 1997; Malikovic et al. 2007). It typically lies posterior to the intersection of the ascending limb of the inferior temporal sulcus and the inferior occipital sulcus. MT cannot be distinguished solely on cytoarchitectural features, but there is agreement on the basic neurochemical properties that differentiate MT. First, cytoarchitecturally MT has a high cell density in layer II/III, a broad layer III, and low cell density in layer V (Malikovic et al. 2007). Second, MT is heavily myelinated and has been localized with gray level index line profiles, with borders that are robustly visible in the deeper layers (Clarke and Miklossy 1990; Sereno and Allman 1991; Zilles and Schleicher 1993). Third, CAT-301 (a proteoglycan at the neuronal

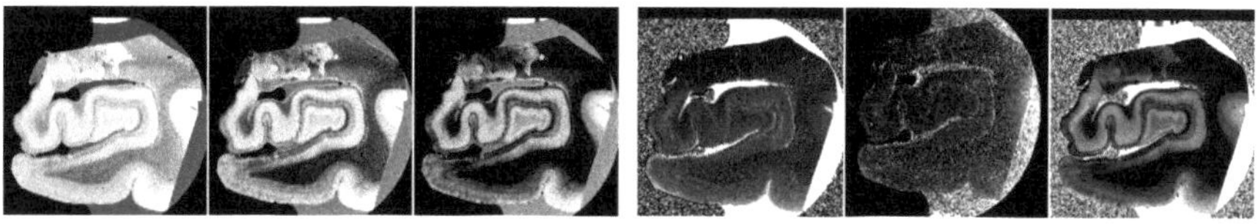

Fig. 4.10 Multi echo flash (images 1–4) with fitted T1 (image 5), PD (image 6) and T2* (rightmost image)

surface) staining also illustrates the intense myelination in MT in its deeper lamina (Tootell and Taylor 1995). Fourth, cytochrome oxidase staining is exceptionally dark in MT (Clarke 1994).

MT has some striking characteristics that make it an excellent candidate for the development of a tool for the segmentation of architectonic areas in the visual stream. It is part of the magnocellular stream (Maunsell et al. 1990) with its larger more heavily myelinated axons, and is lower in the visual hierarchy and hence activates earlier than the surrounding areas (Schmolesky et al. 1998), although some studies have shown timing overlap with the adjacent Medial Superior Temporal area MST. Well-characterized functional localizers exist for demarcating the boundary of MT and surrounding motion sensitive areas (Tootell et al. 1993), collectively known as MT+. In addition, MT is known to contain a retinotopic representation of the visual field and to have smaller receptive fields (RFs) than adjacent areas (Desimone and Ungerleider 1986), properties that have been used to distinguish it functionally from the other areas that make up MT + (Dukelow et al. 2001; Huk et al. 2002; Gardner et al. 2008). MT is directly and heavily connected to V1, again the only area in the region that possesses this type of connectivity pattern (Maunsell and Van Essen 1983; Weller and Kaas 1983). Additional motivation is provided by the fact that recent studies have had some success in detecting MT using in vivo MRI, at least in part (Walters et al. 2003). Finally, it is known to have heavily myelinated deep layers reflecting the large heavily myelinated inputs it receives from V1 (Annese et al. 2005).

Image Analogies for Using Histological Properties to Drive Segmentation

One difficulty with segmenting architectonic boundaries with MRI is the poorly understood relationship between histological and magnetic tissue properties of fixed tissue. This makes it difficult to directly predict what a given histological signature will appear as in an MRI image. For that reason, we propose to borrow technology from the image processing literature that facilitates the synthesis of one type of image from another given some pair training samples (Hertzmann et al. 2001). Specifically, we utilize an histology-to-MR registration that we have developed (Reuter et al. 2012) to align a 100 um MR volume with a photomicrograph of a section stained with Luxol Fast Blue, a myelin stain. We then used the image analogies technique to synthesize a different section only using the ex vivo MR (Fig. 4.12) and compared it to the actual LFB stain of that section (Fig. 4.12).

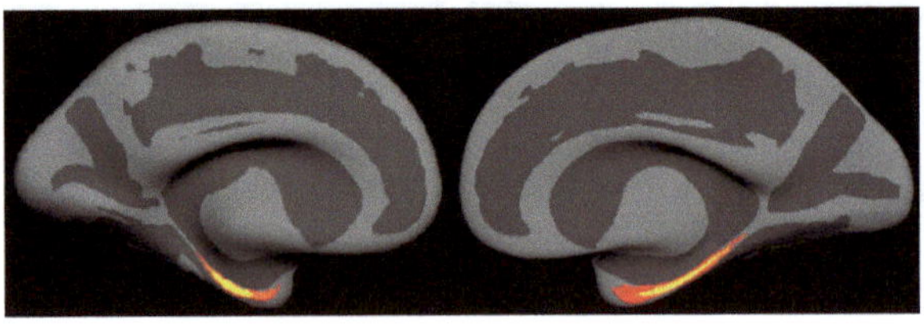

Fig. 4.11 Entorhinal cortex spatial PDFs

The excellent visual similarity between the real (left) and synthesized (right) sections encourage us to develop segmentation algorithms based on known histological features, that we can then apply to synthesized LFB and other stains in a fully 3D fashion, with no cutting, mounting and staining to distort the data.

4.4 The Path to In Vivo Localization

Some attempts have been made to use MRI as a tool to directly image cortical architectonics in vivo (Clark et al. 1992; Barbier et al. 2002; Walters et al. 2003; Clare and Bridge 2005; Duyn et al. 2007). While these examples are encouraging in the sense that they show that currently achievable imaging resolutions can directly image facets of cortical architectonics, they are more proofs of principle than usable techniques, as none of them provide a means for the automated delineation of an entire cortical area from in vivo imaging data.

As noted above, high resolution is difficult to obtain in MRI due to the dramatic decrease in SNR with increasing resolution. Thus, it is unlikely that we will achieve the 100um or so resolution imaging needed to robustly visualize architectonics with MR in the near future. However, as shown above, direct visualization is not required for accurate probabilistic localization. Rather, one needs to find macroscopically visible features the are predictive of the microscopically visible borders. One such feature is cortical folding patterns, as discussed in Sect. 4.2. Here we show another feature and discuss other possibilities, including laminar intensity profiles, and interior distances.

However, before we begin it is worth noting that we frequently assume that areas defined by functional activation in imaging studies are in some sense "the same" as those defined histologically. One such area that serves as a good example is the motion-sensitive Middle Temporal area commonly referred to as MT or V5. Most histological studies localize MT to the crown of a gyrus (e.g. (Clarke and Miklossy 1990)). Most functional imaging studies localize MT in a sulcus (e.g. (Dumoulin et al. 2000)). Here we have access to both histologically defined MT (10 subjects) and functionally defined MT (30 subjects). We averaged these data in spherical coordinates to obtain average histologically-defined MT/V5 maps (see the bottom row, Fig. 4.13) and fMRI-defined MT + (top row, Fig. 4.13). As can be seen, the

Fig. 4.12 *Left*: real LFB stained section, *Right*: synthesized LFB from 100um MRI

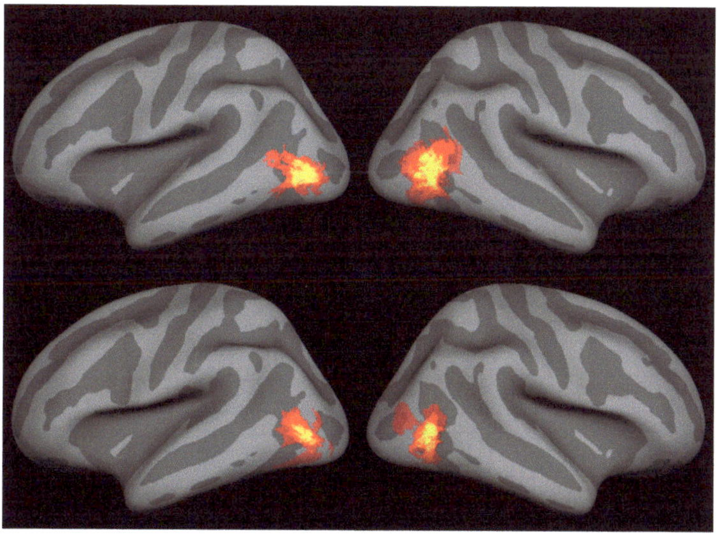

Fig. 4.13 Comparison of fMRI-based MT + (*top row*, 30 subjects, data courtesy of Roger Tootell and Daphne Holt) and histologically defined V5/MT (*bottom row*, 10 subjects, data courtesy of Karl Zilles, Katrin Amunts and Hartmut Mohlberg)

centers of these spatial probability maps are in excellent agreement, giving using confidence that the two techniques are in fact delineating the corresponding cortical area.

While cortical folding patterns appear to be strong, macroscopically visible features that are predictive of microscopic boundaries to varying degrees, there are potentially other derived quantities that can potentially aid in localizing cortical areas. One of these that we have pursued are what we term "interior distances". These represent the shortest path that a fiber bundle could take through the interior of the white matter to connect two distant regions in the cortex. A graphical depiction of what an interior distance might look like is given in Fig. 4.14. The utility of interior distances comes from the hypothesis that brain networks are likely to be formed with some form of minimization of wiring length as a constraint. While the wiring lengths may not be exactly minimal, we at least expect them to be

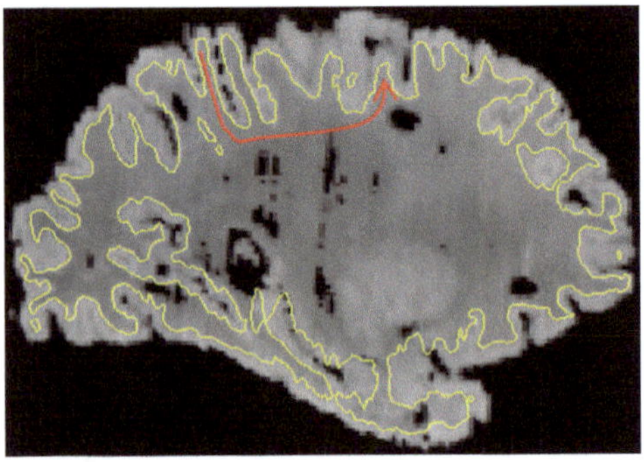

Fig. 4.14 Illustration of an interior distance

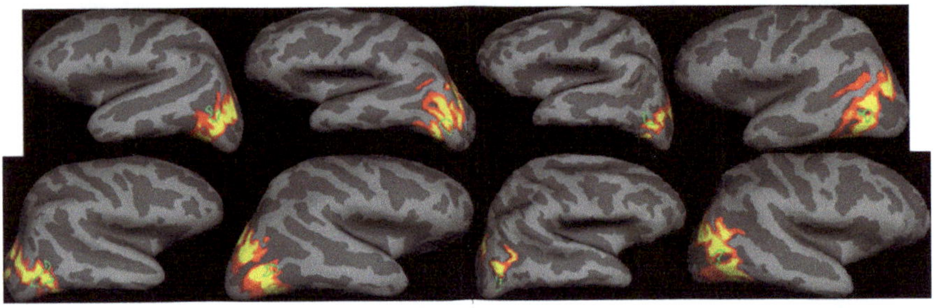

Fig. 4.15 Likelihood p(distances|MT) for 4 subjects distances to posterior callosum, and anterior and posterior estimated V1 and V2 locations (*Green* outline is fMRI-defined MT + .)

relatively consistent across individuals so that propagation delays are reasonably similar from subject to subject for any given pair of areas.

In order to quantify the utility of interior distances we used MT/V5 as a model system that has a set of known connections (Clarke and Miklossy 1990), including callosal projections to the contralateral MT as well as direct projections from primary and secondary visual cortex (Felleman and Van Essen 1991). More specifically, we used a leave-one-out analysis to build probability distributions of the interior distances for the location of MT given interior distances to the boundary of V1, the boundary of V2, and the posterior part of the corpus callosum (through which cross-hemispheric fibers connecting the two MTs course). That is, we compute the mean and variance of the interior distances in order to compute the likelihood of observing the interior distances given the true MT being at each position in the cortex. The results of this analysis are given in Fig. 4.15, which illustrates the likelihood for 4 subjects (shown in heat scale overlay) together with the true location of the histologically defined MT in green.

We can then combine these likelihoods with maps of the prior probability of MT at a given point on the cortical surface (given the folding patterns), shown on the same 4 subjects in Fig. 4.16.

Finally, using Bayes rule, the product of the probability maps in Figs. 4.15 and 4.16 yields a posterior estimate of the probability of MT occurring at each location in the cortex after observing the interior distances, shown in Fig. 4.17. As can be

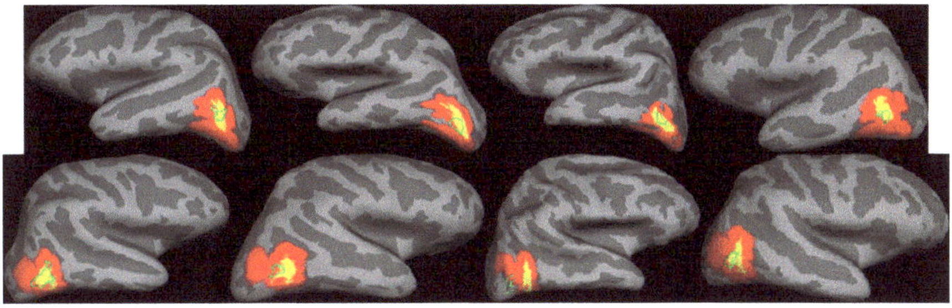

Fig. 4.16 Prior p(MT) of MT location (given geometry and histological MT labels) (*Green* outline is fMRI-defined MT+)

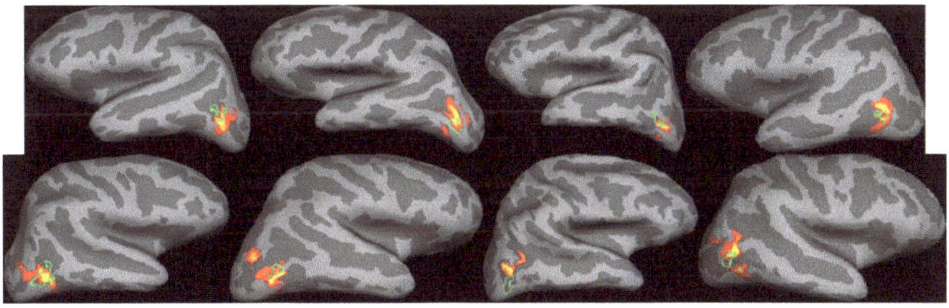

Fig. 4.17 Posterior density p(MT|intensity profiles, distances) for distances to posterior callosum, and anterior and posterior estimated V1 and V2 locations (*Green* outline is fMRI-defined MT+)

seen, even these very simplistic measurements can greatly constrain the location of a cortical area. It is worth stressing again that the small sample sizes that are feasible given the enormous time requirements of whole-brain histological processing greatly limit the types of probabilistic modeling that can be done, and larger sample sizes from automated analysis would potentially greatly enhance our ability to make inferences about architectonic identity from in vivo data (Fig. 4.17).

4.5 Conclusion

The location of architectonic fields in the cortex and nuclear boundaries in subcortical structures is important for our understanding of normal and pathological brain function. From a neuroscientific point of view, these architectonic areas are thought to be the functional modules of the brain, at least in some vague and general sense. Thus being able to assert areal identify across subjects would facilitate our understanding of have various competencies are instantiated by brain anatomy and physiology. From a pathological point of view, it is possible that disease commonalities are obscured by our inability to determine the locations of corresponding areas in different subjects, but within a disease group and in comparison to a control population.

Little is understood about the variability of these boundaries due to the intrinsic difficulty in visualizing them in an undistorted fashion. Historically, the only method for delineating these boundaries involved the massively labor-intensive process of cutting, mounting and staining tissue sections. The success of the Juelich and Dusseldorf groups in this type of analysis on a whole-brain basis has greatly increased our understanding of the nature and variability of these boundaries (Geyer et al. 1997; Roland and Zilles 1998; Schormann and Zilles 1998; Schleicher et al. 1999; Geyer et al. 2000; Morosan et al. 2001; Rademacher et al. 2001; Rademacher et al. 2002; Malikovic et al. 2007), but much work remains to increase the sample size significantly and to remove the unavoidable distortions incurred by these techniques. In recent work, we have shown that aligning cortical geometry greatly improves our ability to infer the locations of architectonic boundaries in the cortex, and that primary and secondary sensorimotor areas are better aligned in this fashion than "higher" cortical areas (Fischl et al. 2008). Further, we have shown in this chapter that the deepest folds are the most stable, implying that procedures for establishing cross-subject homologies should rely more heavily on them than on shallower, more variable folds. In the future, we intend to pursue alternative imaging strategies such as ultra-high-resolution ex vivo MRI and optical coherence tomography (OCT) (Huang et al. 1991) that can produce undistorted, 3D representations of large brain regions. We anticipate that these technologies will facilitate the generation of dozens or hundreds of labeled datasets, allowing us to further out understanding of the complex relationship between macrostructure and microstructure in the human brain.

Acknowledgements Support for this research was provided in part by the National Center for Research Resources (P41-RR14075, and the NCRR BIRN Morphometric Project BIRN002, U24 RR021382), the National Institute for Biomedical Imaging and Bioengineering (R01EB006758), the National Institute on Aging (AG022381, 5R01AG008122-22), the National Center for Alternative Medicine (RC1 AT005728-01), the National Institute for Neurological Disorders and Stroke (R01 NS052585-01, 1R21NS072652-01, 1R01NS070963), and was made possible by the resources provided by Shared Instrumentation Grants 1S10RR023401, 1S10RR019307, and 1S10RR023043. Additional support was provided by The Autism & Dyslexia Project funded by the Ellison Medical Foundation, and by the NIH Blueprint for Neuroscience Research (5U01-MH093765), part of the multi-institutional Human Connectome Project. In addition, I would like to thank Jean Augustinack, Gheorghe Postelnicu, Neda Bernasconi, Daphne Holt, and Thomas Yeo for contributing data or results to this chapter.

References

Allman JM, Kaas JH (1971) A representation of the visual field in the caudal third of the middle temporal gyrus of the owl monkey (Aotus trivirgatus). Brain Res 31:85–105

Amunts K, Schleicher A, Bürgel U, Mohlberg H, Uylings HBM, Zilles K (1999) Broca's region revisited: cytoarchitecture and intersubject variability. J Comp Neurol 412:319–341

Amunts K, Malikovic A, Mohlberg H, Schormann T, Zilles K (2000) Brodmann's areas 17 and 18 brought into stereotaxic space— where and how variable? Neuroimage 11:66–84
Amunts K, Kedo O, Kindler M, Pieperhoff P, Schneider F, Mohlberg H, Habel U, Shah JN, Zilles K (2005) Cytoarchitectonic mapping of the human amygdala, hippocampal region and entorhinal cortex. Anat Embryol 210:343–352
Annese J, Gazzangia M, Toga AW (2005) Localization of the human cortical visual area MT based on computer aided histological analysis. Cereb Cortex 15:1044–1053
Ashburner J, Neelin P, Collins DL, Evans AC, Friston KJ (1997) Incorporating prior knowledge into image registration. Neuroimage 6:344–352
Ashburner J, Hutton C, Frackowiak RSJ, Johnsrude I, Price CJ, Friston KJ (1998) Identifying global anatomical differences: deformation-based morphometry. Hum Brain Mapp 6:348–357
Augustinack J, Avd K, Salat D, Wald L, Blackwell M, Wiggins C, Fischl B (2004) Detection of entorhinal islands using 7 T MRI. Human Brain Mapping Budapest, Hungary
Augustinack J, van der Kouwe AJW, Blackwell ML, Salat DH, Wiggins CJ, Frosch MP, Wiggins GC, Potthast A, Wald LL, Fischl BR (2005) Detection of entorhinal layer II using 7 tesla magnetic resonance imaging. Ann Neurol 57:489–494
Augustinack JC, Helmer K, Huber KE, Kakunoori S, Zollei L, Fischl B (2010) Direct visualization of the perforant pathway in the human brain with ex vivo diffusion tensor imaging. Front Hum Neurosci 4
Augustinack J, Huber K, Postelnicu G, Kakunoori S, Wang R, van der Kouwe A, Wald L, Stein T, Frosch M, Fischl B (2012) Entorhinal verrucae geometry is coincident and correlates with Alzheimer's lesions: a combined neuropathology and high-resolution ex vivo MRI analysis. Acta Neuropathol 123:85–96
Baker JR (1991) Spatially variant tomographic imaging: estimation, Identification, and Optimization. University of California-Berkeley.
Barbier EL, Marrett S, Danek A, Vortmeyer A, Pv G, Duyn J, Bandettini P, Grafman J, Koretsky AP (2002) Imaging cortical anatomy by high-resolution MR at 3.0 T: detection of the stripe of Gennari in visual area 17. Magn Reson Med 48:735–738
Bloch F, Hansen WW, Packard M (1946) The nuclear induction experiment. Phys Rev 70:474–485
Brodmann K (1909) Vergleichende Lokalisationslehre der Großhirnrinde in ihren Prinzipien dargestellt auf Grund des Zellenbaues. Verlag von Johann Ambrosius Barth, Leipzig
Caviness VSJ (1975) Mechanical model of brain convolutional development. Science 189:18–21
Chi JG, Dooling EC, Gilles FH (1977) Gyral development of the human brain. Ann Neurol 1
Clare S, Bridge H (2005) Methodological issues relating to in vivo cortical myelography using MRI. Hum Brain Mapp 26:240–250
Clark VP, Courchesne E, Grafe M (1992) In vivo Myeloarchitectonic analysis of human striate and extrastriate cortex using magnetic resonance imaging. Cereb Cortex 2:417–424
Clarke S (1994) Modular organization of human extrastriate visul cortex: evidence from cytochrome oxidase pattern in normal and macular degeneration cases. Eur J Neurosci 6:725–736
Clarke S, Miklossy JJ (1990) Occipital cortex in man: organization of callosal connections, related myelo- and cytoarchitecture, and putative boundaries of functional visual areas. Comp Neurol 298:188–214
Constable RT, Henkelman RM (1991) Constrast, resolution, and detectibility in MR imaging. J Comput Assist Tomogr 15:297
Constable RT, Skudlarski P, Gore JC (1995) An ROC approach for evaluating functional brain MR imaging and postprocessing protocols. Magn Reson Med 34:57–64
Dale AM (1994) Source localization and spatial discriminant analysis of event-related potentials: linear approaches. University of California, San Diego
Dale AM, Buckner RL (1997) Selective averaging of individual trials using fMRI. Hum Brain Mapp 5:329–340
Dale AM, Fischl B, Sereno MI (1999) Cortical surface-based analysis I: segmentation and surface reconstruction. Neuroimage 9:179–194

Dale AM, Liu AK, Fischl BR, Buckner RL, Belliveau JW, Lewine JD, Halgren E (2000) Dynamic statistical parametric mapping: combining fMRI and MEG for high- resolution imaging of cortical activity. Neuron 26:55–67

Davatzikos C (1997) Spatial transformation and registration of brain images using elastically deformable models. Comput Vis Image Und 66:207–222

Davatzikos C, Bryan RN (1996) Using a deformable surface model to obtain a shape representation of the cortex. IEEE Trans Med Imaging 15:785–795

Davatzikos CA, Prince JL (1995) An active contour model for mapping the cortex. IEEE Trans Med Imaging 14:65–80

Davatzikos CA, Prince JL, Bryan RN (1996) Image registration based on boundary mapping. IEEE Trans Med Imaging 15:112–115

Desimone R, Ungerleider L (1986) Multiple visual areas in the caudal superior temporal sulcus of the macaque. J Comp Neurol 248:164–189

Drury HA, Van Essen DC (1997) Functional specializations in human cerebral cortex analyzed using the visible man surface-based atlas. Hum Brain Mapp 5:233–237

Dukelow S, DeSouza J, Culham J, van den Berg A, Menon R, Vilis T (2001) Distinguishing subregions of the human MT + complex using visual fields and pursuit eye movements. J Neurophysiol 86:1991–2000

Dumoulin S, Bittar R, Kabani N, Baker CJ, Le Goualher G, Pike BG, Evans A (2000) A new anatomical landmark for reliable identification of human area V5/MT: a quantitative analysis of sulcal patterning. Cereb Cortex 10:454–463

Duyn JH, van Gelderen P, Li T-Q, de Zwart JA, Koretsky AP, Fukunaga M (2007) High-field MRI of brain cortical substructure based on signal phase. Proc Natl Acad Sci USA 104 (28):11796–11801

Edelstein WA, Glover GH, Hardy CJ, Redington RW (1986) The intrinsic signal-to-noise ratio in NMR imaging. Magn Reson Med 3:604

Epstein F, Mugler JI, Brookeman J (1994) Optimization of parameter values for complex pulse sequences by simulated annealing: application to 3D MP-RAGE imaging of the brain. Magn Reson Med 31:164–177

Felleman D, Van Essen DC (1991) Distributed hierarchical processing in primate cerebral cortex. Cereb Cortex 1:1–47

Fischl B, Sereno MI, Dale AM (1999a) Cortical surface-based analysis II: inflation, flattening, and a surface-based coordinate system. Neuroimage 9:195–207

Fischl B, Sereno MI, Tootell RB, Dale AM (1999b) High-resolution intersubject averaging and a coordinate system for the cortical surface. Hum Brain Mapp 8:272–284

Fischl B, Liu A, Dale AM (2001) Automated manifold surgery: constructing geometrically accurate and topologically correct models of the human cerebral cortex. IEEE Trans Med Imaging 20:70–80

Fischl B, Salat DH, Albert M, Dieterich M, Haselgrove C, Avd K, Killiany R, Kennedy D, Klaveness S, Montillo A, Makris N, Rosen B, Dale AM (2002) Whole brain segmentation: automated labeling of neuroanatomical structures in the human brain. Neuron 33:341–355

Fischl B, Salat D, van der Kouwe A, Makris N, Ségonne F, Quinn B, Dale AM (2004) Sequence-Independent segmentation of magnetic resonance images. Neuroimage I23:S69–S84

Fischl B, Rajendran N, Busa E, Augustinack J, Hinds O, Yeo BTT, Mohlberg H, Amunts K, Zilles K (2008) Cortical folding patterns and predicting cytoarchitecture. Cereb Cortex 18:1973–1980

Fischl B, Stevens AA, Rajendran N, Yeo BTT, Greve DN, Van Leemput K, Polimeni JR, Kakunoori S, Buckner RL, Pacheco J, Salat DH, Melcher J, Frosch MP, Hyman BT, Grant PE, Rosen BR, van der Kouwe AJW, Wiggins GC, Wald LL, Augustinack JC (2009) Predicting the location of entorhinal cortex from MRI. Neuroimage 47:8–17

Gardner J, Merriam E, Movshon J, Heeger DJ (2008) Maps of visual space in human occipital cortex are retinotopic, not spatiotopic. J Neurosci 28:3988–3999

Gerig G, Kikinis R (1990) Segmentation of 3D magnetic resonance data. In: International conference on image analysis and processing. World Scientific, Singapore
Gerig G, Martin J, Kikinis R, Obler K, Jolesz FA (1991) Automating segmentation of dual-echo MR head data. Proc Inf Process Med Imaging 10:175–187
Gerig G, Kubler O, Kikinis R, Jolesz FA (1992) Nonlinear anisotropic filtering of MRI data. IEEE Trans Med Imaging 11:221–232
Geyer S, Ledberg A (1996) Two different areas within the primary motor cortex of man. Nature 382:805
Geyer S, Schleicher A, Zilles K (1997) The somatosensory cortex of human: cytoarchitecture and regional distributions of receptor-binding sites. Neuroimage 6:27–45
Geyer S, Schormann T, Mohlberg H, Zilles K (2000) Areas 3a, 3b, and 1 of human primary somatosensory cortex. Neuroimage 11:684–696
Grief WL, Buxton RB, Lauffer RB, Saini S, Stark DD, Wedeen VJ, Rosen BR, Brady TJ (1985) Pulse sequence optimization for MR imaging using a paramagnetic hepatoboiliary contrast agent. Radiology 157:461–466
Han X, Pham DL, Tosun D, Rettmann ME, Xu C, Prince JL (2004) CRUISE: cortical reconstruction using implicit surface evolution. Neuroimage 23:997–1012
Hertzmann A, Jacobs C, Oliver N, Curless B, Salesin D (2001) In: SIGGRAPH 2001 conference proceedings. Los Angeles, CA
Hinds OP, Rajendran N, Polimeni JR, Augustinack JC, Wiggins G, Wald LL, Rosas HD, Potthast A, Schwartz EL, Fischl B (2008) Accurate prediction of V1 location from cortical folds in a surface coordinate system. Neuroimage 39:1585–1599
Huang D, Swanson EA, Lin CP, Schuman JS, Stinson WG, Chang W, Hee MR, Flotte T, Gregory K, Puliafito CA, Fujimoto JG (1991) Optical coherence tomography. Science 254
Huk AC, Dougherty RF, Heeger DJ (2002) Retinotopy and functional subdivision of human areas MT and MST. J Neurosci 22:7195–7205
Johnson GA, Thompson MB, Drayer BP, Bone SN (1986) Magnetic resonance microscopy in neurologic models. Acta Radiol Suppl 369:267–268
Johnson GA, Cofer GP, Gewalt SL, Hedlund LW (2002) Morphologic phenotyping with MR microscopy: the visible mouse. Radiology 222:789–793
Kadah YM, Hu X (1998) Algebraic reconstruction for magnetic resonance imaging under B0 inhomogeneity. IEEE Trans Med Imaging 17
Kikinis R, Shenton ME, Iosifescu DV, McCarley RW, Saiviroonporn P, Hokama HH, Robatino A, Metcalf D, Wible CG, Portas CM, Donnino R, Jolesz F (1996) A digital brain atlas for surgical planning, model-driven segmentation, and teaching. IEEE Trans Vis Comput Graph 2:232–241
Lancaster J, Rainey L, Summerlin J, Freitas C, Fox P, Evans A, Toga A, Mazziotta J (1997) Automated labeling of the human brain: a preliminary report on the development and evaluation of a forward-transform method. Hum Brain Mapp 5:238–242
Lancaster J, Woldorff M, Parsons L, Liotti M, Freitas C, Rainey L, Kochunov P, Nickerson D, Mikiten S, Fox P (2000) Automated talairach atlas labels for functional brain mapping. Hum Brain Mapp 10:120–131
Leventon ME, Grimson WEL (1998) Multi-modal volume registration using joint intensity distributions. In: MICCAI. Berlin, Springer-Verlag, pp 1057–1066
Lewis DA (1997) Development of the prefrontal cortex during adolescence: insights into vulnerable neural circuits in schizophrenia. Neuropsychopharmacology 16:385–398
MacDonald D (1998) A method for identifying geometrically simple surfaces from three dimensional images. McGill University, Montreal Neurological Institute, Montreal
Malikovic A, Amunts K, Scheicher A, Mohlberg H, Eickoff S, Wilms M, Palomero-Gallagher N, Armstrong E, Zilles K (2007) Cytoarchitectonic analysis of the human extrastriate cortex in the region of V5/MT+: a probabilistic, stereotaxic map of area hOc5. Cereb Cortex 17:562–574
Marcus DS, Wang TH, Parker J, Csernansky JG, Morris JC, Buckner RL (2007) Open access series of imaging studies: cross-sectional MRI data in young, middle aged, nondemented, and demented older adults. J Cogn Neurosci 19:1498–1507

Maunsell J, Van Essen D (1983) The connections of the middle temporal visual area (MT) and their relationship to a cortical hierarchy in the macaque monkey. J Neurosci 3:2563–2586
Maunsell J, Nealey T, DePriest D (1990) Magnocellular and parvocellular contributions to responses in the middle temporal visual area (MT) of the macaque monkey. J Neurosci 10:3323–3334
Miller MI (2004) Computational anatomy: shape, growth, and atrophy comparison via diffeomorphisms. Neuroimage 23:S161–S169
Miller MI, Christensen GE, Amit Y, Grenander U (1993) A mathematical textbook of deformable neuroanatomies. Proc Natl Acad Sci USA 90:11944–11948
Morosan P, Rademacher J, Schleicher A, Amunts K, Schormann T, Zilles K (2001) Human primary auditory cortex: cytoarchitectonic subdivisions and mapping into a spatial reference system. Neuroimage 13:684–701
Chen N-k, Wyrwicz AM (1999) Correction for EPI Distortions using multi-echo gradient-echo imaging. Magn Reson Med 41:1206–1213
Ogg RJ, Steen RG (1998) Age-related changes in brain T1 are correlated with iron concentration. Magn Reson Med 40:749–753
Prince JL, Tan Q, Pham D (1995) Optimization of MR pulse sequences for bayesian image segmentation. Med Phys 22:1651–1656
Rademacher J, Galaburda AM, Kennedy DN, Filipek PA, Caviness VS (1992) Human cerebral cortex: localization, parcellation and morphometry with magnetic resonance imaging. J Cog Neurosci 4:352–374
Rademacher J, Caviness VS, Steinmetz H, Galaburda AM (1993) Topographic variation of the human primary cortices: implications for neuroimaging, brain mapping and neurobiology. Cereb Cortex 3(4):313–329
Rademacher J, Morosan P, Schormann T, Schleicher A, Werner C, Freund H-J, Zilles K (2001) Probabilistic mapping and volume measurement of human primary auditory cortex. Neuroimage 13:669–683
Rademacher J, Burgel U, Zilles K (2002) Stereotaxic localization, intersubject variability, and interhemispheric differences of the human auditory thalamocortical system. Neuroimage 17:142–160
Rajkowska G, Goldman-Rakic PS (1995a) Cytoarchitectonic definition of prefrontal areas in the normal human cortex: I. remapping of areas 9 and 46 using quantitative criteria. Cereb Cortex 5:307–322
Rajkowska G, Goldman-Rakic PS (1995b) Cytoarchitectonic definition of prefrontal areas in the normal human cortex: II. variability in locations of areas 9 and 46 and relationship to the talairach coordinate system. Cereb Cortex 5:323–337
Reuter M, Sand P, Huber K, Nguyen K, Saygin Z, Rosas HD, Augustinack J, Fischl B (2012) Registration of histology and MRI using blockface as intermediate space. Human Brain Mapping, Beijing
Roland PE, Zilles K (1998) Structural divisions and functional fields in the human cerebral cortex. Brain Res Rev 26:87–105
Sarkissov S, Filimonoff I, Kononova IP, Preobrazenskaja NS, Kukueva L (1955) Atlas of the cytoarchitectonics of the human cerebral cortex. Medgiz, Moscov
Schleicher A, Zilles K (1990) A quantitative approach to cytoarchitectonics: analysis of structural inhomogeneities in nervous tissue using an image analyser. J Microsc 157:367–381
Schleicher A, Amunts K, Geyer S, Morosan P, Zilles K (1999) Observer-independent method for microstructural parcellation of cerebral cortex: a quantitative approach to cytoarchitectonics. Neuroimage 9:165–177
Schmithorst VJ, Dardzinski BJ, Holland SK (2001) Simultaneous correction of host and geometric distortion artifacts in EPI using a multiecho reference scan. IEEE Trans Med Imaging 20
Schmolesky M, Wang Y, Hanes D, Thompson K, Leutgeb S, Schall J, Leventhal AG (1998) Signal timing across the macaque visual system. J Neurophysiol 79:3272–3278

Schormann T, Zilles K (1998) Three-dimensional linear and nonlinear transformations: an integration of light microscopical and MRI data. Hum Brain Mapp 6:339–347

Sereno MI, Allman JM (1991) Cortical visual areas in mammals. In: Leventhal AG (ed) The neural basis of visual function. Macmillan, London, pp 160–172

Smith SM, Jenkinson M, Woolrich MW, Beckmann CF, Behrens TEJ, Johansen-Berg H, Bannister PR, Luca MD, Drobnjak I, Flitney DE, Niazy R, Saunders J, Vickers J, Zhang Y, Stefano ND, Brady JM, Matthews PM (2004) Advances in functional and structural MR image analysis and implementation as FSL Zhang, N De Stefano, JM Brady, and PM Matthews. Neuroimage 23:208–219

Steen RG, Reddick WE, Ogg RJ (2000) More than meets the eye: significant regional heterogeneity in human cortical T1. Magn Reson Imaging 18:361–368

Styner M, Lieberman JA, Pantazis D, Gerig G (2004) Boundary and medial shape analysis of the Hippocampus in Schizophrenia. Med Image Anal 8:197–203

Talairach J, Tournoux P (1988) Co-Planar stereotaxic atlas of the human brain. Thieme Medical, New York

Talairach J, Szikla G, Tournoux P, Prosalentis A, Bordas-Ferrier M, Covello L, Iacob M, Mempel E (1967) Atlas d'Anatomie Stereotaxique du Telencephale. Masson, Paris

Thompson PM, Toga AW (1996) A surface-based technique for warping 3-dimensional images of the brain. IEEE Trans Med Imaging 15:1–16

Thompson P, Schwartz C, Lin RT, Khan AA, Toga AW (1996) Three-dimensional statistical analysis of sulcal variability in the human brain. J Neurosci 16:4261–4274

Thompson PM, MacDonald D, Mega MS, Holmes CJ, Evans AC, Toga AW (1997) Detection and mapping of abnormal brain structure with a probabilistic atlas of cortical surfaces. J Comput Assist Tomogr 21:567–581

Thompson P, Woods R, Mega M, Toga A (2000) Mathematical/computational challenges in creating deformable and probabilistic atlases of the human brain. Hum Brain Mapp 9:81–92

Tootell RB, Taylor JB (1995) Anatomical evidence for MT and additional cortical visual areas in humans. Cereb Cortex 5:39–55

Tootell RBH, Silverman MS, De Valois RL, Jacobs GH (1983) Functional organization of the second cortical visual area in primates. Science 220:737–739

Tootell RBH, Kwong KK, Belliveau JW, Baker JR, Stern CS, Savoy RL, Breiter H, Born R, Benson R, Brady TJ, Rosen BR (1993) Functional MRI (fMRI) evidence for MT/V5 and associated visual cortical areas in Man. In: Society for neuroscience 23rd annual meeting, vol. 19 Washington, D.C.: Society for Neuroscience., Washington, D.C,.p 1500

Tosun D, Rettmann ME, Naiman DQ, Resnick SM, Kraut MA, Prince JL (2004) Cortical reconstruction using implicit surface evolution: a landmark validation study. In: Seventh annual international conference on medical image computing and computer-assisted interventions(MICCAI), vol 3216, Saint-Malo, pp 384–392

Van Essen DC (1997) A tension-based theory of morphogenesis and compact wiring in the central nervous system. Nature 385:313–318

Venkatesan R, Haacke EM (1997) Role of high resolutino in magnetic resonance (MR) imaging: applications to MR angiography, intracranial T1-weighted imaging, and image interpolation. Int J Imaging Syst Tech 8:529

Vogt O (1911) Die Myeloarchitektonik des Isocortex parietalis. J Psychol Neurol 18:107–118

von Economo C (1929) The cytoarchitectonics of the human cerebral cortex. Oxford University Press, London

Walters NB, Egan GF, Kril JJ, Kean M, Waley P, Jenkinson M, Watson JD (2003) In vivo identification of human cortical areas using high-resolution MRI: an approach to cerebral structure-function correlation. Proc Natl Acad Sci USA 100:2981–2986

Weller R, Kaas J (1983) Retinotopic patterns of connections of area 17 with visual areas V-II and MT in macaque monkeys. J Comp Neurol 220:253–279

Wells W, Grimson W, Kikinis R, Jolesz F (1996) Adaptive segmentation of MRI data. IEEE Trans Med Imaging 15:429–442

Yeo B, Sabuncu M, Vercauteren T, Holt D, Amunts K, Zilles K, Golland P, Fischl B (2010) Learning task-optimal registration cost functions for localizing cytoarchitecture and function in the cerebral cortex. IEEE Trans Med Imaging 29:1424–1441

Yu P, Grant PE, Qi Y, Han X, Ségonne F, Pienaar R, Busa E, Pacheco J, Makris N, Buckner RL, Golland P, Fischl B (2007) Cortical surface shape analysis based on spherical wavelets. IEEE Trans Med Imaging 26:582–597

Yushkevich PA, Avants BB, Pluta J, Das SR, Minkoff D, Mechanic-Hamilton D, Glynn S, Pickup S, Liu W, Gee JC, Grossman M, Detre JA (2009) A high-resolution computational atlas of the human hippocampus from postmortem magnetic resonance imaging at 9.4 T. Neuroimage 44:385–398

Zeng X, Staib LH, Schultz RT, Duncan JS (1999) Segmentation and measurement of the cortex from3-d MR images using coupled-surfaces propagation. IEEE Trans Med Imaging 18:927–937

Zhang Y, Brady M, Smith S (2001) Segmentation of brain MR images through hidden markov random field model and the expectation maximization algorithm. IEEE Trans Med Imaging 20:45–57

Zilles K, Clarke S (1997) Architecture, connectivity and transmitter receptors of human extrastriate cortex. Comparison with non-human primates. In: Rockland K et al (eds) Cerebral Cortex, 12th edn. Plenum, New York, pp 673–742

Zilles K, Schleicher A et al (1993) Cyto- and myelarchitecture of human visual cortex and periodical GABA-a recptor distribution. In: Gulyas B et al (eds) Functional organization of the human visual cortex. Pergamon Press, Oxford, pp 111–121

Zilles K, Schleicher A, Palomero-Gallagher N, Amunts K (2002) Quantitative analysis of cyto- and receptor architecture of the human brain. In: Mazziotta JC, Toga A (eds) Brain Mapping: the methods. Elsevier, Amsterdam, pp 573–602

Chapter 5
Database-Driven Identification of Functional Modules in the Cerebral Cortex

Simon B. Eickhoff and Danilo Bzdok

Abstract The organization of the cerebral cortex into distinct modules may be described along several dimensions, most importantly structure, connectivity and function. Functional neuroimaging provides a powerful tool for the localization of function, which allows testing hypotheses about structure-function relationships. This method is, however, intrinsically less well suited to delineate the organization of a particular brain region. While neuroimaging studies may thus test hypotheses about a functional differentiation between cortical modules, their potential for delineating those in a particular region of interest is limited. Identification of cortical modules by differences in whole-brain connectivity profiles derived from diffusion tensor imaging or resting state correlations have therefore raised much recent interest. As these approaches, however, do not carry task-related information, the functional relevance of the obtained parcellation have so far remained largely elusive.

The emergence of comprehensive databases for functional neuroimaging results provides a novel basis for delineating cortical modules by co-activation networks. Importantly, such approaches are data-driven in that they do not rely on a classification of tasks or paradigms, but merely rely on the spatial pattern of whole brain co-activation profiles. The key idea behind database-informed cortical parcellation is computing the whole-brain co-activation patterns of each individual voxel within a seed region, regardless of the ontological classification of the original

S.B. Eickhoff (✉)
Cognitive Neuroscience Group, Institute of Clinical Neuroscience and Medical Psychology, Heinrich-Heine University, Düsseldorf, Germany

Brain Network Modelling Group, Institute for Neuroscience and Medicine (INM-1), Research Center, Jülich, Germany
e-mail: Simon.Eickhoff@uni-duesseldorf.de; S.Eickhoff@fz-juelich.de

D. Bzdok
Brain Network Modelling Group, Institute for Neuroscience and Medicine (INM-1), Research Center, Jülich, Germany
e-mail: Danilo.Bzdok@rwth-aachen.de

S. Geyer and R. Turner (eds.), *Microstructural Parcellation of the Human Cerebral Cortex*, DOI 10.1007/978-3-642-37824-9_5,

experiments. Recording the co-activation likelihoods of all grey-matter voxels outside the region of interest then yields a functional co-activation matrix. This connectivity matrix may then be used to group the seed voxels in such manner, that voxels showing similar co-activation are clustered together and separated from those showing different co-activation profiles. Hereby functional modules may be identified in a data-driven fashion using task-based neuroimaging information. By assessing the functional characteristics and spatial response patterns of those experiments associated with the ensuing clusters, the derived parcellation may be directly related to network properties and task properties.

5.1 Cortical Units and How to Define Them

The organization of the cerebral cortex into distinct modules may be described along several dimensions, most importantly, structure, connectivity and function. This conclusion was mainly derived from concurrent invasive examination of microstructure (histological preparation), connectivity (axonal tracing) and functional properties (single cell recordings) in individual animals, in particular non-human primates. Given that this approach is not feasible in man, delineating modules of the human cortex has been a longstanding challenge.

Regarding the (micro-)structural dimension, early histological investigations into the microscopic heterogeneity of the human cerebral cortex have resulted in several detailed, though partially incongruent, anatomical maps (Brodmann 1909; Vogt and Vogt 1919). More recent advances in structural brain mapping have led to the development of observer-independent probabilistic cytoarchitectonic maps in stereotaxic space (Zilles and Amunts 2010). Moreover, the constantly growing field strength of fMRI scanners yields increasingly fine-grained microstructural images of the living human brain (Walters et al. 2007).

Regarding the connectional dimension, each cortical area is assumed to possess a unique set of input and output connections. Axonal connectivity between areas can be revealed by injection of a tracing dye that is transported to interconnected brain regions in animals. Axonal tracing studies in primates suggest that cortical areas tend to be connected hierarchically and reciprocally by feed-forward and feed-backward connections (Maunsell and van Essen 1983). While operating at a considerably more coarse level, fiber tracking approaches based on diffusion-weighted imaging in humans likewise provide information about the anatomical connections of a particular cortical location that are consistent with those detected by invasive tracing approaches in animals. Mapping connectivity patterns of a particular part of the cortex may thus allow identifying cortical areas by demonstrating differences between the connections of neighbouring grey matter locations.

Regarding the functional dimension, Franz and Lashley (1917) were probably the first to induce focal lesions in circumscribed cortical areas in rats and measure the ensuing effects on the performance of behavioral tasks. Accidental lesion in the human brain likewise provided hints regarding the cortical localization of psychological processes (Harlow 1869). The emergence of neuroimaging methods has later enabled the precise localisation of functional responses across the whole brain throughout various psychological tasks.

Slightly reframing the concept of structure, connectivity and function as three pillars of brain organization, regional specialization of computational processes (i.e., function) can be conceived as being a consequence of both the local (micro-) structural and global connectional properties, as suggested by research in animals. That is, specialization of a particular function is not regarded as an intrinsic property of a brain region that is independent of its connectivity. Rather, input and output *connectivity* of an area in combination with the local "infrastructure" provided, e.g., by cyto- and chemoarchitecture, crucially determine what particular functions that area can perform (Passingham et al. 2002). Conversely, each particular *cortical module* is probably characterized by a unique set of input and output connections, which is supported by statistical analyses of cortical connectivity in the primate (Young 1993) and feline (Scannell et al. 1995) cortex. More generally, cortical modules and connections between those actually reflect *functional segregation* and *functional integration*, respectively (Friston 2002). In sum, a cortical module of functional specialization is likely to be defined by the intersection of regionally specific microstructure and connectivity patterns. Given that microstructural borders are interindividually highly variable and difficult to discern in vivo, a promising avenue to distinguish functionally specialized modules in the human brain is by examining the regional heterogeneity of brain-wide connections.

This notion prompted the development of approaches for connectivity-based parcellation (CBP) of the human cerebral cortex into distinct modules. The key idea behind connectivity-based parcellation is to first analyse the connectivity of each individual voxel in a particular seed region of interest (ROI) with the rest of the brain. By comparing the connectivity profiles of the individual seed voxels with each other, these may then be grouped into distinct clusters of homogeneous connectivity. This approach has first been introduced for the analysis of anatomical connectivity using diffusion tensor imaging (DTI; Johansen-Berg et al. 2004). More recently, a similar approach has also been successfully applied to resting state correlations, i.e., functional connectivity (Kim et al. 2010).

The *exploratory* character of CBP approaches is of particular importance because the majority of current methods in the neuroimaging field are predominantly *confirmatory* in nature and thus rely on a priori hypotheses may be derived from animal research or clinical neurophysiology. Using appropriate experimental designs, functional neuroimaging is an extremely powerful tool for testing hypotheses about, e.g., a functional differentiation between two regions or a dichotomy between the neural correlates of two processes. While many hypotheses derived in particular from primate work and lesion mapping studies could be explicitly tested using this technique, neuroimaging is intrinsically less well suited to delineate the organization of a particular brain region. A prime reason for this drawback is that in most instances the tasks that would allow differentiating different modules in a region of interest are unknown so that experiments cannot be specifically designed to reveal the functional distinction. Consequently, connectivity-based parcellation is a precious addition to the repertoire of neuroscientific tools, given its potential for data-led identification of biological modules in the human cerebral cortex. The obtained functional compartments can subsequently serve as a stepping stone for hypothesis-led methods, such as neuroimaging and connectivity methods.

Connectivity-based parcellation based on DTI and resting-state functional connectivity allow for the definition of individual modules of the cerebral cortex, but also share drawbacks with respect to providing functional hypotheses. In particular, DTI provides information about anatomical connectivity (though not in the strict sense of axonal connections) but does not hold any functional implications. That is, examining anatomical connection patterns does not provide clear hypotheses about the functional differences that may be expected between two clusters. Resting-state connectivity, on the other hand, is founded on the (in the clausal sense) not yet well understood neurophysiological correlations in the absence of a structured task and hence the cognitive processes corresponding to the observed correlated activity remain open. Importantly thus, neither DTI-based nor resting-state-based parcellation of cortical areas may functionally characterize the ensuing clusters. Providing such relations to functional properties, in turn, represents a key advantage of recently developed task-based functional connectivity approaches such as meta-analytic connectivity mapping (MACM).

5.2 MACM as a Measure of Functional Connectivity

There are different methods and theoretical concepts to delineate the functional connectivity of a particular brain area (Eickhoff and Grefkes 2011), that have developed from the basic definition of functional connectivity as the "temporal correlation of spatially distant neurophysiological events" (Friston et al. 1996) and have preceded spike coincidence analyses. Hence, a straightforward assumption would be that coincidences in metabolic changes describe functional connectedness. More specifically, brain regions that tend to increase or decrease neural activity in parallel with the target area *consistently across many subjects* and *throughout various experimental paradigms* can be expected to have some functional relationship with that target area. This approach to task-based functional connectivity analysis has been coined "meta-analytic connectivity modelling" (MACM) (Eickhoff et al. 2011b; Robinson et al. 2010). Put differently, MACM refers to the structure-based (i.e., seed region driven) computation of between-paradigm co-occurrence probabilities. In contrast to functional connectivity analyses on (resting-state) fMRI time-series, occurrences of activation *across many different experiments* rather than changes in the voxel-specific BOLD signal *over scans* represent the unit of observation in MACM analyses. This is achieved by assessing the brain-wide co-activation patterns of a particular seed region across a large array of diverse neuroimaging studies, such as found in comprehensive databases for neuroimaging results.

The emerging MACM approach to functional connectivity analysis has recently been juxtaposed with other established connectivity measures. First, Robinson and collaborators (2010) compared MACM connectivity in humans and axonal white-matter connectivity in monkeys of the amygdala using the non-human primate literature database CoCoMac. These authors reported high consistency of MACM results with tract-traced connections in monkeys. Second, Eickhoff and collaborators (2010) compared MACM connectivity and DTI-based anatomical connectivity of cytoarchitectonically constrained areas of the human parietal operculum, which yielded close correspondence. Third, Cauda and collaborators (2011) compared MACM-derived and resting-state-derived functional connectivity of the nucleus accumbens, again with widely converging results. On the one hand side, MACM thus provides a robust approach to measuring functional relationships between brain regions in humans, given the large correspondence with traditional connectivity techniques. On the other hand side, MACM is based on different assumptions and may thus reflect complementary information of task-based functional connectivity. This potentially renders previously untapped facets of brain connectivity accessible.

As noted above, MACM relies on the integration of a large array of neuroimaging results. This requirement can be met by existing neuroimaging databases, such as AMAT, BrainMap, Brede, and SumsDB (Derrfuss and Mar 2009). Despite many idiosyncrasies, they have in common what is the crucial prerequisite for MACM: They provide coordinates of peak metabolic changes that were observed in various neuroimaging experiments. The localization information is provided by reference to common stereotactic coordinate systems. In fact, from the very beginning most neuroimaging articles report significant activation foci according to either the Talairach-Tournoux (1988) or Montreal Neurological Institute (MNI; Evans et al. 1992) 3D reference space, which can be easily converted into each other (Lancaster et al. 2007). A common coordinate system is crucial in mapping neuroimaging results from heterogeneous sources to a same reference space, which enables testing for across-experiment co-activation in MACM.

The BrainMap database (http://brainmap.org) is the currently most exhaustive database, in particular with respect to the amount of coded meta-data. This open-access repository of functional neuroimaging studies was developed by Fox, Lancaster and colleagues (Fox and Lancaster 2002). It currently contains about 21 % of the entire neuroimaging literature. By virtue of this wealth of neuroimaging data, it contains fMRI and PET experiments resulting from diverse experimental designs with different paradigm classes, stimulus types, and response modalities. BrainMap classifies each experiment according to a rigorous taxonomy that is composed chiefly of structured keywords. More specifically, each database entry includes descriptions on the scanned subjects and experimental conditions, including the presented stimuli, instructions, and responses.

Based on this database, the MACM algorithm comprises the following steps. First, the experiments in the database that reported activation closest to the currently considered seed voxel are identified. This is done by computing for each databased experiment the distance between the current seed voxel and the nearest activation of that particular experiment. This first step thus determines the set of experiments from the database that activate the respective seed voxel. That is, each particular seed voxel in the seed region is associated with an individual experiment pool of neurotopographically most related activation patterns.

The focus of MACM is then to assess the consistency of activation foci reported in the retrieved experiments. As these were defined by activation in the seed, this region will evidently feature the highest convergence. Significant convergence outside the seed, however, should reflect robust co-activation, i.e., functional connectivity. In MACM, the brain-wide co-activation pattern for each individual seed voxel is therefore computed by means of an ALE meta-analysis (activation likelihood estimaton; Eickhoff et al. 2009; Eickhoff and Bzdok 2011) over the experiments that were associated with a particular seed. ALE is the currently most widely used method for coordinate-based meta-analysis, which enables the quantitative localization of above-chance convergence across multiple neuroimaging experiments in a 3D reference space. This integration and synthesis of neuroimaging data thus permits statistically defensible inference on the convergence across a set of experiments, e.g., those activating a particular seed. The key idea behind ALE is to treat the foci reported in associated experiments not as single points, but as centres for 3D Gaussian probability distributions that reflect the spatial uncertainty associated with neuroimaging results. The voxel-wise union of the Gaussian-modelled activation maps of all experiments associated with a particular seed voxel then yields an ALE value for each voxel of the brain that describes the co-activation probability of that particular location with the current seed voxel.

In sum, the described procedure enables the statistical mapping of the co-activation pattern for each individual voxel of the seed region, in spite of the variable and usually low number of foci located precisely at any particular voxel. Experiments are defined as relevant by activation at or close to a particular voxel. Quantitative coordinate-based meta-analysis over all foci reported in these experiments then determines the probability of any other voxel throughout the brain to co-activate with that seed voxel. In this way, MACM follows exactly the definition of meaningful functional connectivity by testing for global coincidences of spatially distant neurophysiological events.

In MACM-based connectivity analyses, selection of relevant experiments in the database is only constrained by featuring at least one focus of activation in the seed region and not by any pre-selection of taxonomic categories (e.g., paradigms, instructions or psychological processes). It is instructive to contemplate why it is this precondition that allows conducting a genuinely data-driven analysis. Only considering a sub-selection of BrainMap experiments would constitute a strong a priori hypothesis about how brain networks are organized. Restricting MACM to specific taxonomic labels would fail to appreciate the fact that it is not known how well psychological constructs, such as motor function, emotion, memory or cognition, map onto neuronal systems or how these are realized in the human brain. This is illustrated by the fact that, up to now, there is no widely accepted cognitive taxonomy. For example, the motor cortex is not only involved in mere action planning and execution, but also in perception of facial emotion, representation of other's motor action and probably even inference of the underlying intentions (Leslie et al. 2004; Mukamel et al. 2010; Carr et al. 2003). Consequently, restricting MACM to experiments from strictly motor-related paradigms would hardly capture the full scope of functional roles likely assumed by the motor cortex. Taken together, MACM determines the coupling strength between a seed region and the rest of the brain across multiple behavioral domains acknowledging the repertoire of operations brain networks can perform.

5.3 Connectivity-Based Parcellation Using MACM

As eluded to above, research in non-human primates using invasive methods strongly suggests that functionally specialized modules in the brain are defined by and can be identified with the regional heterogeneity of brain-wide connections. As described in the last section, a given seed region's task-based functional connectivity may be assessed by performing seed-voxel-wise MACM connectivity analysis to obtain a connectional fingerprint for each seed voxel independently. The connectivity profiles of the seed voxels can then be compared against each other to "blindly" discriminate functional modules in the seed region. For this purpose, a co-activation matrix is created, where the seed voxels are pitted against the entirety of the voxels in the reference brain volume. This co-activation matrix thus summarizes, how likely each seed-voxel co-activates with each voxel in the rest of the brain. The similarity of brain-wide co-activation profiles between seed voxels is then quantified, e.g., by performing cross-correlation, yielding a symmetric square matrix providing the similarity in functional connectivity between any two seed voxels. This cross-correlation- (more general, distance-) matrix is then subjected to a clustering approach in order to identify groups of seed voxels that have similar connectivity and in turn differentiate themselves from other seed-voxels by differences in connectivity patterns.

Given the dimensional nature of regional heterogeneity, more than one distinct scheme for clustering the seed region is often conceivable and feasible. Arguments for a biologically plausible number of clusters can be derived from spectral reordering algorithms (Barnard et al. 1995; Ng et al. 2002). More specifically, the cross-correlation matrix may be rearranged by a spectral reordering algorithm to minimize the cross-correlation values off the diagonal, hereby forcing strongly correlated voxels close to each other. In doing so, sets of seed voxels emerge that are strongly correlated with each other and weakly correlated with the rest of the matrix. Importantly, these sets of seed voxels can then be mapped back on the seed region to obtain topographically defined clusters of homogeneous connectivity (Johansen-Berg et al. 2004). In sum, the spectral reordering algorithm constitutes a useful semi-automated procedure to segregate the seed region into a specific number of clusters. Spectral reordering thus provides an important representation of the data for the identification of a meaningful cluster number in the seed region but does not per se provide a "hard" attribution of voxels to individual clusters.

Alternatively, the co-activation matrix can be fed into a hierarchical cluster analysis to group sets of voxels that feature similar brain-wide co-activation profiles (Timm 2002; Eickhoff et al. 2007). Hierarchical clustering is a multivariate method for solving classification problems by revealing similarities and dissimilarities between elements in a multidimensional feature space. More specifically, individual voxels initially represent separate clusters which are then successively included into a hierarchy merging the least dissimilar cluster to derive progressively larger clusters of voxels. Correlation between the brain-wide co-activation profiles of seed voxels can for instance be used as a similarity measure and average linkage criterion for cluster merging (Timm 2002). The seed voxels are therefore successively merged into clusters as a function of similarity between their co-activation profiles, i.e., into clusters of convergent functional connectivity. That is, there is no need for an a priori definition of expected cluster number. Instead, an organisational hierarchy is generated that allows a multi-layered delineation of cortical fields from individual modules to larger regions. In sum, multi-level hierarchical clustering analysis of the seed-voxel-wise connectivity profiles allows delineating *tiled* functional-connectional patterns of the seed region, while spectral reordering segregates the seed region into *one* "optimal" solution.

K-means clustering is yet another tool to identify clusters in the seed region. This recursive algorithm can be used to parcellate a seed region into a pre-selected number of k non-overlapping clusters (Hartigan and Wong 1979; Lloyd 1982). Initially, k voxels in the seed region are randomly chosen to represent the centers of the k emerging clusters. Two steps are then iterated multiple times. First, the individual seed voxels are assigned to the closest candidate center, which equates with partitioning the seed region into k clusters. Second, the k candidate centers are readjusted to each cluster's current center. As soon as the center needs to be shifted by only a slight pre-set difference, the iterative process stops. In contrast to the multi-layered hierarchical clustering, each seed voxel is thus assigned to only one of the k clusters in this non-hierarchical approach. Given that the ensuing assignments of voxels to particular clusters can vary with the starting point chosen initially, the algorithm is conventionally applied numerous times and the optimal solution is retained (Nanetti et al. 2009). In contrast to spectral reordering and hierarchical clustering, k-means clustering crucially depends on a priori specified parameters, namely the cluster number k and distance threshold for the center shifts. Moreover, it is important to remember, that using k-means clustering with multiple k's may not emulate a hierarchical approach, as the solution at each level (k) is independent of the other ones, which makes parent-children situations possible but by no means necessary.

Taken together, there are several approaches that may be used to derive a segregation of a seed region into multiple clusters based on the cross-correlation matrix reflecting the similarity of whole-brain co-activation profiles between any pair of seed-voxels. Spectral reordering is useful in suggesting a biologically meaningful number of clusters, while hierarchical clustering allows for a multi-level stratification of various cluster solutions. Repeated k-means clustering can yield robust assignments of seed voxels to the resulting clusters. The described approaches compose a methodological family to reveal functional cortical units based on task-based co-activations in the human brain.

While very similar arguments may be made for connectivity-based parcellation on the basis of anatomical connectivity patterns or resting-state correlations, task-based functional connectivity analyses and reference to the underlying databases may provide important additional information about the derived clusters. For one, following the co-activation based parcellation of the seed region into separate clusters, MACM can be performed on each of the ensuing clusters to characterize their co-activation profiles. In this context, "clusters" refers to sets of voxels within the seed region that were identified by the co-activation based parcellation outlined as having similar co-activation patterns to each other but distinct to the rest of the seed voxels. The co-activation profiles of the different clusters can be obtained by first identifying all experiments in the BrainMap database that featured at least one focus of activation in the cluster derived from the co-activation based hierarchical cluster analysis. Then, an ALE meta-analysis is performed on these experiments. To establish which regions are significantly co-activated with a cluster, random and "true" convergence is distinguished by comparing the computed ALE map against a null-distribution that reflects a random spatial association between the experiments. That is, the null-distribution indicates the voxel-wise ALE values that would be observed if the considered experiments converged entirely by chance (Eickhoff et al. 2011a). The observed ALE values from the actual meta-analysis of experiments activating within a particular cluster are then tested against the ALE values obtained under the null-distribution. This yields a p-value based on the proportion of equal or higher random values. Finally, spatial inference can be drawn by identifying those voxels where the experiments converged more robustly (reflected by the ALE values) than expected if the results were independently distributed (reflected by the null-distribution). Taken together, task-based whole-brain connectivity of each cluster can be delineated by performing an ALE meta-analysis across all experiments featuring at least one peak activation in that region. While these analyses reflect the task-based functional connectivity of the derived clusters, contrasting the MACM analyses between different derived clusters in turn reflects those differences that drove the distinction in the clustering process.

Apart from parcellation and connectivity analysis, functional characterization of the co-activation based clusters can provide a link between the derived parcellation and the corresponding functional differentiation. Using MACM (rather than resting-state fMRI) for functional connectivity analysis entails the availability of functional meta-data on the experiments associated with a particular cluster. By using this information, functional characteristics of identified cortical modules may be determined by quantitative correspondence with cognitive and experimental descriptions of the BrainMap taxonomy. In fact, the BrainMap taxonomy has been developed and refined throughout the last 20 years by leaders of the neuroimaging community (Laird et al. 2009b). It describes the category of mental processes isolated by the statistical contrast of each experiment stored in the database. More specifically, behavioral domains include the main categories cognition, action, perception, emotion, interoception, as well as their sub-categories. The respective paradigm classes categorize the specific task employed. It is thus beneficial to analyze the behavioral domain and paradigm class meta-data of BrainMap experiments associated with clusters to determine the frequency of domain 'hits' relative to its likelihood across the entire database. Put differently, the functional roles of the obtained clusters can be identified by determining those behavioral domains and paradigm classes that showed a significant over-representation in the experiments activating within the clusters relative to the entire BrainMap database (Laird et al. 2009a; Eickhoff et al. 2011b). That is, a cluster can be thought of as related to a taxonomic label if more experiments activating the cluster carry that label than it would be expected if such experiments were evenly distributed throughout the whole brain. Mathematically, the test for above-chance occurrence of labels is performed by a binomial test. Taken together, meta-data profiling is valuable in quantifying the correspondence of the connectivity-derived clusters with a comprehensive taxonomy of behavioural domains and paradigm classes. This analysis may then provide a crucial link to functional properties that is not (easily) feasible by any other approach for connectivity-based parcellation.

5.4 Applying Connectivity-Based Parcellation

An exemplary application on a seed region in the medial premotor cortex will illustrate how MACM can be used to identify cortical modules, their connectivity and function in a data-driven fashion. The seed region was drawn from two activation sites of a neuroimaging study on speeded motor responses (Fig. 5.1a). The posterior activation was consistently observed during left, right and bilateral responses, whereas the anterior showed increased activation when subjects responded to (randomly) bilateral as compared to unilateral visual stimuli. Both clusters were merged into a single seed region. We then tested whether the two original regions could be recovered from this combined seed region in a model-free analysis based on similarities between co-activation patterns of the individual seed voxels across neuroimaging experiments (Fig. 5.1b, c).

As described above, the brain-wide co-activation pattern of each seed voxel was first computed using the BrainMap database. Hierarchical cluster analysis was performed on the seed region's co-activation matrix to create a stratification of cluster solutions and similarity measures (Fig. 5.2). This revealed a clear-cut separation of the seed region into two co-activation defined clusters. Hierarchical clustering of the seed region was reiterated for a variety of different parameters as an acid test for the robustness of the co-activation derived sub-regions. This assignment of individual seed voxels to specific clusters was stable across all explored analysis parameters such as the similarity-measure between two co-activation profiles or the criterion for cluster-merger. Importantly, the MACM connectivity derived segregation in the seed region corresponded exactly to the two activation clusters obtained from the two different contrasts of the original fMRI study. Cluster formation driven by dissimilarity in co-activation patterns thus

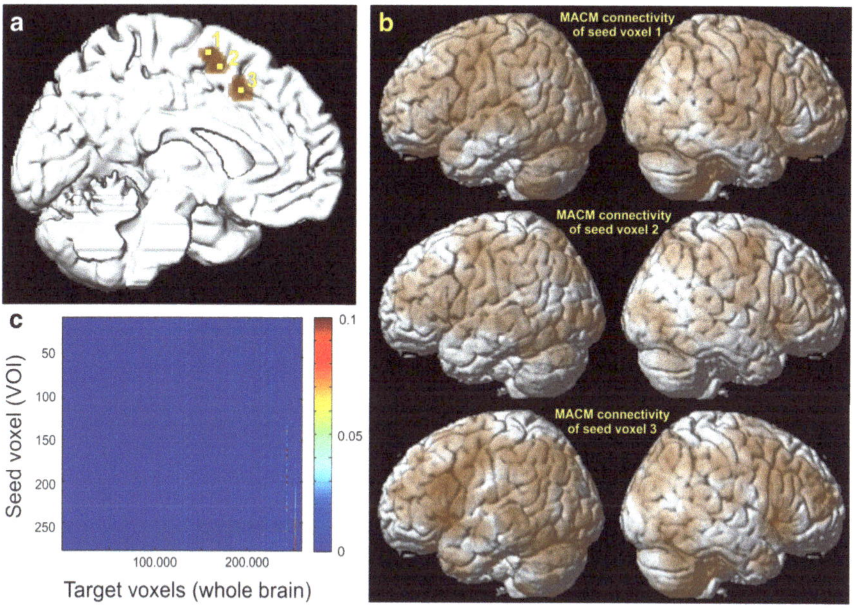

Fig. 5.1 (**a**) Location of the seed region (*brown*) and the three exemplary voxels for which co-activation maps are illustrated in *panel B*, displayed on a surface rendering of the MNI single subject template. The *yellow* colored exemplary seed voxel 1 is located at −4/−6/+68, seed voxel 2 is located at −2/0/+60, and seed voxel 3 at −6/+12/+48 (all coordinates in MNI space). (**b**) Brain-wide co-activation maps of three voxels indicated by the *yellow* numbers in *panel A* as revealed by meta-analytical connectivity modelling using ALE meta-analysis on the brain-wide foci reported in those experiments in BrainMap that featured the closest activation peaks to the respective seed voxels. (**c**) Co-activation matrix summarising the co-activation likelihood (ALE values) of all seed voxels to the rest of the grey matter. The grey matter mask is based on at least 10 % probability according to the ICBM (International Consortium on Brain Mapping). This matrix containing the brain wide co-activation pattern of each individual seed voxel served as the basis for co-activation based parcellation of the medial premotor seed region (Reprinted from Neuroimage, 57, Eickhoff, S.B., Bzdok, D., Laird, A.R., Roski, C., Caspers, S., Zilles, K., Fox, P. T., "Co-activation patterns distinguish cortical modules, their connectivity and functional differentiation", 938–949, 2011, with permission from Elsevier)

recovered the two original activation blobs from the combined seed region in a reliable manner and without a priori constraints imposed on the analysis.

Task-based co-activations of each cluster were then delineated by performing an ALE meta-analysis across all experiments featuring at least one activation in that region, i.e., cluster. The conjunction analysis (Fig. 5.3a) revealed an overlap between the individual co-activation maps of the two clusters in dorsal and ventral lateral premotor cortex, BA 44, primary motor and somatosensory cortices, anterior insula, basal ganglia (particularly putamen), thalamus, superior cerebellum as well as intraparietal sulcus and adjacent inferior parietal lobule. That is, those brain areas are functionally connected to both the posterior and anterior cluster. The difference

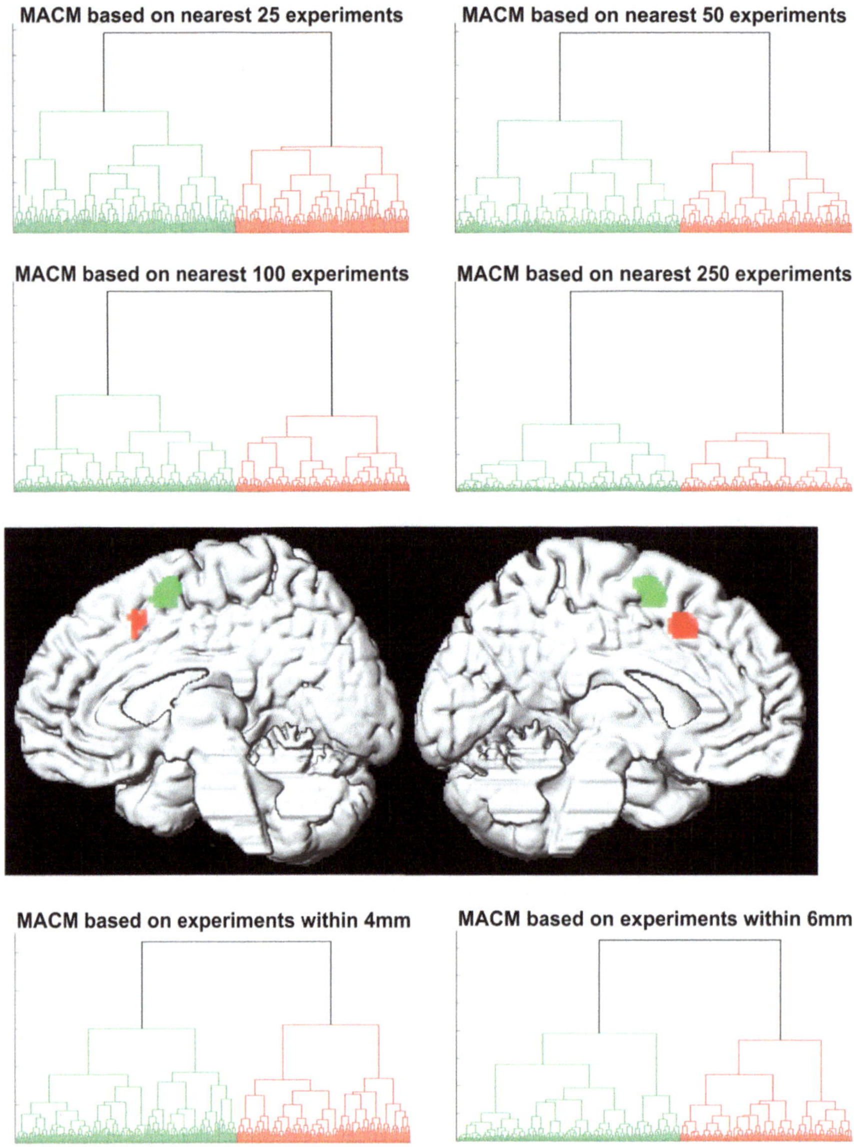

Fig. 5.2 Hierarchical cluster analysis of the co-activation profile matrix (cf. Fig. 5.1c) revealed a highly reliable separation of the seed voxels into two distinct clusters independent of filter criterion and cluster parameters. Projecting the voxels back onto their brain location revealed that these clusters were spatially continuous and corresponded to an anterior and posterior cluster in the medial premotor cortex (Reprinted from Neuroimage, 57, Eickhoff, S.B., Bzdok, D., Laird, A.R., Roski, C., Caspers, S., Zilles, K., Fox, P.T., "Co-activation patterns distinguish cortical modules, their connectivity and functional differentiation", 938–949, 2011, with permission from Elsevier)

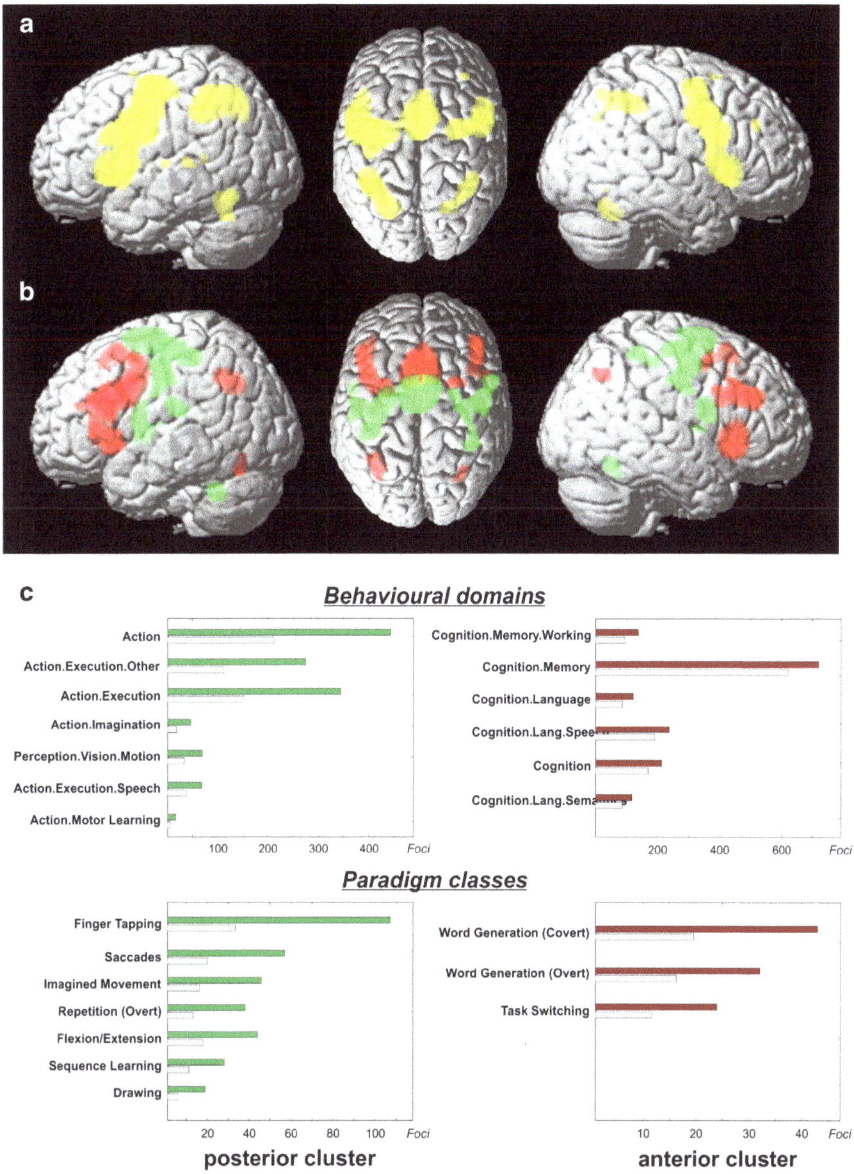

Fig. 5.3 (**a**) Conjunction analysis over the MACM maps for the two main clusters indicates that several fronto-parietal regions show significant co-activation with both medial premotor regions. (**b**) Contrasting the MACM maps revealed that the anterior cluster showed significantly higher co-activation probabilities with ventral premotor, inferior frontal and posterior parietal cortices. The posterior cluster showed significantly higher co-activation probabilities with dorsal premotor cortex, primary sensory-motor cortices, cerebellum and basal ganglia. It should be noted that, at the given threshold, many brain regions appear both in the conjunction as well as the contrast analysis. The reason behind this observation is that the connectivity maps of the two clusters differ

analysis (Fig. 5.3b) of the co-activation maps for the posterior and anterior cluster showed the clear distinction in functional connectivity pattern that actually drove parcellation of the seed region. The posterior cluster showed higher co-activation probabilities in dorsal premotor cortex, primary motor and somatosensory cortices, cerebellum and basal ganglia (putamen). In contrast, the anterior cluster showed significantly higher co-activation probabilities in ventral premotor cortex, middle frontal gyrus and BA 44, anterior insula and intraparietal sulcus/inferior parietal cortex. The stable distinction of the two clusters in the above analyses thus seems to originate from different co-activation likelihood of the posterior and anterior seed region with areas implicated in sensory-motor functions and cognitive control, respectively.

Interestingly, there seemed to be a profound topographical correspondence between the conjunction (Fig. 5.3a) and the difference analysis (Fig. 5.3b) of the task-based functional connectivity for the derived clusters. This is related to the fact that the connectivity maps of the two clusters differ mainly quantitatively (i.e., the connectivity likelihood different for the individual target-voxels) and less qualitatively (i.e., the topographical pattern of the target voxels). Put differently, although the two clusters interact with a similar set of brain regions, the connection strengths with those target brain areas were significantly different for the two cluster. In fact, the difference analysis revealed a quantitatively higher connectivity likelihood of the anterior cluster with anterior brain regions, and vice versa, in spite of highly similar network nodes. Importantly, this differentiation was additionally supported by a resting-state functional connectivity analysis in an independent sample of subjects. This analysis revealed, that all voxels indicated as local maxima in the MACM difference analysis also showed significantly stronger resting-state functional connectivity with the respective cluster.

After connectivity-based parcellation and MACM connectivity analysis, the ensuing clusters were functionally characterized by quantitative analysis of the underlying experiments' meta-data (Fig. 5.3c). This step introduced a first link between the obtained clusters to functional differentiation, i.e., task-relatedness or

---

Fig. 5.3 (continued) mainly quantitatively (i.e., the target-voxel-wise connectivity likelihood) and hardly qualitatively (i.e., the topographical pattern of the target voxels). That is, although the two clusters interact with a number of same brain regions, the connection strengths with those target brain areas are different for each cluster. (**c**) Functional characterisation by behavioural domain and paradigm class metadata from BrainMap. The *red*/*green* bars denote the number of foci for that particular behavioural domain and paradigm class within the anterior/posterior cluster. The *grey* bars represent the number of foci that would be expected to hit the particular cluster if all foci with the respective behavioural domain or paradigm class were randomly distributed throughout the cerebral cortex. That is, the *grey* bars denote the by-chance frequency of that particular label given the size of the cluster. This analysis indicated that the posterior cluster was strongly related to motor functions whereas the anterior cluster showed lower specificity but was activated predominantly by more cognitive processes, such as language, working memory, and task switching (Reprinted from Neuroimage, 57, Eickhoff, S.B., Bzdok, D., Laird, A.R., Roski, C., Caspers, S., Zilles, K., Fox, P.T., "Co-activation patterns distinguish cortical modules, their connectivity and functional differentiation", 938–949, 2011, with permission from Elsevier)

task-specificity. All behavioural domains (BDs) and paradigm classes (PCs) that were significantly over-represented in experiments activating the posterior cluster were related to motor functions (execution, imagery and learning, overt speech and saccades). The only exception was a significant over-representation of visual motion experiments that may be attributable to the high prevalence of (reflexive) eye movements in these tasks. In contrast to the posterior cluster, conspicuously fewer PCs were over-represented in the anterior cluster indicating lower specificity to particular cognitive processes. Notably, none of the BDs and PCs over-represented in the anterior cluster related to motor behaviour. Rather this region was primarily activated in experiments assessing "higher" cognitive processes, such as working memory, language and task switching. In sum, functional characterisation of the connectivity-derived sub-regions linked the posterior cluster to motor function and the anterior cluster to higher associative processes.

This example thus showed a highly robust distinction in task-based functional connectivity between two medial premotor regions that could be linked to differences in functional properties. The posterior cluster co-activated with (pre-) motor areas and was activated by action-related tasks. The anterior cluster in turn co-activated with pre-frontal and parietal cortices and was associated with cognitive functions. Indeed, the pre-SMA is conceptualised to be more strongly involved than the SMA in more complex, "cognitive" aspects of motor behaviour, such as motor selection or inhibition (Picard and Strick 1996; Rizzolatti and Luppino 2001; Vogt et al. 2007). Supporting this view, connectivity tracing in non-human primates revealed that the pre-SMA receives afferences from the inferior parietal lobule (Luppino et al. 1993) and the prefrontal cortex. The described functional characterisation and co-activation pattern of the anterior cluster relate very well to these pre-SMA properties. Invasive tracing, moreover, provided no evidence for direct connections towards the primary motor cortex from pre-SMA but only from SMA proper (Luppino et al. 1993; Rizzolatti and Wolpert 2005). Taken together, earlier observations converge with the presented findings, and allow identification of the anterior cluster as pre-SMA and the posterior one as SMA proper.

5.5 Perspectives and Future Directions

Research in animals suggested that functional specialization in the cerebral cortex is strongly driven by local differences in brain-wide connectivity patterns. This relation between borders of (functional) cortical modules and brain connectivity is exploited by recent additions to the neuroscientist's toolbox. Connectivity-based parcellation allows for "blind" discrimination of cortical modules by regional connectional heterogeneity in a given seed region without any prior knowledge.

From a neurobiological perspective, it is possible that more than one clustering solution for a given region of interest can be biologically meaningful. That is because a region of interest may feature a multi-layered functional hierarchy, which can be captured and described at more than one level. To illustrate this idea by an example: The amygdala is usually discussed as a monolithic functional unit in the neuroimaging literature. However, this brain region has been divided into three main nuclei groups based on cytoarchitectonic assessment (Amunts et al. 2005). It is therefore likely to be informative to study the amygdala at both the macroscopically defined and cytoarchitectonically defined hierarchical levels. Indeed, some in vivo connectivity studies focused on the human amygdala as a whole (Roy et al. 2009; Robinson et al. 2010), whereas other authors studied its connectional properties on a sub-regional scale (Saygin et al. 2011; Solano-Castiella et al. 2011). From a global perspective, limitation to a single parcellation scheme of a given target region might thus entail potential loss of biologically plausible information. Even more so, it would mitigate one of the actual strengths of connectivity-based parcellation, namely, the segregation of brain regions at different neurobiological levels.

Indeed, connectivity-based parcellation relying on task-based measures of functional connectivity enables the identification of cortical functional modules at coarse and finer grained scales, ultimately methodologically constrained by the spatial resolution of fMRI and PET measurements. Although different algorithms are conceivable to determine an "optimal" number of parcellation clusters, applying a single *automated* method seems sub-optimal in the authors' opinion. Technically, there is no unequivocally accepted approach for this matter. Automated algorithms represent up to now mere attempts that often rely on subjectively pre-specified parameters and are therefore still far from representing reliable all-purpose solutions for clustering similarity matrices. In particular, determination of the "optimal" number of clusters (cluster validity problem) is generally acknowledged to be an unresolved issue in bioinformatics and pattern recognition (Handl et al. 2005; Jain et al. 1999). There is no universal clustering method that works excellent for all types of data and any number of clusters. Existing automated clustering algorithms thus tend to produce unstable parcellation schemes across similarity measures and other pre-specified parameters. Consequently, an informed and confident choice about the most reasonable cluster number in a seed region should build upon consistency across different clustering mechanisms and methodological modalities.

A variety of such clustering methods have already been employed to segment brain regions into functional modules. Spectral reordering lends itself as a semi-automated approach that suggests a specific number of clusters contained in the dataset, i.e., the seed-voxel-brain-connectivity matrix. This approach differs from the hierarchical clustering and k-means clustering by primarily providing a representation of the dataset that allows visually identifying the number of contained distinct components, i.e., potential functional modules. It is noteworthy that spectral reordering does not depend on any choices of parameters or association-measures, in contrast to other approaches. Hierarchical cluster analysis has the advantage of allowing parcellation of the seed regions at various coarseness levels. A set of clustering solutions can thus be obtained from a dendrogram that represents the nested seed voxel assignments to clusters and the similarity levels at which groupings change. K-means clustering, in turn, represents a non-hierarchical approach that is computationally attractive and provides an allocation of all voxels to a pre-specified number of clusters. Both hierarchical and k-means clustering depend on pre-existing domain knowledge, as they require (subjective) input parameters chosen by the investigator.

Taken together, there is currently no gold standard for the clustering mechanism in connectivity-based parcellation of brain regions as each approach comes with its own advantages and inconveniences and the very mechanics of a given clustering approach inherently bias the grouping process. Combing different clustering methods can improve the reliability of the results by consecutive application or by cross-validation. Up to now, the majority of CBP implementations in the neuroimaging field allocated each seed voxel definitively to one, and only one, cluster in a particular cluster solution. Future CBP approaches should shed more light on probabilistic versus absolute (in technical terms, "fuzzy" versus "hard") clustering methods, which can reveal membership degrees of seed voxels to multiple clusters. Moreover, graph theory might prove to be useful addition to the methodological arsenal. Ultimately, the development of cluster validity measures for post-hoc assessment of the clusters' meaningfulness (Handl et al. 2005) deserve further improvement as this was neglected so far in neuroimaging, comparing to other bioinformatic disciplines, such as broad-scale gene expression analysis.

Connectivity-based parcellation has so far been performed based on three different modalities of connectivity analysis, namely, MACM, resting-state data and DTI. In this chapter, we have detailed how Meta-analytic connectivity-modelling (MACM), which recently emerged as a measure of functional connectivity between brain areas in a task-based setting by assessing co-activation across hundreds of neuroimaging experiments, may be used to identify functional modules. After that, these modules may be functionally characterized through quantitative correspondence with cognitive and experimental descriptions of the BrainMap taxonomy. This latter step introduced a link between the obtained network nodes or clusters with task-relatedness respectively task-specificity. MACM may hence provide a crucial link between connectivity-derived cortical modules and functional properties. This advantage of MACM allows formulation of hypotheses for targeted experiments on functional activation properties and inter-regional connections.

Comparing MACM, RS and DTI reveals several idiosyncrasies and commonalities. MACM is *task-constrained* respectively *task-dependent* as opposed to RS and DTI being *task-unconstrained* respectively *task-independent*. That is, MACM builds on fMRI and PET studies that impose task settings and thus ensuing cognitive sets on the subjects participating in the scanner session, whereas subjects are simply asked to lie still during RS and DTI measurements. The highly parameterized, hypothesis-based fMRI and PET experiments underlying MACM analyses might thus be thought of as *interventional*, i.e. capturing metabolic changes in the brain caused by manipulation of environmental variables. The heuristic, data-descriptive RS and DTI analyses, on the other hand, would be *observational*, i.e. capturing baseline brain activity without controlled environmental modulation. Hence, many more assumptions need to be made when drawing conclusions from *observational* as opposed to *interventional* findings (Steyvers et al. 2003). Furthermore, MACM and RS are measures of *functional connectivity* as both identify temporal correlations between spatially distant neurophysiological events. Contrarily, DTI is a measure of *anatomical* or *structural connectivity* as it depicts the trajectories of white-matter tracts. On a different note on MACM and RS, it should be stressed that synchronic metabolic change in two brain regions may be mediated by either supplementary intermediary structures or a third area could induce correlated activation in unconnected regions. This is for instance known to be the case for visual-stimulus-driven activity in early sensory areas that can be simultaneously fed to somatosensory areas for perceptual analysis and to premotor cortex for response preparation. This may entail significant functional correlation between the somatosensory and premotor cortex, although the underlying neural processes are not functionally related to each other. This property partly explains why MACM and RS are generally more susceptible to false positive results, while DTI is more susceptible to false negative results (Smith 2012). Taken together, MACM, RS and DTI analyses all reflect different features of brain-network organization. As a consequence, an important question for future research will be, whether connectivity-based parcellation informed by MACM, RS and DTI engenders identical segregation of a particular seed region.

Summing up, connectivity-based parcellation is a powerful tool to distinguish biological modules in the human cerebral cortex in an unsupervised manner. Connectivity-based parcellation using task-based MACM capitalizes on the increasing trend for large-scale data aggregation in neuroimaging science. This variant of CBP is special in enabling experimental and psychological characterization of the obtained cortical segments, in contrast to CBP based on resting-state or DTI data. The ensuing topographically delineated and functionally qualified compartments may provide a basis for generating novel hypotheses that can subsequently guide targeted experimentation.

References

Amunts K, Kedo O, Kindler M, Pieperhoff P, Mohlberg H, Shah NJ, Habel U, Schneider F, Zilles K (2005) Cytoarchitectonic mapping of the human amygdala, hippocampal region and entorhinal cortex: intersubject variability and probability maps. Anat Embryol (Berl) 210 (5–6):343–352. doi:10.1007/s00429-005-0025-5

Barnard ST, Pothen A, Simon HD (1995) A spectral algorithm for envelope reduction of sparse matrices. Numer Linear Algebra Appl 2:317–334

Brodmann K (1909) Vergleichende Lokalisationslehre der Großhirnrinde. Barth, Leipzig

Carr L, Iacoboni M, Dubeau MC, Mazziotta JC, Lenzi GL (2003) Neural mechanisms of empathy in humans: a relay from neural systems for imitation to limbic areas. Proc Natl Acad Sci USA 100(9):5497–5502. doi:10.1073/pnas.0935845100

Cauda F, Cavanna AE, D'Agata F, Sacco K, Duca S, Geminiani GC (2011) Functional connectivity and coactivation of the nucleus accumbens: a combined functional connectivity and structure-based meta-analysis. J Cogn Neurosci 23(10):2864–2877. doi:10.1162/jocn.2011.21624

Derrfuss J, Mar RA (2009) Lost in localization: the need for a universal coordinate database. Neuroimage 48(1):1–7. doi:10.1016/j.neuroimage.2009.01.053
Eickhoff SB, Bzdok D (2011) Meta-analyses in basic and clinical neuroscience: state of the art and perspective. In: Ulmer S, Jansen O (eds) fMRI – Basics and clinical applications. 2nd edn. Springer, Heidelberg, Germany
Eickhoff SB, Grefkes C (2011) Approaches for the integrated analysis of structure, function and connectivity of the human brain. Clin EEG Neurosci 42(2):107–121
Eickhoff SB, Rottschy C, Zilles K (2007) Laminar distribution and co-distribution of neurotransmitter receptors in early human visual cortex. Brain Struct Funct 212(3–4):255–267. doi:10.1007/s00429-007-0156-y
Eickhoff SB, Laird AR, Grefkes C, Wang LE, Zilles K, Fox PT (2009) Coordinate-based activation likelihood estimation meta-analysis of neuroimaging data: a random-effects approach based on empirical estimates of spatial uncertainty. Hum Brain Mapp 30(9):2907–2926. doi:10.1002/hbm.20718
Eickhoff SB, Jbabdi S, Caspers S, Laird AR, Fox PT, Zilles K, Behrens TE (2010) Anatomical and functional connectivity of cytoarchitectonic areas within the human parietal operculum. J Neurosci 30(18):6409–6421. doi:30/18/6409 [pii]
Eickhoff SB, Bzdok D, Laird AR, Kurth F, Fox PT (2011a) Activation likelihood estimation meta-analysis revisited. Neuroimage. (in press) doi:S1053-8119(11)01062-7 [pii]
Eickhoff SB, Bzdok D, Laird AR, Roski C, Caspers S, Zilles K, Fox PT (2011b) Co-activation patterns distinguish cortical modules, their connectivity and functional differentiation. Neuroimage 57(3):938–949. doi:S1053-8119(11)00509-X [pii]
Evans AC, Collins DL, Milner B (1992) An MRI-based stereotactic atlas from 250 young normal subjects. Soc Neurosci Abstr 18(408)
Fox PT, Lancaster JL (2002) Opinion: mapping context and content: the BrainMap model. Nat Rev Neurosci 3(4):319–321. doi:10.1038/nrn789
Franz SI, Lashley KS (1917) The retention of habits by the rat after destruction of the frontal parts of the cerebrum. Psycholobiology 1:3–18
Friston K (2002) Beyond phrenology: what can neuroimaging tell us about distributed circuitry? Ann Rev Neurosci 25:221–250
Friston KJ, Frith CD, Fletcher P, Liddle PF, Frackowiak RS (1996) Functional topography: multidimensional scaling and functional connectivity in the brain. Cereb Cortex 6(2):156–164
Handl J, Knowles J, Kell DB (2005) Computational cluster validation in post-genomic data analysis. Bioinformatics 21(15):3201–3212. doi:bti517 [pii]
Harlow MJ (1869) Recovery from the passage of an iron bar through the head. Boston Medical and Surgical Journal 3(7)
Hartigan JA, Wong MA (1979) A k-means clustering algorithm. Appl Stat 28:100–108
Jain AK, Murty MN, Flynn PJ (1999) Data clustering: a review. ACN Computing Surveys 31 (3):264–323
Johansen-Berg H, Behrens TE, Robson MD, Drobnjak I, Rushworth MF, Brady JM, Smith SM, Higham DJ, Matthews PM (2004) Changes in connectivity profiles define functionally distinct regions in human medial frontal cortex. Proc Natl Acad Sci USA 101(36):13335–13340. doi:10.1073/pnas.0403743101
Kim JH, Lee JM, Jo HJ, Kim SH, Lee JH, Kim ST, Seo SW, Cox RW, Na DL, Kim SI, Saad ZS (2010) Defining functional SMA and pre-SMA subregions in human MFC using resting state fMRI: functional connectivity-based parcellation method. Neuroimage 49(3):2375–2386. doi: S1053-8119(09)01087-8 [pii]
Laird AR, Eickhoff SB, Li K, Robin DA, Glahn DC, Fox PT (2009a) Investigating the functional heterogeneity of the default mode network using coordinate-based meta-analytic modeling. J Neurosci 29(46):14496–14505. doi:29/46/14496 [pii]
Laird AR, Lancaster JL, Fox PT (2009b) Lost in localization? The focus is meta-analysis. Neuroimage 48(1):18–20. doi:S1053-8119(09)00673-9 [pii]
Lancaster JL, Tordesillas-Gutierrez D, Martinez M, Salinas F, Evans A, Zilles K, Mazziotta JC, Fox PT (2007) Bias between MNI and Talairach coordinates analyzed using the ICBM-152 brain template. Hum Brain Mapp 28:1194–1205

Leslie KR, Johnson-Frey SH, Grafton ST (2004) Functional imaging of face and hand imitation: towards a motor theory of empathy. Neuroimage 21(2):601–607. doi:10.1016/j.neuroimage.2003.09.038
Lloyd SP (1982) Quantization in PCM. IEEE Trans Inf Theory 28:129–137
Luppino G, Matelli M, Camarda R, Rizzolatti G (1993) Corticocortical connections of area F3 (SMA-proper) and area F6 (pre-SMA) in the macaque monkey. J Comp Neurol 338:114–140
Maunsell JH, van Essen DC (1983) The connections of the middle temporal visual area (MT) and their relationship to a cortical hierarchy in the macaque monkey. J Neurosci 3(12):2563–2586
Mukamel R, Ekstrom AD, Kaplan J, Iacoboni M, Fried I (2010) Single-neuron responses in humans during execution and observation of actions. Curr Biol 20(8):750–756. doi:S0960-9822(10)00233-2 [pii]
Nanetti L, Cerliani L, Gazzola V, Renken R, Keysers C (2009) Group analyses of connectivity-based cortical parcellation using repeated k-means clustering. Neuroimage 47(4):1666–1677. doi:S1053-8119(09)00626-0 [pii]
Ng AY, Jordan MI, Weiss Y (2002) On spectral clustering: analysis and an algorithm. Proc Neur Inform Proc Syst 14:849–856
Passingham RE, Stephan KE, Kotter R (2002) The anatomical basis of functional localization in the cortex. Nat Rev Neurosci 3(8):606–616. doi:10.1038/nrn893
Picard N, Strick PL (1996) Motor areas of the medial wall: a review of their location and functional activation. Cereb Cortex 6:342–353
Rizzolatti G, Luppino G (2001) The cortical motor system. Neuron 31:889–901
Rizzolatti G, Wolpert DM (2005) Motor systems. Curr Opin Neurobiol 15:623–625
Robinson JL, Laird AR, Glahn DC, Lovallo WR, Fox PT (2010) Metaanalytic connectivity modeling: delineating the functional connectivity of the human amygdala. Hum Brain Mapp 31(2):173–184. doi:10.1002/hbm.20854
Roy AK, Shehzad Z, Margulies DS, Kelly AM, Uddin LQ, Gotimer K, Biswal BB, Castellanos FX, Milham MP (2009) Functional connectivity of the human amygdala using resting state fMRI. Neuroimage 45(2):614–626. doi:S1053-8119(08)01221-4 [pii]
Saygin ZM, Osher DE, Augustinack J, Fischl B, Gabrieli JD (2011) Connectivity-based segmentation of human amygdala nuclei using probabilistic tractography. Neuroimage 56 (3):1353–1361. doi:S1053-8119(11)00265-5 [pii]
Scannell JW, Blakemore C, Young MP (1995) Analysis of connectivity in the cat cerebral cortex. J Neurosci 15(2):1463–1483
Smith SM (2012) The future of FMRI connectivity. Neuroimage. doi:S1053-8119(12)00039-0 [pii]
Solano-Castiella E, Schafer A, Reimer E, Turke E, Proger T, Lohmann G, Trampel R, Turner R (2011) Parcellation of human amygdala in vivo using ultra high field structural MRI. Neuroimage 58(3):741–748. doi:S1053-8119(11)00692-6 [pii]
Steyvers M, Tenenbaum JB, Wagenmakers E-J, Blum B (2003) Inferring causal networks from observations and interventions. Cogn Sci 27(3):453–489. doi:10.1207/s15516709cog2703_6
Talairach J, Tournoux P (1988) Co-planar stereotaxic atlas of the human brain. Thieme, New York
Timm NH (2002) Applied multivariate analysis. Springer, New York
Vogt C, Vogt O (1919) Allgemeinere Ergebnisse unserer Hirnforschung. Journal für Psychologie und Neurologie 25:279–461
Vogt S, Buccino G, Wohlschlager AM, Canessa N, Shah NJ, Zilles K, Eickhoff SB, Freund HJ, Rizzolatti G, Fink GR (2007) Prefrontal involvement in imitation learning of hand actions: effects of practice and expertise. NeuroImage 37:1371–1383
Walters NB, Eickhoff SB, Schleicher A, Zilles K, Amunts K, Egan GF, Watson JD (2007) Observer-independent analysis of high-resolution MR images of the human cerebral cortex: in vivo delineation of cortical areas. Hum Brain Mapp 28(1):1–8. doi:10.1002/hbm.20267
Young MP (1993) The organization of neural systems in the primate cerebral cortex. Proc Biol Sci 252(1333):13–18. doi:10.1098/rspb.1993.0040
Zilles K, Amunts K (2010) Centenary of Brodmann's map–conception and fate. Nat Rev Neurosci 11(2):139–145. doi:nrn2776 [pii]

Part III
"In Vivo Brodmann Mapping" with High-Field Magnetic Resonance Imaging

Chapter 6
Where Matters: New Approaches to Brain Analysis

Robert Turner

Abstract Emerging MRI and fMRI methods, most effective at high magnetic field, provide details of brain cortical architecture and function that are very difficult to include in analysis methods established in the 1990s. For essentially pragmatic reasons, these early methods for analysing spatial maps of functional brain activity have incorporated several assumptions that fail to approximate actual neural operations in any living brain. Some of these assumptions have been made explicit, but others remain implicit and unexamined. However, the improved data quality now available allows the more unrealistic assumptions to be discarded, opening a way forward to far more realistic methods for brain functional analysis. Principles of neural organization and function that should be respected by analysis techniques are listed, and their implications for the formulation of improved methods are explored. Methods based on in-vivo cortical parcellation will have much greater sensitivity and spatial specificity than existing techniques, and will allow much deeper understanding of the co-operative action of neurons across the brain in task accomplishment. Neuroscientific understanding of the relationships between structure, function and connectivity in anatomically distinct brain areas may at last begin to catch up with the achievements of nephrology.

6.1 Introduction

It is instructive to compare our understanding of the functional systems anatomy of the brain with that of the kidney. In the kidney, the cellular and subcellular microarchitecture is well known from histological studies, the systems anatomy is well known, and the relationship between the multiple functions of the kidney, as

R. Turner (✉)
Max-Planck Institute for Human Cognitive and Brain Sciences, Stephanstraße 1A, 04103 Leipzig, Germany
e-mail: turner@cbs.mpg.de

S. Geyer and R. Turner (eds.), *Microstructural Parcellation of the Human Cerebral Cortex*, DOI 10.1007/978-3-642-37824-9_6, © Springer-Verlag Berlin Heidelberg 2013

part of the body, and its component parts, such as the nephrons, is very well understood (Smith 1951). By contrast, despite two centuries of neuroanatomy and the genius of such pioneers as Ramon y Cajal, we are still very unsure of the nature and function of the component parts of the brain. It is clear that brain grey matter can be categorized as cortical and subcortical, and subcortical regions such as the amygdala, the basal ganglia, striatum and thalamus can be subdivided into nuclei with specific connections and well-understood developmental pathways (Swanson 2012). We can even make some testable inferences regarding the role that each nucleus plays in the coordinated activity of the brain.

We are on weaker ground, however, in regard to the white matter and the cortical grey matter. The typical poverty of our understanding of the organization of white matter fibres in human brain is revealed, for instance, by the widespread neuroimaging assumption over the decade 2000–2010 that most of this tissue can be considered to comprise a small number of highly coherent axonal fascicles, with infrequent fibre crossings. A few minutes spent examining histological samples of brain tissue stained for myelin with a good optical microscope reveals that this assumption is obviously incorrect. The 'small world' connectivity (Hilgetag et al. 2000) of a brain indeed implies a very large number of short connections between brain areas, and a smaller number of longer connections (Schüz and Braitenberg 1998), which may well be quite coherent, but are inevitably crossed frequently by connections between other brain regions (Jeurissen et al. 2012).

The situation is worse still when it comes to cortical grey matter. It has been known for more than a century that the 0.23 m^2 area of grey matter in human brain shows many compact subregions (e.g. Brodmann areas) defined by their distinctive cytoarchitecture and myeloarchitecture (Brodmann 1909; Elliot Smith 1907; Vogt and Vogt 1919). In order to explain scientifically how any mechanism works, it is necessary to be able to define the component parts, their specific functional roles, and how these sub-functions are integrated into the operation of the mechanism as a whole. It is still unclear at what critical spatial scale it is necessary to identify the relevant component parts of human brain, but a reasonable starting point would be to consider such cortical areas that have shared structure, together with anatomically definable subcortical nuclei, as units of analysis and modelling.

However, there is no consensus (see the chapter by Nieuwenhuys, this book) regarding how many such cortical regions can and should be distinguished, and no useful concordance atlas between myeloarchitecture and cytoarchitecture in the same cadaver human brains has ever been systematically compiled. Research in the myeloarchitecture of human cortical grey matter has stagnated for about a century, and even though myeloarchitectural details, such as the stria of Gennari in primary visual cortex, may be more easily visible than cytoarchitecture even with the naked eye in cadaver brain, there has been almost no speculation or research regarding the functional role that they play in each cortical area. Some important details of cortical connections have been carefully researched, such as the respective inputs and output to and from other brain regions and the spinal cord, although far too little is still known about the reasons for variability of the horizontal bands of myelinated cortical fibres comprising the Exner stripe and the the two Bands of Baillarger. With few exceptions, the arrangement of cytoarchitectural features within the cortex has also gone largely unexplained in terms of function. Terms such as 'granular' and 'agranular', 'allocortex' and 'isocortex' create an illusion that the mode of operation of specific types of cortex is already well understood.

Even more problematic, although it is trivially obvious that brain areas with different microarchitecture have different neural processing competences, present-day traditions of data analysis in functional neuroimaging have grown up that pay scant regard to the precise division of labour that this seems to imply. Such a division of labour is probably intrinsic to the brain's efficient strategy of parallel processing, that compensates remarkably well for the slowness and electrochemical noise of its wetware components.

When imaging neuroscience in human brain became readily feasible with the demonstration of functional BOLD contrast in 1991 by Kwong et al. (1992) it was quickly recognized that gross sulcal and gyral brain anatomy could usually provide only approximate indications of the location of specific cortical areas, because these are defined cytoarchitecturally only in cadaver brains, and human brains naturally show considerable variability, both in the pattern of sulcal folding and in the relative locations of cortical areas on the sulci and gyri. (Exceptionally, certain areas such as primary visual cortex, primary motor cortex and the supplementary motor area are reliably defined by their sulcal location.) The spatial localization of brain activity was initially defined instead by reference to the work of Talairach and Tournoux (1988), who outlined the cytoarchitecturally defined Brodmann areas on the external cortex of the brain of an elderly French woman. The usefulness of this work lay mainly in the definition of a 3D coordinate system, which could be used to suggest locations of Brodmann area in other brains. The invention of MRI by Mansfield and Lauterbur in the early 1970s made it feasible to project the Talairach map onto any desired living brain, which could be scanned with a spatial resolution of about 1 mm by the time of Kwong's discovery in the early 1990s. The coordinate system defined by Talairach was subsequently replaced by one based on MRI scans of many, much more typical brains, known as "MNI coordinates" (Le Goualher et al. 1999).

Thus a methodology was rapidly developed by leading imaging neuroscience laboratories, such as the FIL in London, which attempted to link brain location, neuroanatomy and function at a spatial scale of about 8 mm, about as close as anyone dared to expect that corresponding cortical areas could be located across brains. A strategy of spatial smoothing, using a Gaussian smoothing kernel of typically about 8 mm, was fundamental to this approach. This had the following very important benefits: (a) it often considerably improved the signal to noise (SNR) of functional data; (b) after structural brain images had been spatially normalized into a standard template brain registered within MNI space, it allowed for the residual mismatch of actual cortical areas, so that positive results could be anticipated from group averaging across normalized brains; and (c) it enabled very simple analytic equations (Worsley et al. 1992) to be used for assessing the statistical significance of measured brain activity, and thus for thresholding the resulting group images to provide spatial activation maps.

Such maps could then be interpreted in relation to the psychological task paradigm experienced by the human subjects, and discussed in the context of the rapidly growing body of scientific results obtained using a similar strategy.

From this perspective it is clear that at the time of discovery of BOLD fMRI in the early 1990s, the methodology just outlined was driven by necessity. MRI scanner resolution for structural in-vivo brain images was about 1 mm, unable to resolve details within the 3 mm thickness of the cortex, and BOLD functional imaging could provide adequate quality images with no better than 3 mm isotropic resolution. MRI using any type of sequence can barely depict variations in cortical cytoarchitecture, and very few researchers were aware in 1992 that it could be made quite sensitive to myeloarchitectural details (Clark et al. 1992). However, with the advent of ultra high field (UHF) MRI scanners, with static magnetic field strength of 7 T and greater, the situation is radically changed.

This chapter will continue with a summary of current UHF-MRI acquisition capabilities, described in fuller detail in Chap. 7. Next will be described the difficulties that traditional imaging neuroscience analysis methods are likely to encounter in dealing with the data quality and data size generated by UHF-MRI. To illustrate the weakness of such traditional methods, the implicit and explicit assumptions embedded within them will be outlined and critiqued, and a more realistic description of important human brain characteristics will be provided. The extent to which UHF-MRI is likely to be able to capture such information is then considered. In particular, the prospects are outlined for relatively detailed parcellation of the cortex into regions comparable to those identified by Brodmann, which can then plausibly be taken as brain components with definable processing competences. The crucial features of more suitable analysis and richer modelling strategies based on such components are outlined, and finally the multiple benefits of such a scientific paradigm shift for understanding human brain function are briefly discussed.

6.2 Summary of Current In-Vivo UHF-MRI Capabilities

Neuroscientific applications of MRI at lower field strengths generally require three types of data: anatomical information ('structural images'); functional information ('BOLD data'); and connectivity information ('DWI') (see Chap. 7). These types of information are of course available with UHF-MRI, but the increased sensitivity and contrast that become available open up useful developments of each imaging modality.

Using 7 T MRI and high sensitivity RF receive coils, the current state of the art allows structural images of entire in vivo human brains to be obtained with 0.4 mm isotropic resolution, or better. Such images mainly derive their contrast from the tissue distributions of myelin or iron, or both. Maps of the longitudinal relaxation time T1 effectively indicate the presence of myelin, and closely resemble myelin-stained histological sections. Maps of the transverse relaxation time T2* are especially sensitive to the presence of iron, although they also indicate myelin features.

Functional images at 7 T, showing BOLD contrast changes of blood oxygenation, can be obtained with an isotropic spatial resolution much better than 1 mm using a range of techniques (Chen and Ugurbil 1999; Barth and Norris 2007; Poser et al. 2010; Polimeni et al. 2010; Zimmerman et al. 2011; Heidemann et al. 2012a).

Diffusion weighted imaging (DWI) allows evaluation of the distribution of oriented fibres in each voxel. An early version of this technique, diffusion tensor imaging (DTI), which assumes a single dominant fibre orientation in each voxel, clearly fails to correspond to the fibre neuroanatomy within human brain, where each of the up to 200 cortical areas typically project to at least ten other areas, resulting in fibre crossings in almost every voxel (Jeurissen et al. 2012). However, good quality tractography can now be achieved using techniques that sample fibre directions in each voxel reasonably well, and at 7 T there is sufficient SNR to allow isotropic voxels of 0.8 mm (Heidemann et al. 2012b).

Thus the remarkable improvements in RF coil sensitivity for 3 T MRI and maturation of the technology of 7 T MRI now provide enough sensitivity and contrast, with spatial resolution of better than 0.5 mm, to depict submillimeter-scale variations in image intensity corresponding to intra-cortical structures. This easily observable intracortical contrast arises very largely from myeloarchitecture. Cortical areas defined by myeloarchitecture (largely neglected in the last 50 years) show good congruence with those mapped using cytoarchitecture, such as Brodmann areas, and may provide even better native structural maps, with more conspicuous contrast ex-vivo and in-vivo. Deposits of iron that are found within cortical myelin further enhance MRI contrast.

6.3 Structural and Functional Characteristics of Brains That Should Be Included in Mechanistic Modelling

For any type of modelling, the research question itself inevitably drives the level of detail that is considered to be adequate. But all present-day hypotheses relating to cognitive science and to neuropsychology make some reference to functional localization in the human brain. For such hypotheses to be meaningfully testable, they should refer to a realistic description of the brain, rather than to a poorly defined abstract conception of what a brain might be like.

It is now well established that a human brain has about 10^{11} neurons, about 10^{11} astrocytes, and on the order of 10^{14} synaptic connections. The cerebral cortex can be parcellated into probably more than 100 areas of recognizably different cyto and myeloarchitecture. Each of these areas is axonally connected to at least ten other areas, and to several thalamic nuclei, basal ganglia, and specific cerebellar areas. Many cortical areas have sharp boundaries, and since brain areas and even sparsely-encoding separate neurons can be selectively activated by appropriate combinations of stimuli, brain activity cannot be assumed in general to be spatially smooth. However, for most tasks many brain areas are jointly in operation, because of the intense connectivity. No neurons are more than about eight synapses apart, and most are within five synapses (Schüz and Braitenberg 1998).

Significant local activity takes place on time scales from 1 ms to minutes and days, and inter-area transit times for action potentials are about 10–50 ms. The brain itself is non-stationary: its connections are dynamic. The strengths of synaptic connections are continually changing and the concentration patterns of modulatory neurotransmitters, neuropeptides, and corticosteroids, which strongly modulate brain network activity, vary diurnally and with situation. New dendrites and terminal axons are continually growing and new synapses are being formed, while at the same time other synapses are lost, dendrites are reabsorbed, and neurons die.

Thus the facts that stable, reproducible neurovascular responses to carefully controlled tasks can be observed, and that these can be replicated across subjects, should be considered somewhat surprising, and in themselves worthy of scientific curiosity. That many tasks require concerted activity among large assemblages of neurons is clearly evident, but this should not exclude the possibility that increased (or decreased) task-related activity of small numbers of neurons, perhaps spatially distributed, may have enormous impact on how a task is performed.

6.4 Typical In-Vivo MRI and fMRI Image Parameters, as Used in Standard Functional Studies at Magnetic Fields of 3T and Below

Since the beginnings (in the 1980s) of imaging neuroscience, the predominant approach to localization of functional brain activity has relied on the approximate similarity of individual human brains. While it has been well accepted that gyrification differs from brain to brain, and that identifiable neuronal territories may not closely respect the folding pattern, the absence of reliable cortical landmarks has largely obviated the use of well-defined regions of interest, both for analysis of individual brain activity and comparison across subjects. Only a few important brain areas, such as V1 and M1, can be clearly associated with sulcal features, and the boundaries of even these areas are hard to define without the use of functional localization, which is vulnerable to circular reasoning.

In the late 1990s most labs studying brain function using MRI converged on a standard protocol for imaging. This was especially helpful for cognitive neuroscience studies of perception, cognition and motor control, because it made reasonably good use of the current MR scanner capabilities, and allowed straightforward comparisons to be made between studies performed at different laboratories. While some variations on this theme continued to be used, in relation to specialized hypotheses and specific brain areas (such as early visual areas), a convention was rapidly agreed upon. In broad terms, this consists of acquisition of structural images of entire volunteer subject brains with 1 mm isotropic resolution, using the T1-weighted MP-RAGE MRI sequence, followed by acquisition of functional data using gradient-echo echo-planar imaging sensitive to BOLD contrast, at about 3 mm isotropic resolution.

The structural images are often used for studies of comparative morphometry (Ashburner and Friston 2000), and recently especially for exploring neural plasticity. Segmented grey matter images are usually smoothed using a Gaussian kernel of up to 10 mm, in order to construct a fictitious parameter described as 'grey matter density', which can be compared across the appropriately normalized brains of a group of subjects using standard statistical tests relying on Gaussian Random Field theory.

The functional images are analysed using the apparatus of the Design Matrix and the General Linear Model (Friston et al. 1995). As described above, spatial smoothing is an integral feature of this analysis strategy.

6.5 Current Analysis Methods Are Unable to Make Effective Use of UHF-MRI Data

Since 1995 the number of cognitive neuroscience studies that utilize neuroimaging techniques has increased exponentially. Between 1995 and 2000 a very widespread consensus emerged regarding acceptable strategies for analysis of fMRI data, led by the work of Friston at the Institute of Neurology in London. These strategies were embodied in software packages such as SPM, AFNI, FSL and others. It was clearly of great benefit to this young research field to have such consensus, but it can now be argued that this was a case of premature closure, given the continual improvement in the brain image data available.

To make this explicit, I will consider the stages of data analysis that are recommended in using SPM8 for a standard neurocognitive study. The first step is of course to acquire the fMRI data, which has typically about 3 mm spatial resolution, and is obtained as a series of 300–600 brain volumes, each taking about 2 s to acquire, while a well-controlled brain task is performed. It is considered that results that can be generalized to a subject group or an entire population require scanning about 16 subjects, a number just large enough to enable estimation of inter-subject variance.

The next steps in analysis immediately involve a set of explicit and implicit assumptions regarding brain mechanisms and brain spatial organization. This set of assumptions is commonly adopted by users of all data analysis packages, such as SPM, FSL, that use spatial smoothing and probabilistic atlases.

6.5.1 Problematic Assumption 1: Smoothing Is Required for Statistical Inference

The first assumption is that statistical inference using fMRI data is only possible if the image is smoothed by a kernel three times the size of the acquired voxels. This is required for the applicability of Gaussian Random Field theory for estimating statistical significance of brain activations. With the standard acquisition voxels of about 3 mm, this means that a smoothing of 8 mm or more is normally applied.

One implication of this assumption is the tacit acceptance of the idea that Neural Mass Modelling is valid everywhere in the brain, at a scale of 8 mm or more. Friston (2008) states that

> the basis of these models rests on modelling, not on the behaviour of individual nerve cells or neurons, but on the probability density over ensembles or populations of similar neurons. The Fokker-Planck formalism becomes central here and can be harnessed using neuronal models that are cast in terms of differential equations, with or without discrete behaviours (e.g., neuronal spiking or firing). From the density dynamics afforded by the Fokker-Planck equation, we then pursue various simplifications and special cases. An important example is when the density becomes a point-mass over the expected states of a population. These are referred to as neural-mass models and predominate in the computational neuroscience literature. A key generalisation of these neural-mass models is to neural-field models, where the location of the mass or expected state of a population becomes a function of both time and position on the brain's cortical surface or subcortical structures. These models generate all sorts of interesting and neuronally plausible patterns and self-organising phenomena, which can be inferred through invasive or non-invasive electrophysiological recordings of real brains.

This raises the important neurophysiological question: Is there a spatial scale for which Neural Mass Modelling might be reliably valid? Spatial smoothing lumps together the highly specific and well-localized task-related changes of blood oxygenation observable with fMRI, such that only the location of the centroids of the thresholded smoothed map of signal changes retains any physical meaning. If Neural Mass Modelling were valid on the spatial scale of 8 mm, smoothing on this scale might be appropriate, but there is compelling evidence that this assumption is generally incorrect. To take but a few examples, columnar structure with a scale of 1 mm or less is found in primary visual and somatosensory cortices, where the perceptually relevant 'neural fields' are certainly smaller than 1 mm in extent; and the brain is equipped with many spatially organized maps, again with a scale of 1 mm, such as the retinotopic, tonotopic, and somatotopic maps in the cortex and thalamus. Smoothing to 8 mm would destroy all of this functionally crucial detail.

When data are smooth enough that the residuals after fitting to some model can be described as a Gaussian Random Field, the Worsley formula (1992) allows a very robust and computationally quick evaluation of significance. Given the vast amount of data to be analyzed in a typical fMRI experiment, the speed of this operation was very appealing in the early 1990s.

But can statistical inference be performed without smoothing? The answer is, yes, of course. The basis of all statistical inference is repeated measures, which allows the computation of the mean and the variance of each measured variable. A normal fMRI experiment, consisting of a time course of hundreds of brain volumes, provides many samples of the signal in each image voxel of the brain for each condition, and thus in principle contains all the information needed to assess significance. Spatial smoothing might help by effectively performing a local average of the functional signal, which would improve the signal to noise ratio (SNR), and hence sensitivity, if the noise were random and uncorrelated across voxels. But most of the apparently randomly fluctuating component of the BOLD signal at typical spatial resolution arises from more-or-less spatially correlated variations in neuronal activity (Bianciardi 2009), not from random stochastic thermal noise.

As already mentioned, it is quick and easy to make whole-brain statistical inferences on spatially smooth data. However, Moore's Law and the remarkable reduction in the cost of computing hardware mean that much more appropriate, albeit more elaborate methods for assessing significance in functional data, without smoothing, are now feasible (Lee et al. 2012). Based mainly on the concept of False Discovery Rate (Benjamini and Hochberg 1995), these methods still have to account for the fact that in comparing the intensities of images voxel by voxel, many statistical inferences are made simultaneously. Significance estimates must therefore be corrected for multiple comparisons, which is not straightforward when data are not smooth.

6.5.2 Problematic Assumption 2: Brain Architecture Is Conserved Across Brains at a Scale of 8 mm, and Is Defined by a Probabilistic Brain Atlas Valid for All Adult Brains

The variability across brains of the relationship between cytoarchitectonically different cortical areas and their gross anatomical location has been well known for more than a century (Elliot Smith 1907). Even when brain image volumes are normalized quite precisely using non-linear warping techniques, corresponding brain areas may fail to overlap by as much as 10 mm (Amunts 2004). One increasingly popular strategy is to use the probabilistic atlas by Eickhoff and the Jülich group (2005), which defines a set of brain areas on an internationally agreed template brain using data from ten cadaver brains which have been parcellated by cytoarchitecture into the equivalent of Brodmann areas. The probabilistic atlas depicts boundaries between areas on the template brain corresponding to the maximum probability that a given voxel will belong to a given brain area. However, examination of the source overlap maps which show the data from all ten brains reveals that some Brodmann areas show very few concordant voxels.

Thus it is invalid to assign a Brodmann location to the centroid of a smoothed activation – in the general sense that it is quite improbable that in any particular brain the peak activation actually lies within the specified Brodmann area.

6.5.3 Problematic Assumption 3: The Importance of a Given Brain Area for a Specific Task Is Indicated Only by a Positive Mean Activity, on a Scale of 8 mm

Most BOLD fMRI cognitive studies show thresholded maps of positive BOLD signal with tables of activation centroids in MNI space. But pattern classification techniques can decode reliably from patterns of scale 1 mm (e.g. Bode 2011), which include decreases of BOLD signal as well as increases. It should not be surprising that localized decreases of neuronal activity in a given cortical area should have informational relevance for other brain areas.

Lateral inhibition (Von Békésy 1967) is a well known neural strategy, that enhances edge detection in the visual system, for instance. It is not known how often a brain typically uses such a means of attentional focusing. For a small area of activation, smoothing by 8 mm is likely to include an area of negative BOLD corresponding to lateral inhibition, and thus reduce the apparent amplitude of the activation. For instance, in primary visual cortex (Shmuel et al. 2006) regions of positive BOLD activation are usually surrounded by a ring of negative BOLD signal, over cortical distance scales of several millimetres. Obscuration of this ring by smoothing incurs the obvious risk that a vital component of the mechanism of vision will remain undiscovered.

6.5.4 Problematic Assumption 4: Activity in Any Given Brain Voxel Is Statistically Independent from Any Other Brain Voxel

Many fMRI analyses, whether using SPM, FSL, AFNI or Brain Voyager, are conducted using the General Linear Model, which treats each image voxel as independent, except insofar as the image is spatially smooth. The solution of the GLM entails formulation of a Design Matrix, which includes time series of all covariates of interest and of no interest. Computation of the pseudo-inverse of the design matrix then provides least-squares-fit best estimates of the linear dependences (betas) of the data time course in each voxel upon the covariates of interest and of no interest. To estimate the statistical significance of the coefficients of variation thus obtained, spatial smoothness must be taken into account. In SPM this is estimated statistically from the residuals after the GLM fit, and its value is used in correcting the final voxel-by-voxel significance estimates for multiple comparisons. Once this has been performed, cluster size can be used to obtain an additional estimation of significance, which clearly allows the use of lower values for the z-score or t-score in computing significance thresholds.

The GLM has proven to be a very powerful tool (Monti 2011) in objectively identifying brain regions in which significant changes in BOLD contrast take place in controlled experimental conditions. But at a conceptual level its validity is fundamentally compromised by the well-established fact that a brain (of any species) is intrinsically multivariate in operation. It is the very dependence of activity in one group of neurons on the activity in a different group of neurons that enables a brain to function. Even the brain of a fly is intrinsically multivariate in operation. Any neuron projects to several other neurons, often in a variety of areas. Synchronized synaptic input at any neuron from many other neurons is required to trigger delivery of an action potential. Analyses that ignore these trivially obvious facts will never be capable of providing insight into brain mechanisms.

Attempts to include spatial covariance using graph theory with an artifactually small number of nodes, such as Dynamic Causal Modelling (critiqued by Lohmann 2012) or Granger Causality (Goebel 2003), are unlikely to do justice to the complex detail of neural performance. Recent work (Gonzalez-Castillo 2012) using unusually long averaging times per volunteer subject in a simple fMRI study have shown that time-locked activity can be detected 95 % of grey matter voxels, when the data are not smoothed. This result suggests that conventional fMRI analyses are likely to incur a very large number of false negative findings. The appearance of simplicity provided by spatial smoothing and conservative thresholding of fMRI data may be highly misleading.

Fortunately, as mentioned above, multivariate analytic techniques – brain decoding–have recently come to the forefront (e.g. Formisano and Kriegeskorte 2012; Kriegeskorte et al. 2006) which offer access to co-ordinated spatial patterns of brain activity, that manifest themselves most particularly in the numerous cortical maps (e.g. retinotopic, tonotopic, somatotopic maps) in which variations in perceptual parameters become spatially encoded in brain networks. In such analyses spatial smoothing is rarely explicitly performed, though an effective smoothing is implicit in 'searchlight' methods for pattern classification (Kriegeskorte et al. 2006). It must be pointed out that such techniques would produce no results if the activity in each voxel was uncorrelated.

6.5.5 Problematic Assumption 5: Grey Matter, White Matter and CSF Are Considered to Be Equally Likely, as Possible Locations of Changed Brain Activity

After BOLD fMRI data have been spatially smoothed by 8 mm, analyzed and the resulting t-statistic map has been thresholded (typically at $p < 0.05$ corrected for multiple comparisons) it is usually overlaid on a high resolution grey-scale T1 weighted image of the brain, inviting assignment of the spatial localization of the activations found. Sadly, the activated region usually then has the topology of a distorted sphere, rather than the ~3 mm thick curved surface corresponding to the cortical grey matter. Thus it is inevitable that many voxels in the high resolution map that are either CSF, bone or white matter are then labelled as part of the activated region. In point of fact, all such voxels are false positives: there can be no activation in them.

6.5.6 Problematic Assumption 6: If a Given Voxel's Task-Related Variance Is Smaller Than a Statistical Threshold of $p < 0.05$, After Correction for Multiple Comparisons, It Can Play No Role in the Task Performance

This assumption infringes one of the fundamental principles of experimental science: absence of evidence does not constitute evidence of absence. The just-cited paper of Gonzalez-Castillo (2012) shows that at least for some experimental paradigms, longer data collection and averaging of more samples reveals activation in a much greater number of voxels. The critical question arises, by what criteria can it be decided which of these voxels are most important for the performance of the given experimental task? Should an activated region of limited extent, but strong BOLD contrast, so that when spatially smoothed it expands over a wide area, be considered more vital to some task performance than a much larger area of voxels with activations that fail to reach the arbitrary theshold of $p < 0.5$ corrected?

It can be strongly argued that attempting model fits to data with many false negatives is just as pointless as trying to fit data with many false positives (as with Assumption 5).

6.5.7 Problematic Assumption 7: The Centroid of an Thresholded Activated Area Completely Defines Its Localization

It has become customary practice to display results in tabular form from cognitive neuroscience experiments using functional brain imaging, giving coordinates of maximal activation in MNI space. With the typical 8 mm spatial smoothing, this maximum will be located close to the centroid of the activated volume. Any other spatial parameters of the activated area are very difficult to interpret anatomically, particularly when much of the thresholded volume fails even to lie within grey matter. However, especially when group average data are considered, the centroid itself may lie in a part of the brain that cannot be identified with any particular cortical area. For example, if the crown of a gyrus and each of its banks are activated in some particular study, the centroid will lie in white matter somewhere within the gyrus, where there is in fact no change of BOLD contrast. If one of the goals of functional neuroimaging is to associate function with specific brain architecture, the MNI coordinates cited in such a study can provide no useful information whatever.

It should be noted that the strategies criticized here have generally been restricted to imaging studies related to cognitive neuroscience and human subjects. Even in humans, there is an extensive body of fMRI research, beginning with the retinotopic studies of Engel et al. (1994), that has explored the functional anatomy of occipital cortex using little or no smoothing, and group data were combined after separate detailed analysis of the data from each subject. The same is true for many fMRI studies of animal brain, ranging from rat (e.g. Silva et al. 1999) to cat (e.g. Zhao et al. 2006), to macaque brain (e.g. Logothetis et al. 1999). In such studies it has been vital to respect the distinction between spatial extent and amplitude, and the underlying neuroanatomy is typically sufficiently well understood that there is no likelihood of conflating distinct brain regions. Thus the difficulties that have encouraged reckless smoothing have mainly arisen when the so-called 'higher' brain functions of humans–cognition, emotion, attention, language–have been the object of study.

6.6 Technical Issues Regarding High Resolution Data Sets

We have argued that use of the much higher spatial resolution image data available at high fields such as 7 T can result in greatly improved modelling of human brain function. But before discussing promising strategies for this, there are specific technical issues that relate to the acquisition and processing of such data: (a) motion artifact and (b) data size.

(a) **Motion Artifact**. This arises from the simple fact that high resolution images take longer to acquire. Human volunteer subjects cannot voluntarily keep their heads immobile, even when restrained using a bite-bar, to better than 0.5 mm over the period of up to an hour that may be required to obtain a high SNR volume image of their entire brain with 0.4 mm isotropic resolution. This motion is enough to cause distributed motion artifact that completely obscures the details of cortical layer structure.

The best solution so far devised for this problem is the use of "prospective motion correction". This methodology comprises a means for independently and continuously tracking all six degrees of freedom of the head motion (three orthogonal displacements, and rotations about all three axes), and a method for inputting these motion parameters to the MRI scanner. The scanner is equipped with software that can continuously update the parameters that control the slice position using these input data. Current state-of-the-art hardware uses high resolution near infrared cameras, which can measure changes in head position with an accuracy of 10 μm, and provide updates to the scanner within 20 ms (Schulz 2012). This level of performance essentially removes motion artifact from structural brain images which have little distortion due to magnetic field inhomogeneity. Applying the the same technique for high resolution functional MR images, which tend to have greater image distortion, some of which does not move with the subject's head, is more complicated, although solutions are in place (Ooi 2012).

(b) **Data Size.** The second critical technical issue arises from the sheer size of the MRI data sets. These can be many hundreds of gigabytes in extent, for an entire brain. Image analysis software packages such as SPM and FreeSurfer were developed at a time (1995–2000) when 400 megabytes was considered to be a large data set. Perhaps part of the rationale for spatial smoothing was that this strategy could further reduce the size and complexity of the data, thus allowing data analysis in acceptable times, such as a few hours per data set. Platforms such as MATLAB offered the opportunity of easy-to-read code, and subroutines well suited for the analysis of four-dimensional data sets and implementation of the General Linear Model.

However, data sets have increased in size, computer memory has become much less expensive, and CPU's have dramatically gained in speed since that time. Radical changes in processing strategies have become attractive, which can cope with the data size and do not require oversimplification of the brain's complex architecture. Some of these methods are already available in efficiently coded and publicly available software: the package MIPAV www.cbs.mpg.de/institute/software/cbs-hrt/index.html, which handles brain image segmentation and cortical contouring, and the package LIPSIA www.cbs.mpg.de/institute/software/lipsia/download.html which provides algorithms for evaluating the significance of fMRI results without spatial smoothing, and a range of options for multivariate image analysis.

6.7 Keys to New Analysis Methods

A more powerful strategy is clearly desirable for neuroimaging studies of human cognition. One important concept might be the idea of 'cortical competence', that is, the description of the relationship between input and output streams of action potentials for any given clearly identifiable area of cortex – and for completeness, the linked concept of 'nuclear competence', in regard to the deep brain nuclei such as the thalamic nuclei and the basal ganglia. The explicit assumption is made, following the great twentieth century neuroanatomists, that myeloarchitecture and cytoarchitecture are not epiphenomenal, as they are commonly treated at the time of writing, but are closely related to the transformational properties of specific areas. Careful study of cyto- and myeloarchitecture may afford important clues regarding such competences. The fact that some details of myeloarchitecture can be observed in vivo, and correlated with task-related activation, invites attempts at interpretation of why it varies across cortical areas.

The fundamental procedures that would enable a more powerful routine strategy for analyzing and modelling human brain function are these: (i) in-vivo brain imaging at submillimeter resolution of structure, function and connectivity, (ii) in-vivo cortical parcellation, (iii) multivariate analysis, (iv) avoidance of spatial smoothing, and (v) across-subject averaging by anatomically defined regions-of-interest.

It is too early to specify comprehensive details of such a strategy, since no complete examples can be found in the literature. However, the process might develop somewhat as follows:

(a) Starting with a behaviourally-validated research question, design a fMRI experiment that respects the technical capabilities of MRI and fMRI, and the time constants involved in neurovascular coupling.
(b) Recruit a suitable number of appropriately characterized human subjects (as usual).
(c) With 7 T MRI, scan each subject with (i) high resolution structural scans designed to be maximally sensitive to myeloarchitecture (Geyer 2011) and (ii) a high resolution DWI scan with as many diffusion gradient directions as possible, but at least 60 (Jones 2012).
(d) Analyze these scans, for each individual subject, to obtain (by cortical parcellation) a "native cortical atlas" of definable cortical areas, a "native nuclear atlas" of deep brain regions, and a submillimeter volume image of fiber orientations in each voxel.
(e) Perform the fMRI scan, with isotropic voxels of 2 mm or less.
(f) Without applying spatial smoothing, analyze the fMRI data, both with the GLM technique and with multivoxel pattern analysis.
(g) Explore the GLM data using False Discovery Rate statistics, and the multivoxel pattern analysis data with prediction accuracy methods.
(h) Identify by subject the spatial locations of activations and deactivations above suitable significance thresholds using their specific native cortical atlas and native nuclear atlas.
(i) Identify by subject the major white matter connections associated with the most important areas of differential brain activity, using MRtrix (Tournier et al. 2012) or similar software to deal with crossing fibres.
(j) Create a mean native cortical atlas by non-linearly warping the boundaries of the cortical areas in each native cortical atlas onto a common reference brain.
(k) Use the transformations thus defined to warp the individual thresholded activation maps onto the reference brain.
(l) Create overlap maps of activations across all subjects.
(m) Assess commonalities of connectivity across subjects using a connectivity matrix approach, by cortical area.

This approach clearly allows a much closer association of function and connectivity with each distinct cortical and deep brain area. Obvious advantages are the greatly increased statistical power provided by the region-of-interest analysis across subjects, and the continued separation of spatial extent and amplitude as independent variables. Two potentially groundbreaking opportunities for hypothesis building emerge from such data, regarding the specific competences of cortical and nuclear architecture, and the structure and integration of brain networks related to specific cognitive tasks.

It is important to recognize that cognitive neuroscientific analysis of human brain function could be productive at a much coarser spatial scale. However, such understanding can only be validated once it has become clear what approximations a coarse-grain approach entails, and what justifies the ignoring of neuroanatomical details. So far, high resolution functional studies of animal brain offer little support for the sweeping generalizations inherent in Neural Mass Modelling.

Beyond this, fundamental research on the neuroscience of cognition should progress far more rapidly when cortical areas supporting specific psychological tasks can be more unambiguously identified. Currently, standard functional image analysis methodology is unable even to assign brain activity to a particular bank of a sulcus, once averaging has been performed across human brains. The great power of spatial mapping of brain activity for understanding when differently labelled tasks are in fact the same, and when apparently similar tasks are actually dissociated, can only be used to its fullest when the neuronal substrate of brain functional activity has been properly identified. Meta-analysis, already showing great promise, should really take off when such correlations have even been partly established.

Acknowledgments I would like to thank David van Essen for useful discussions on human brain neuroanatomy, and Geraint Rees for his perceptive comments on the manuscript.

References

Amunts K, Schleicher A, Zilles K (2004) Outstanding language competence and cytoarchitecture in Broca's speech region. Brain Lang 89(2):346–353

Ashburner J, Friston KJ (2000) Voxel-based morphometry – the methods. Neuroimage 11 (6 Pt 1):805–821

Barth M, Norris DG (2007) Very high-resolution three-dimensional functional MRI of the human visual cortex with elimination of large venous vessels. NMR Biomed 20(5):477–484

Benjamini Y, Hochberg Y (1995) Controlling the false discovery rate: a practical and powerful approach to multiple testing. J R Stat Soc Ser B (Methodol) 57(1):289–300

Bianciardi M, Fukunaga M, van Gelderen P, Horovitz SG, de Zwart JA, Shmueli K, Duyn JH (2009) Sources of functional magnetic resonance imaging signal fluctuations in the human brain at rest: a 7 T study. Magn Reson Imaging 27(8):1019–1029

Bode S, He AH, Soon CS, Trampel R, Turner R, Haynes JD (2011) Tracking the unconscious generation of free decisions using ultra-high field fMRI. PLoS One 6(6):e21612

Braitenberg V, Schüz A (1998) Cortex: statistics and geometry of neuronal connectivity. Springer, Berlin

Brodmann K (1909) Vergleichende Lokalisationslehre der Grosshirnrinde in ihren Prinzipien dargestellt auf Grund des Zellenbaues. J.A. Barth, Leipzig

Chen W, Ugurbil K (1999) High spatial resolution functional magnetic resonance imaging at very-high-magnetic field. Top Magn Reson Imaging 10(1):63–78

Clark VP, Courchesne E, Grafe M (1992) In vivo myeloarchitectonic analysis of human striate and extrastriate cortex using magnetic resonance imaging. Cereb Cortex 2(5):417–424

Eickhoff SB, Stephan KE, Mohlberg H, Grefkes C, Fink GR, Amunts K, Zilles K (2005) A new SPM toolbox for combining probabilistic cytoarchitectonic maps and functional imaging data. Neuroimage 25:1325–1335

Elliot Smith G (1907) A new topographical survey of the human cerebral cortex, being an account of the distribution of the anatomically distinct cortical areas and their relationship to the cerebral sulci. J Anat Physiol 41(Pt 4):237–254

Engel SA, Rumelhart DE, Wandell BA, Lee AT, Glover GH, Chichilnisky EJ, Shadlen MN (1994) fMRI of human visual cortex. Nature 369:525

Formisano E, Kriegeskorte N (2012) Seeing patterns through the hemodynamic veil – the future of pattern-information fMRI. Neuroimage 62(2):1249–1256

Friston K (2008) Mean-fields and neural masses. PLoS Comput Biol 4(8):e1000081. doi:10.1371/journal.pcbi.1000081

Friston KJ, Holmes AP, Worsley KJ, Poline J-B, Frith CD, Frackowiak RSJ (1995) Statistical parametric maps in functional imaging: a general linear approach. Hum Brain Mapp 2:189–210

Geyer S, Weiss M, Reimann K, Lohmann G, Turner R (2011) Microstructural parcellation of the human cerebral cortex-from Broadmann's post-mortem map to invivo mapping with high-field magnetic resonance imaging. Front Hum Neurosci 5:19

Goebel R, Roebroeck A, Kim DS, Formisano E (2003) Investigating directed cortical interactions in time-resolved FMRI data using vector autoregressive modeling and sranger causality mapping. Magn Reson Imaging 21(10):1251–1261

Gonzalez-Castillo J, Saad ZS, Handweker DA, Inati SJ, Brenowitz N, Bandettini PA (2012) Whole-brain, time-locked activation with simple tasks reveated using massive averaging and model-free analysis. Proc Natl Acad Sci USA 109(4):5487–5492

Heidemann RM, Ivanov D, Trampel R, Fasano F, Meyer H, Pfeuffer J, Turner R (2012a) Isotropic submillimeter fMRI in the human brain at 7 T: combining reduced field-of-view imaging and partially parallel acquisitions. Magn Reson Med. doi:10.1002/mrm.24156, Jan 9

Heidemann RM, Anwander A, Feiweier T, Knösche TR, Turner R (2012b) k-space and q-space: combining ultra-high spatial and angular resolution in diffusion imaging using ZOOPPA at 7 T. Neuroimage 60:967–978

Hilgetag CC, Burns GAPC, O'Neill MA, Scannell JW, Young MP (2000) Anatomical connectivity defines the organization of clusters of cortical areas in the macaque monkey and the cat. Philos Trans R Soc Lond B 355:91–110

Jeurissen B, Leemans A, Tournier JD, Jones DK, Sijbers J (2012) Investigating the prevalence of complex fiber configurations in white matter tissue with diffusion magnetic resonance imaging. Hum Brain Mapp. doi:10.1002/hbm.22099, May 19

Jones DK, Knosche TR, Turner R (2013) White matter integrity, fiber count and other fallacies: the do's and don'ts of diffusion MRI. Neuroimage 73:239–254

Kriegeskorte N, Goebel R, Bandettini P (2006) Information-based functional brain mapping. Proc Natl Acad Sci USA 103:3863–3868, Epub 2006 Feb 28

Kwong KK, Belliveau JW, Chesler DA, Goldberg IE, Weisskoff RM, Poncelet BP, Kennedy DN, Hoppel BE, Cohen MS, Turner R, Cheng H-M, Brady TJ, Rosen BR (1992) Dynamic magnetic resonance imaging of human brain activity during primary sensory stimulation. Proc Natl Acad USA 89:5675–5679

Le Goualher G, Procyk E, Collins DL, Venugopal R, Barillot C, Evans AC (1999) Automated extraction and variability analysis of sulcal neuroanatomy. IEEE Trans Med Imaging 18:206–217

Lee SH, Lim J, Park D, Biswal BB, Petkova E (2012) Input permutation method to detect active voxels in fMRI study. Magn Reson Imaging 30:1495–1504, Jul 19 [Epub ahead of print]

Logothetis NK, Guggenberger H, Peled S, Pauls J (1999) Functional imaging of the monkey brain. Nat Neurosci 2:555–562

Lohmann G, Erfruth K, Muller K, Turner R (2012) Critical comments on dynamic causal modeling. NeuroImage 59(3):2322–2329

Monti MM (2011) Statistical analysis of fMRI time-series: a critical review of the GLM approach. Front Hum Neurosci 18:5–28

Ooi MB, Krueger S, Thomas WJ, Swaminathan SV, Brown TR (2009) Prospective real-time correction for arbitrary head motion using active markers. Magn Reson Med 62(4):943–954

Polimeni JR, Fischl B, Greve DN, Wald LL (2010) Laminar analysis of 7 T BOLD using an imposed spatial activation pattern in human V1. Neuroimage 52:1334–1346

Poser BA, Koopmans PJ, Witzel T, Wald LL, Barth M (2010) Three dimensional echo-planar imaging at 7 Tesla. Neuroimage 51:261–266

Schüz A, Braitenberg V (2002) The human cortical white matter: quantitative aspects of cortico-cortical long-range connectivity. In: Cortical areas: unity and diversity, Conceptual advances in brain research. Taylor & Francis, London, pp 377–386

Schulz J, Siegert T, Reimer E, Labadie C, Maclaren J, Herbst M, Zaitsev M, Turner R (2012) An embedded optical tracking system for motion-corrected magnetic resonanace as 7T. MAGMA 25(6):443–453

Shmuel A, Augath M, Oeltermann A, Logothetis NK (2006) Negative functional MRI response correlates with decreases in neuronal activity in monkey visual area V1. Nat Neurosci 9:569–577

Silva AC, Lee SP, Yang G, Iadecola C, Kim SG (1999) Simultaneous blood oxygenation level-dependent and cerebral blood flow functional magnetic resonance imaging during forepaw stimulation in the rat. J Cereb Blood Flow Metab 19:871–879

Smith WH (1951) The kidney: structure and function in health and disease. Oxford University Press, USA

Swanson LW (2012) Brain architecture: understanding the basic plan. Oxford University Press, New York

Talairach J, Tournoux P (1988) Co-planar stereotaxic atlas of the human brain: 3-dimensional proportional system – an approach to cerebral imaging. Thieme Medical, New York

Tournier JD, Calamante F, Connelly A (2012) MRtrix: diffusion tractography in crossing fibre regions. Int J Imaging Syst Technol 22:53–66

Vogt C, Vogt O (1919) Allgemeinere Ergebnisse unserer Hirnforschung. J Psychol Neurol 25:279–468

Von Békésy G (1967) Sensory inhibition. Princeton University Press, Princeton

Worsley KJ, Evans AC, Marrett S, Neelin P (1992) A three -dimensional statistical analysis for CBF activation studies in human brain. J Cereb Blood Flow Metab 12(6):900–918

Zhao F, Wang P, Hendrich K, Ugurbil K, Kim SG (2006) Cortical layer-dependent BOLD and CBV responses measured by spin-echo and gradient-echo fMRI: insights into hemodynamic regulation. Neuroimage 30:1149–1160, Epub 2006 Jan 18

Zimmermann J, Goebel R, De Martino F, van de Moortele PF, Feinberg D, Adriany G, Chaimow D, Shmuel A, Uğurbil K, Yacoub E (2011) Mapping the organization of axis of motion selective features in human area MT using high-field fMRI. PLoS One 6(12):e28716

Chapter 7
MRI Methods for In-Vivo Cortical Parcellation

Robert Turner

Abstract The advent of whole-body MRI scanners at field strengths as high as 7 T has enabled a dramatic improvement of the spatial resolution of human brain images, in all three major attributes: structure, function and neural connectivity. Structural imaging of entire living human brains with an isotropic resolution of 300–400 μm is now feasible. Such images allow the discrimination of discrete cortical areas based largely on their distinctive myeloarchitecture. This chapter describes the challenges that have to be overcome in creating such images. Sources of contrast in structural MR brain images are summarized. Rationales are then provided for the currently preferred acquisition techniques, which give good signal-to-noise ratios, and relative freedom from the effects of the non-uniformity of the radiofrequency magnetic fields relating to spin excitation and MR signal reception.

7.1 Histology Versus Magnetic Resonance Microscopy

A major goal of imaging neuroscience is precisely, routinely and objectively to identify the different brain areas associated with specific observable operations for individual human subjects. This enables questions regarding the computational competence of specific cortical architectures to be empirically addressed. This goal would be easier to achieve if MRI-based structural parcellation of the living human brain could be automatically performed.

The variety of possible MRI tissue contrasts and the capability of MRI for volumetric imaging have made it indispensable for both ex vivo and in vivo investigations. As will discussed in further detail below, use of the ultra-high field strength of 7 T gives dramatically improved image quality, as compared

R. Turner (✉)
Max Planck Institute for Human Cognitive and Brain Sciences, Stephanstraße 1A, 04103 Leipzig, Germany
e-mail: turner@cbs.mpg.de

S. Geyer and R. Turner (eds.), *Microstructural Parcellation of the Human Cerebral Cortex*, DOI 10.1007/978-3-642-37824-9_7, © Springer-Verlag Berlin Heidelberg 2013

with the more commonly used field strength of 3 T and 1.5 T. The human brain can thus be investigated with a resolution of a few 100 μm in vivo with adequate signal-to-noise ratio (SNR) in a reasonable scanning time. Ex vivo, using fixed brain, the much longer scan time that becomes feasible allows resolution of 70 μ or even higher, closing the gap between in vivo MRI and cadaver brain histology.

It is interesting to compare these two methodologies for the study of cadaver brain. Techniques that use visible light to observe differences in tissue constituents, whether with histological stains, autoradiography of neuroreceptor-labelled sections (Schleicher et al. 1999), polarized light imaging (PLI, Axer et al. 2011), electron microscopy (Denk et al. 2012) or even particle-induced X-ray emission (PIXE, Duflou et al. 1989) all require the brain tissue to be sliced into sections much thinner than 1 mm before examination. Accurate reconstruction of entire volumes of tissue, which is vital for understanding the topology of the cortex and white matter, is then difficult, tedious, prone to error, and extremely time-consuming. MRI, although much more limited in spatial resolution, has the unique ability to produce perfectly registered three-dimensional maps of water-containing tissue. Such maps allow a far firmer grasp on the topology and connectivity of brain substructures than can be achieved by any other means. Furthermore, the technique offers a wide range of relevant contrasts that can explore spatial variations in tissue orientation, myeloarchitecture, cortical thickness, vascularity and area-dependent axonal connectivity.

Beyond these considerations, the overwhelming advantage of MRI is that it can also probe the structure, function and connectivity of living human brain tissue, without any harmful side effects.

7.2 Sources of MRI Contrast in Brain Tissue

Variations of cytoarchitecture are difficult to detect using MRI, whereas myelin is very easy to pick out. It is the major source of contrast in T1-weighted and proton density images, and is responsible for T2 and T2* contrast in most of the brain, except for the small deep brain structures naturally containing sequestered ferritin, such as the basal ganglia and the subthalamic nucleus, where the magnetic effects of iron obviously dominate the contrast. Oriented bundles of myelinated axons determine most of the anisotropy of water mobility detected using diffusion tensor MRI. Where myelinated fibres are found within the cortex, in the form of pyramidal cell axons and the bands of Baillarger, they also give excellent MRI contrast.

While the use of MRI to map cortical thickness has already shown success in cortical parcellation (Meyer et al. 1996; MacDonald et al. 2000), at the present time the most mature MR techniques for exploring cortical areas in-vivo and ex-vivo are intended to visualize patterns of cortical myelination. The results can then be compared with the scanty literature on human brain myeloarchitecture, derived from myelin stained cadaver brain tissue, excellently summarized in the chapter by Nieuwenhuys (this volume), to identify corresponding Brodmann areas.

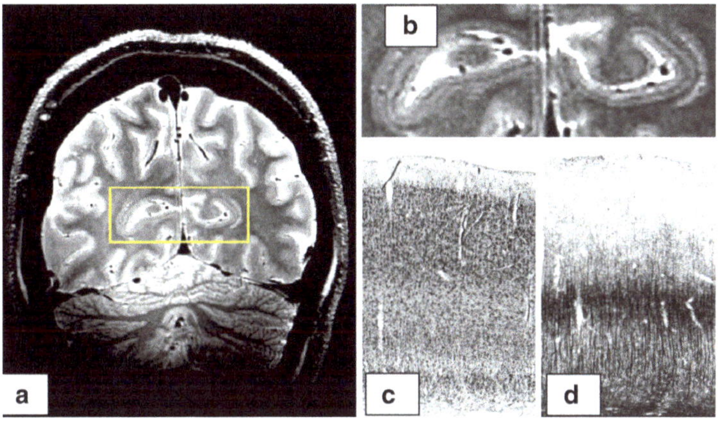

Fig. 7.1 (**a**) Coronal MR image through occipital lobe, including calcarine sulcus. Normal human subject. MT-weighted Turbo Spin Echo (TSE) sequence, 7 T MR scanner (MAGNETOM 7 T, Siemens). Voxel size is $(0.50\ \text{mm})^3$. Acquisition time was 14 min 33 s for 20 slices. (**b**) Expanded view of *box* in Fig. 7.1a. (**c**) Comparable section of lower bank of calcarine sulcus in cadaver human brain. Merker stained (cell body). (**d**) Adjacent section using Gallyas myelin stain, showing Stria of Gennari in cortical layer IV, corresponding closely in appearance and location to the in-vivo MR images in (**a**) and (**b**)

Comparison of appropriate quantitative MR images of brain tissue with histological sections stained for cell bodies and for myelin shows that it is overwhelmingly myelin content that dominates MRI contrast (Fig. 7.1).

It is thus important to map MR-accessible parameters that correlate well with the cortical myelin content (Hopf 1954, 1955, 1956) measured in cadaver brain sections. Among suggested candidates are proton density (Shah et al. 2011), myelin water fraction (MWF, Mackay et al. 2006), magnetization transfer rate (MTR, Wolff and Balaban 1989; Koenig 1991), transverse relaxation rate (R2*, Duyn 2011), phase maps formed using gradient echo imaging (Duyn et al. 2007), and longitudinal relaxation rate (R1) (Fischl et al. 2004). The challenge is to obtain a high enough spatial resolution to distinguish intra-cortical details that can assist in realistic parcellation, within a scanning time short enough for in-vivo studies of humans.

At this point it must be mentioned that currently there is no technique in existence that can quantitatively assay the density of myelin at high spatial resolution in cadaver brain tissue. The optical density of myelin stained sections of cadaver brain clearly reflects the amount of myelin, but this does not correspond to actual myelin density, because the microscope images are essentially shadowgrams, like flat X-rays, giving no information about the distribution of myelin in the third dimension perpendicular to the plane of the section. Chemical assays, such as those pioneered in the nineteenth century by Thudichum (1884), require volumes of tissue, at least 30 mm^3, that preclude intracortical spatial resolution (Dasgupta and Hogan 2001).

Because the water content of grey and white brain tissue is about 80 % and 70 % respectively, it is possible to obtain useful contrast simply using maps of proton density (Shah et al. 2011), which demonstrate this very basic physical distinction. These can be obtained by means of very short echo time (TE) imaging, or by extrapolating sets of images with a range of echo times to zero TE. However, high signal to noise ratio (SNR) is required because of the relatively small grey-white difference in water content, and this method has not been widely used for distinguishing intra-cortical structures.

Myelin water fraction (MWF) is defined as the fraction of the NMR signal from water protons that can be unambiguously assigned to the thin layers of water that separate the layers of myelin membrane that wrap myelinated axons. There is good evidence (Kalantari et al. 2011) that this water exchanges very slowly with the other major pools of water in brain tissue, those within cell bodies and between cells. MWF can be estimated by determining the distribution of T_2 relaxation times (Whittall et al. 1997). In brain tissue, at 3 T magnetic field strength, T_2 distributions show three peaks, empirically assigned to compartmentalized spin populations: a short T_2 peak around 20 ms (between 10 and 50 ms) representing the water trapped between the layers of myelin, a second peak around 70–90 ms assigned to intra- and extracellular water, and a third peak with $T_2 > 2$ s usually assigned to cerebrospinal fluid signal. The myelin water fraction can be computed as the fraction of signal in the T_2 distribution below about 40 ms, typically representing about 15 % of the total water content. T_2 relaxation times can be obtained using the well-known Carr-Purcell-Meiboom-Gill technique (Kolind et al. 2009).

There are two practical limitations regarding the use of MWF for investigation of cortical myeloarchitecture. Firstly, as just mentioned, because there is no simple quantitative assay for myelin content in living human brain, it has not yet been demonstrated that MWF provides a linear measure of myelin density. The second limitation is the comparatively low signal-to-noise ratio (SNR) of accurate MWF methods, which means that either very large voxels or very long averaging times must be used. Although a faster sequence (mcDESPOT) based on steady-state free precession and fast gradient echo acquisition has been developed (Deoni et al. 2008), this method for quantifying MWF is very approximate, due to the large number of variables involved in the fitting procedure employed (Lankford and Does 2012), and the requirement of uniform transmit RF fields. For such reasons MWF mapping has not yet been used in cortical parcellation studies.

Measurement of magnetization transfer rate (MTR) has also been proposed as a method for myelin density quantification (e.g. Vavasour et al. 1998; Alexander et al. 2011). This method (Wolff 1989) relies on the fact that the spin–lattice relaxation of longitudinal magnetization in tissue arises largely from transfer of the spin magnetization of protons on free water molecules to protons bound to large relatively immobile biological molecules, mostly embedded in membranes. Transfer of spin magnetization in the opposite direction also occurs, from protons bound to the large molecules to the free water protons. This can be enhanced by applying an RF magnetic field at a frequency offset from the free water proton Larmor frequency, which excites the protons of hydrogen atoms attached to large

Table 7.1 Concentrations of lipids in gray matter, white matter, and myelin of human brains (Reprinted with permission from Siegel et al. 1999). Protein and lipid figures in percent dry weight; all others in percent total lipid weight

	Myelin		White matter				
Substance	Human	Bovine	Rat	Human	Bovine	Gray matter (human)	Whole brain (rat)
Protein	30.0	24.7	29.5	39.0	39.5	55.3	56.9
Lipid	70.0	75.3	70.5	54.9	55.0	32.7	37.0
Cholesterol	27.7	28.1	27.3	27.5	23.6	22.0	23.0
Cerebroside	22.7	24.0	23.7	19.8	22.5	5.4	14.6
Sulfatide	3.8	3.6	7.1	5.4	5.0	1.7	4.8
Total galactolipid	27.5	29.3	31.5	26.4	28.6	7.3	21.3
Ethanolamine phosphatides	15.6	17.4	16.7	14.9	13.6	22.7	19.8
Lecithin	11.2	10.9	11.3	12.8	12.9	26.7	22.0
Sphingomyelin	7.9	7.1	3.2	7.7	6.7	6.9	3.8
Phosphatidylserine	4.8	6.5	7.0	7.9	11.4	8.7	7.2
Phosphatidylinositol	0.6	0.8	1.2	0.9	0.9	2.7	2.4
Plasmalogens	12.3	14.1	14.1	11.2	12.2	8.8	11.6
Total phospholipid	43.1	43.0	44.0	45.9	46.3	69.5	57.6

molecules, having a broad resonance spectrum. The coupling of this spin magnetization to the free water protons accelerates their relaxation, and transfer of the magnetization saturates them, reducing the net magnetization. These effects on T1 and spin magnetization M_0 thus provide an indication of the local molecular environment. In much of brain tissue, especially white matter, myelin dominates the local molecular environment, and so it has been considered that MTR also provides a map of brain myelin (e.g. Gringel et al. 2009; Helms et al. 2010). However, while there is a rough concordance between maps generated using MWF and MTR methods, they do not match in detail (Vavasour et al. 1998), giving rise to a concern that neither of these methods attain this goal. MTR imaging also tends to provide a relatively low SNR per unit time, because it compares the signal from images with rather similar intensity, with and without a magnetization saturation pulse.

Given the ubiquity and concentration of myelin in brain tissue, it is expected to affect MR image contrast in many ways. Table 7.1 (Siegel et al. 1999) provides the dry weight composition for several important constituents of brain tissue.

It is clear from this table that myelin concentration levels are high enough to affect proton density contrast. Myelin also has a strong effect on the transverse relaxation times T2 and T2*. T2 is primarily affected because the water between the myelin layers wrapping axons has a particularly short T2, as described above. The T2 of the non-myelin water in white matter is surprisingly similar to T2 in grey matter (Whittall et al. 1997).

It is still poorly recognized how little T2 contrast there really is in the brain, outside of deep brain structures containing iron (see below). Because its volume

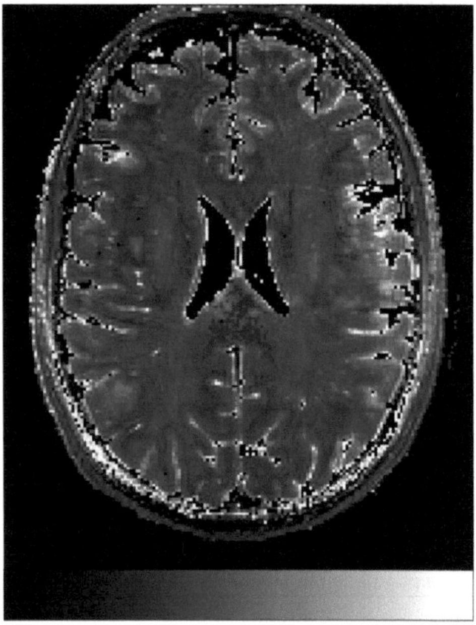

Fig. 7.2 Quantitative map of T2, measured at 7 T in an axial slice through normal human brain. Spatial resolution was 1 mm isotropic. Data obtained at five echo times were fitted to a single exponential decay curve. Grey scale gives T2 in milliseconds. Note the minimal *grey-white* contrast (Courtesy of Zaheer Abbas)

fraction is small (less than 15 %), and its T2 is short ($<$ 15 ms), myelin water contributes little to the value of T2 measured conventionally by comparing the signal at two echo times, or by measuring with the Carr-Purcell sequence. Instead, so-called "T2-weighted images" provided by MRI sequences such as turbo-spin-echo (TSE) typically gain much of their grey-white matter contrast from magnetization transfer (see Thomas et al. 2004; Turner et al. 2008), differences in T1, or proton density variations. Optimal contrast-to-noise for TSE sequences, which typically have echo trains extending much longer than T2, is obtained at the shortest possible echo times, which would not be the case if T2 difference were the major contrast mechanism. When T1 was measured using an inversion-recovery prepared turbo-spin-echo sequence, it was found that MTC effects reduced its value by about 30 % (Turner et al. 2008). It has been historically unusual to create genuine maps of either the short or long T2 component of the bi-exponential transverse relaxation decay curve. A very careful study was performed by Oros-Peusquens et al. (2008). Her quantitative maps of T2 in normal brain show very little overall contrast between grey and white matter, with a large overlap in their T2 distributions. See also Fig. 7.2, which shows a quantitative map of T2 in normal brain obtained at the Leipzig MPI at 7 T. The futility of attempting cortical parcellation using T2 maps alone is underscored by West et al. (2012) who show the enormous overlap of the T2 distributions in WM and GM.

In practice, when magnetization transfer effects are carefully avoided, the grey-white contrast obtained by "T2-weighted sequences", such as TSE, is dominated by variations in T1 and MTC at echo times longer than 100 ms. The main reason for this is that the echo train generated by the repeated radiofrequency refocussing pulses of the TSE sequence consists largely of stimulated echoes, for echo times longer than about 100 ms. Stimulated echoes decay with a time constant of T1, which is much longer than T2 in tissue, and thus the signal in these later echoes is heavily T1 weighted. Even with echo times as short as 26 ms (e.g. Trampel et al. 2011) TSE contrast is clearly dominated by magnetization transfer effects.

T2* weighted images and T2* maps of the brain, on the other hand, show highly significant signal variations, both between grey and white matter and within each of these tissues. This reflects the presence of myelin, but also another very important source of MRI contrast in brain, iron (Drayer et al. 1986). It is now becoming clear that the main origin of T2* contrast is that the magnetic susceptibility of lipid molecules, and of commonly found iron-containing molecules, particularly ferritin, is different from that of water. (Duyn 2011; Lee et al. 2011) When the tissue experiences a static magnetic field, this difference causes microscopic field gradients to arise in water associated with such molecules. This is especially the case in WM, with a high density of lipid membranes forming myelin sheaths around the axons, and in deep brain nuclei, such as the substantia nigra, with high densities of iron in the form of ferretin and neuromelanin. The net result of the dephasing of the spin magnetization caused by these magnetic field gradients is a reduction in T2*. This susceptibility difference also partly explains the strong phase contrast between GM and WM (Li et al. 2011). T2* weighted images and phase images of the brain can be obtained with very high spatial resolution. Cortex often contains a significant quantity of myelinated axons (the basis of myeloarchitecture, indeed), and iron deposits are often associated with such cortical myelin (Fukunaga et al. 2010) (see Fig. 7.3).

Thus T2*-weighted images show excellent detail in cortical grey matter (Duyn 2011) (see Fig. 7.2), which has been shown to depend on the angle between B_0 and the local axonal orientation (Cohen-Adad et al. 2012) (Fig. 7.4).

But probably the most striking impact of myelin on MR images lies in its effect on T1. For precisely the same reason that it affects MTR, the presence of myelin substantially accelerates T1 relaxation, by the transfer of magnetization from slowly relaxing free water protons to rapidly relaxing protons bound to large molecules in the cell membranes. Those specific large molecules that are particularly responsible for this powerful relaxation mechanism have been discovered by careful study of model membrane systems comprised of phospholipids and water (Koenig 1991; Kucharczyk et al. 1994). When either of two important components of typical cell membranes, cholesterol and cerebroside, were added in biologically typical concentrations, they were found to dramatically shorten T1. These molecules have hydrophilic hydrogen-containing head groups that sit at the membrane-water interface and provide a conduit for the transfer of magnetization from the free water protons to the bulk of the membrane.

Thus so-called “T1-weighted” images have come to be regarded as the standard for depicting grey-white contrast in human brain, and used in numerous applications which require image segmentation of grey and white matter, such as voxel-based morphometry (Ashburner and Friston 2000). Generally low flip-angle gradient-echo (FLASH) sequences (Frahm et al. 1986) are used to provide such

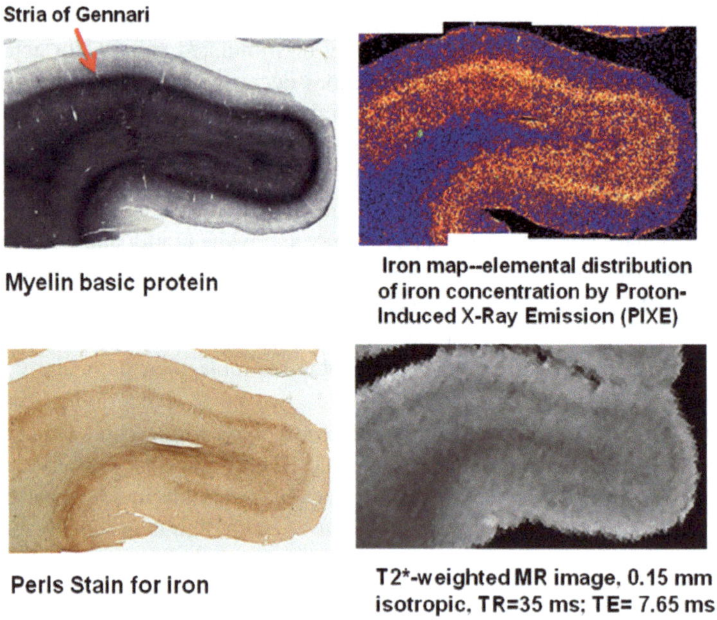

Fig. 7.3 Image data from a block of human cadaver brain tissue containing primary visual cortex, V1 (Courtesy of Carsten Stüber, Leipzig). The Stria of Gennari is prominent in this block. Ex-vivo MRI scanning was performed in Leipzig at 7 T (Siemens Magnetom). The brain block was then sliced. Adjacent slices were given various histological stains. One slice was analyzed using Proton-Induced X-ray Emission provided by a proton beam accelerator (Lipsion) to map elementary iron content within the slice. Note that cortical myelin is much richer in iron than white matter myelin, but both iron and myelin contribute to T2* contrast

images. Here the contrast arises because white matter protons recover faster than grey matter protons to alignment with the static field, providing a larger equilibrium signal. The contrast can be further enhanced using a preparatory inversion pulse, such that the grey matter signal is close to being nulled at the time of image data acquisition, at which time the white matter signal has already partially recovered. The resultant sequence (Mugler and Brookeman 1990) is named MP-RAGE (magnetization-prepared rapid-acquisition gradient echo). However, it should be pointed out that neither FLASH nor MP-RAGE sequences actually produce maps of the value of T1. They have higher intensity where T1 is shorter, and their intensity is also weighted by proton density and T2* values.

In order to obtain quantitative maps of T1 and thus to provide a better guide to tissue myelin, each of these sequences must be modified. Adding a second excitation pulse with a different flip angle to the FLASH sequence (Preibisch and Deichmann 2009), or sampling the MP-RAGE signal at two different inversion times, allows the longitudinal relaxation time T1 to be mapped with reasonable precision. The second of these sequences is known as MP2RAGE (Marques et al. 2010). Maps of T1 may also be generated by acquiring a series of inversion-recovery T1-weighted images, and fitting the inversion time dependence in each

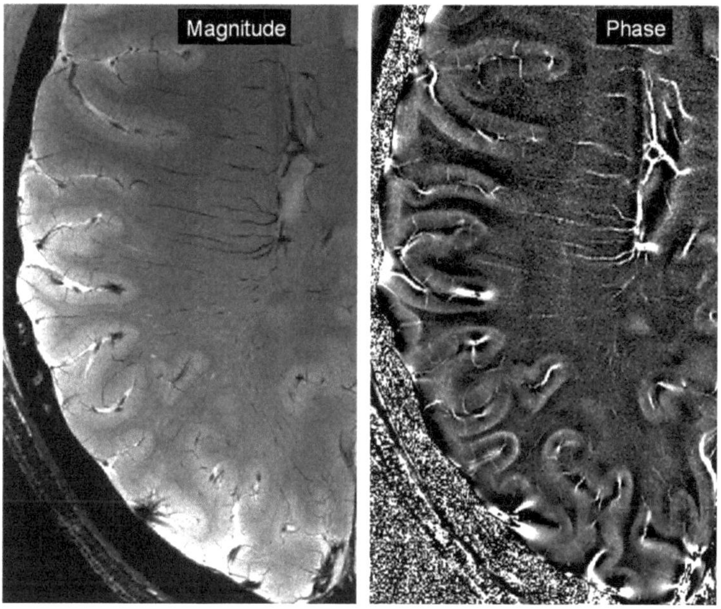

Fig. 7.4 Axial images of normal human brain acquired on a Siemens Magnetom 7 T scanner at Leipzig by Andreas Schaefer. FLASH sequence. In-plane resolution $0.23 \times 0.23\ \text{mm}^2$. Slice thickness 1 mm. 24-channel NOVA RF coil

voxel to an exponential function. Importantly, these maps show little effect of the inhomogeneity of image intensity (bias-field) normally caused by spatial variations in RF excitation and reception.

In conclusion, comparing MRI methods for detecting the presence of myelin, it can be hypothesized that maps of relaxation rate R1 (= 1/T1) and MTR maps provide essentially identical information, proportional to the area of membranes in good contact with the free water pool (i.e. the outer and inner circumferences of myelinated axons), while MWF maps provide a good quantification of total myelin density. This concept merits much deeper exploration, including quantitative chemical assays of myelin density, in future research.

7.3 In-Vivo MRI Microscopy: Issues of Signal, Sequence and Field Strength

Whatever the source of contrast and the structures to be investigated, the most important consideration for useful cortical mapping is to have a sufficiently large contrast to noise ratio (CNR). Although fundamental limits exist, related to the density of water protons in brain tissue and the fact that MRI is performed at body temperature, achievable CNR can be improved by maximizing raw NMR signal, minimizing noise, optimizing the MRI sequence strategy, maximizing acquisition time, minimizing effects of brain motion, and optimizing the static magnetic field strength.

7.3.1 Signal

For a given water proton density, flip angle, static magnetic field uniformity, sampling rate, and relaxation times, the NMR signal depends linearly on the voxel volume, and the RF receiver coil sensitivity. Intracortical detail exists at a scale of less than 500 μm, and the cortical surface is always multiply curved. Thus cubic voxels of less than 500 μm are essential. At this resolution, using a very high performance MR sequence, Turner et al. (2008) showed that at least 30 min of scan time was required at 3 T to give enough CNR to detect the Stria of Gennari reliably in multiple slices with resolution $0.4 \times 0.4 \times 0.5$ mm^3, with the 12-channel RF receiver coil then available.

The receiver coil sensitivity itself depends largely on how close the coil elements are to the tissue to be imaged. Thus, other things being equal, to obtain good signal from an entire head it is desirable to have a large number of small coil elements, mounted on a coil former only just large enough to accommodate the head. These are configured in the form of a "phased array" (Roemer et al. 1990; Wiggins et al. 2009). With a 64 channel array an improvement in SNR of a factor about 2–3 is claimed in cortical areas relatively close to the skull, which should shorten the acquisition time for the isotropic detection of the Stria of Gennari at 3 T, as just described (Turner et al. 2008) to about 8 min.

7.3.2 Noise

Random noise in MRI is primarily white noise, arising from thermal fluctuations in the object to be imaged, in the conductive elements of the coil itself, and in the receiver preamplifiers. The present generation of receiver preamplifiers adds well below 1 dB of noise to the signal. Thermal noise can be reduced by making the receiver bandwidth as small as possible, since the noise is proportional to the square root of the bandwidth. There are two limits to this minimization. The first is that the NMR signal decays over a time of about 100 ms, so there is never any point in having a receiver bandwidth less than 10 Hz. Secondly, and more important for imaging, the acquisition readout window for a given phase-encoded echo is never greater than 100 ms, and within this window up to 512 data points must be obtained to sample the signal which has been spatially encoded by the read gradient. This means that the acquisition bandwidth is at least a few kilohertz. This has the advantage, for typical multi-pulse imaging, that the imaging magnetic field gradients are much larger than the endogenous field gradients produced by inhomogeneities of magnetic susceptibility within the head, so there is little spatial distortion in the final images.

7.3.3 MRI Sequence Choice

MR microscopy has been performed for several decades (reviewed in Blackband et al. 1999). With a short echo-time spin echo sequence, a field strength of 14 T, and an hour of scan time, spatial resolution of 1 μm in-plane was achieved ex vivo (Lee et al. 2001). Fresh interest in MRI neuroanatomy was generated by the publication by Fatterpekar et al. (2002) of high resolution images of cadaver brain at 9.4 T with a spatial resolution of $78 \times 78 \times 500\ \mu m^3$. These images required 14 h of scan time. No attempt was apparently made to maximise the CNR per unit time, and the authors fail to state what type of sequence they used. From the contrast that they found, it is fair to assume that they used a single-slice gradient-echo sequence, with a stated echo time of 45 ms and a very long TR of 2.4 s. Such a sequence provides good grey-white matter contrast due to the shorter T2* in white matter, and good intracortical contrast due both to myelinated cortical layers and the associated iron deposition. However, owing to the lack of technical detail it is quite impossible to deduce a quantitative relationship between the MR contrast and the optical density of the corresponding stained sections which they show.

Fortunately, MRI physics is a science, and not a black art. It is possible to relate measured image intensity to the basic parameters of the Bloch equation, and in the present context it is highly desirable to do so (see also Fischl et al. 2004). The duration of an MRI scan in vivo is basically limited by the capacity of a human volunteer subject to remain immobile for an extended period of time. Thus the ideal MRI sequence for in-vivo cortical parcellation makes maximum use over time of the available spin magnetization, with minimum time spent preparing the spin magnetization, when data are not acquired. In addition, the highest possible contrast between grey matter and white matter is desired. Further requirements are that the image intensity should simply reflect an NMR parameter that can be used as an index of myelination, and that the RF power deposition should be the minimum required to generate sufficient contrast. This final consideration is associated with the need to use high magnetic field strength (see next section), and the fact that RF power deposition (SAR, specific absorption rate) increases as the square of the field strength.

It is an unfortunate historical fact that so far there has been no systematic investigation as to which sequence delivers the closest approximation to these requirements. The two main classes of sequence are (i) gradient echo sequences, lacking refocusing pulses, in which low flip angle pulses are repeated at short intervals, each followed with read gradients applied while data are acquired; and (ii) spin echo sequences, which typically use higher flip angle excitation pulses, with refocusing pulses to recover the spin magnetization dephased by local magnetic field gradients.

(i) Gradient echo sequences. Here the three main contenders for high SNR and good grey matter/white matter contrast are the MP-RAGE sequence (Mugler), the modified version using a second inversion pulse, MP2-RAGE (Marques) and the Modified Driven Equilibrium Fourier Transform (MDEFT) sequence, optimized by Deichmann (2006). Acquisition parameters for the MP-RAGE sequence have recently been optimized by Bock et al. (2013), giving a substantial improvement in CNR per unit time.

(ii) Spin echo sequences. Until the invention by Hennig et al. (1986) of the RARE sequence, gradient echo sequences (which include SSFP, FISP, PSIF, etc.) made more efficient use of the steady-state magnetization that was available, but the string of refocusing pulses following a single excitation pulse that form the essence of RARE (turbo-spin-echo (TSE), fast spin-echo, etc.) greatly increased the efficiency of this sequence. They also greatly increase the SAR, making standard TSE sequences difficult to use at field strength above 3 T.

Fortunately an alternative sequence to TSE exists, which gives nearly identical contrast, somewhat higher signal to noise per unit time, and has a much smaller RF power deposition. This is the GRASE sequence (GRAdient echo and Spin Echo) first described by Oshio and Feinberg (1991). This sequence is similar to TSE, in that it uses a train of refocussing 180° RF pulses, but most of these pulses are replaced by alternating gradients, as in echo-planar imaging, so that the signal forms a train of gradient echoes, which are periodically refocused by the 180° pulses. As with TSE, this echo train decays with a time constant closer to T1, which enables a long acquisition window, but the radiofrequency power deposition is drastically reduced. Trampel et al. (2010) has shown that for the same effective echo time, voxel size and TR time, the TSE and GRASE images of a brain are almost identical, so that image subtraction reveals only a few edges, and mostly noise. The GRASE sequence has great promise for ultra-high resolution scanning of human brain at very high field.

7.3.4 *Maximizing Acquisition Time*

Signal to noise can obviously be increased by longer averaging of image data. However, SNR improves only with the square root of the imaging time, so that scanning for an hour gives only double the SNR obtained in 15 min. And this apparent gain can easily be cancelled out by the increasing likelihood with scan duration that the subject will move significantly. One solution (Turner et al. 2008) is to average a number of separate scans after they have been co-registered with each other. Keeping one's head still for 20 min is considerably easier than trying to remain immobile for an hour or more. In this approach, it is advisable to average in the image domain, but to retain the full complex data, so that the real and imaginary components of the image are separately averaged, and then recombined to form a magnitude image (Oros-Peusquens et al. 2010). In this way, data from as many as ten imaging sessions with a given human subject can be combined with very little loss of spatial resolution.

7.3.5 *Minimizing Effects of Brain Motion*

Subject motion presents severe difficulty for ultra-high resolution in vivo studies. Even with the use of a bite-bar, motions of as much as 1 mm over a scan duration of 20 min are hard to avoid. It is remarkable that an occasional image with better than 0.5 mm isotropic resolution can appear sharp and precise. This can occur when any head motion takes place while the scan trajectory is within a region of k-space where there is little signal. Obviously, the effects of head motion on a typical structural scan acquired using multiple spin excitations are blurring and formation of multiple spatially translated images. These are often accompanied by an artefact distributed across the image that results from discontinuities in the k-space data arising from motion between successively acquired lines in k-space. Post-hoc methods for correction of such artefacts (Bourgeois et al. 2003) have had little success.

Effective measures to overcome these problems depend mainly on prospective adjustment of the parameters determining the MRI slice position. Measured head movement parameters are passed to the scanner, and converted into a corresponding adjustment of the imaging gradients. The imaged volume therefore always moves with the subject's head, and the resulting images require far less post-acquisition correction for head motion artefacts. Methods include the use of navigator echoes and several strategies for independent tracking of the head position.

Navigator echo techniques (Lee et al. 1996) rely on sampling the subject's head position using low flip angle rf pulses and appropriate magnetic field gradients between actual image data acquisitions. While these methods have shown effectiveness, they have limited accuracy and update rate, and some signal is inevitably lost. Active NMR marker techniques (Ooi et al. 2009) can be much more accurate and faster. A headband integrated with three active markers is secured to the forehead. Prospective correction is achieved by interleaving a rapid track-and-update module into the imaging sequence. For every repetition of this module, a short tracking pulse-sequence re-measures the marker positions, and during head motion, the rigid-body transformation that realigns the markers to their initial positions is fed back to adaptively update the image-plane. The slowness of nuclear spin dynamics ultimately limits the update rate of this technique to a few tens of milliseconds.

Several labs have developed prospective motion correction systems based on optical tracking with infrared cameras (Speck et al. 2006; Andrews-Shigaki et al. 2011; Maclaren et al. 2013; Schulz et al. 2012). Reflective markers are attached to the subject's head, enabling measurement of head movement within an accuracy of a few tens of micrometres.

7.3.6 *Optimizing the Static Magnetic Field Strength*

The most expensive but ultimately the most satisfactory means of improving the resolution of MR images is to increase the strength of the static magnetic field. This linearly increases the spin magnetization of the water protons. The SNR of MR images is also roughly proportional to field strength, although this depends quite strongly on the type of sequence used, and on the quality of the RF coils. Ideally the transmit RF field should be uniform across the head, but as the field strength increases, the wavelength of the Larmor frequency RF magnetic field decreases, which makes it much more difficult to assure field uniformity. Considerable effort has been made in recent years to provide uniform transmit fields, using techniques often involving multiple channel RF transmit systems, such as RF shimming (Mao et al. 2006), transmit SENSE (Katscher et al. 2003), and improved RF coil design (Kozlov and Turner 2011).

However, there is a range of MRI sequences that suffer rather little from transmit RF field nonuniformity. These include MDEFT (Thomas et al. 2004), fast spin-echo (TSE) (De Vita et al. 2003; Thomas et al. 2005), and gradient-echo echo-planar imaging. Nonuniformity corrections can be made for images obtained using certain other sequences, by computing ratio images in which the image intensity variations (bias field) are in common between two acquisitions having different contrast (van de Moortele et al. 2009; Marques et al. 2010).

One may ask whether increasing the field strength can give an unlimited improvement in CNR, and thus in spatial resolution. Unfortunately, at fields above 10 T the RF fields at the Larmor frequency become extremely non-uniform inside an object as large as the human brain, and the RF power deposition continues to grow as the square of the field strength, causing severe restrictions on the type of MRI sequence and scanning rate that can be used. While MRI scanners for human subjects with field strength of 11.7 T are under development, it is by no means certain that these will be useful in the quest for in-vivo cortical parcellation of human brain. It is already clear that considerable engineering effort will be required to obtain acceptable image quality at this spatial scale, although much is already being learned regarding neuroanatomy and neural function in studies of small animal brains at 11.7 T and greater (e.g. Marques et al. 2009).

7.4 MRI Studies of Cortical Myeloarchitecture

The history of attempts to visualize cortical myeloarchitecture using in-vivo MRI tracks closely the progressive increase in available magnetic field strength. This is easily understandable, given the dependence of SNR in MRI on static magnetic field, and the extreme need for high SNR in the visualization of submillimetre-sized structures such as cortical layers.

The earliest published study was that of Clark et al. (1992), working at 1.5 T, who used proton density-weighted images with 0.39 mm in-plane resolution, and 3 mm slice thickness. At 1.5 T, adequate SNR was achieved by averaging four acquisitions, resulting in about 40 min per slice. This study constituted a proof of principle, but the extreme slowness of the method and the highly anisotropic resolution needed to attain enough SNR at 1.5 T make it impractical for mapping of cortical areas.

A gap of 10 years in the pursuit of in vivo cortical parcellation then followed, until the group of Koretsky at NIH published a 3 T study, also of primary visual cortex (Barbier et al. 2002), which used T1-weighted contrast and voxels of $0.35 \times 0.35 \times 0.6$ mm^3, with an averaging time of 45 min for 62 slices (see Fig. 7.5). In these images the Stria of Gennari, which indicates the cortical area V1, was significantly easier to pick out, but the relatively thick slice used made it hard to trace this intracortical structure in three dimensions in the folded cortex. This work was followed by that of Walters et al. (2003) at 1.5 T, using T1-weighted FLASH images to study both V1 and the visual motion area V5/MT in human brain, at a resolution of $0.556 \times 0.556 \times 0.5$ mm^3. These images had insufficient SNR to allow depiction of the complete V1 area, but the authors were able to show that their T1-weighted contrast gave an indication of myeloarchitecture, rather than cytoarchitecture. Unfortunately, their analysis of the MRI contrast mechanism was not deep enough to provide more insight.

The next important work on this topic came from Bridge et al. (2005), working at 3 T. These researchers used an inversion-recovery sequence to provide good T1 contrast with voxels of dimensions $0.3 \times 0.3 \times 1.5$ mm^3. Where the imaging plane was normal to the cortical surface, this allowed the stria of Gennari to be very clearly visualized, although this was impossible elsewhere. Retinotopic fMRI in the same subjects was performed, and the spatial locations of V1 determined structurally and functionally were confirmed to be congruent, within the limitations of the technique.

Similar more recent work has been done by Sanchez-Panchuelo, working at 7 T (2012), where the use of isotropic voxels of 0.4 mm size enabled a much more complete depiction of V1 to be obtained. In this work, comparison was made between the use of T2*-weighted FLASH imaging and MP-RAGE data, with the conclusion that FLASH provides a significantly higher contrast-to-noise ratio, allowing reliable visualization of the stria of Gennari in every slice of a volume covering the occipital cortex. The independently derived boundary of V1, identified in the same subjects using retinotopic mapping by fMRI, closely matched the border of anatomically defined striate cortex in the human brain.

The stria of Gennari was also depicted very clearly using the TSE sequence at 4.7 T field strength by Carmichael et al. (2006). The sequence was slightly modified to improve the point-spread function (Thomas et al. 2004). Voxels in this study were $0.352 \times 0.352 \times 2.0$ mm^3, too anisotropic (for reasons of SNR) to allow tracing of the entire visual cortex.

Understanding of significant MRI contrast within the cerebral cortex was further enhanced by the work of Sigalovsky et al. (2006). Working at the low field strength

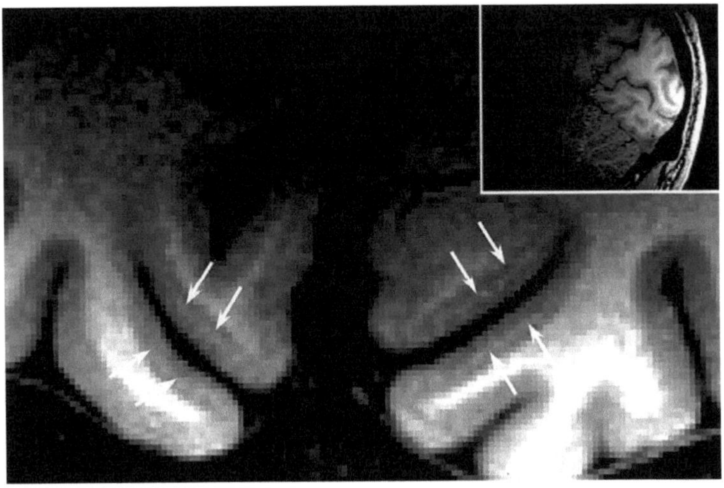

Fig. 7.5 T1-weighted MRI (350 × 350 × 600 μm^3 spatial resolution) in the visual cortex, acquired at 3 T. The *white arrows* point to a *thin white line*, identified as the stripe of Gennari. The insert shows the surface coil coverage on a *T*1-weighted sagittal MR image (Barbier et al. 2002, with permission)

of 1.5 T, these researchers used a 3D FLASH sequence, scanning four times with different flip angles α, set at 5°, 10°, 20° and 40°, so that the relaxation rate R1 (= 1/T1) could be computed for each voxel, using the equation (Nishimura 2010)

$$S\,(TR, \alpha, R1) = K \sin\alpha\ (1 - e^{-TR_* R1})/(1 - \cos\alpha\ e^{-TR_* R1})$$

Here the voxel proportions were close to isotropic, 1.3 × 1.0 × 1.3 mm^3, but too large to resolve intracortical detail. However, the R1 maps so obtained (see Fig. 7.6) revealed well-defined cortical regions of high R1 (low T1) situated on the anterior Heschl's Gyrus and corresponding closely to the known increased myelin density in primary auditory cortex (Hopf 1954).

These displays show all high R1 regions on the superior temporal lobe.

Further work from the laboratory of Fischl (see Fischl chapter) using an optimized 3D-FLASH MRI sequence giving excellent visualization of the stria of Gennari via T2* contrast at 7 T in cadaver brain (Hinds et al. 2008) has shown that the boundaries of primary visual cortex V1, as defined by the presence of this cortical feature, are accurately predicted by the pattern of sulcal folding in occipital cortex. Here the isotropic voxel size ranged between 180 and 200 μm, requiring typically more than 10 h of acquisition time. Similar findings have been shown for other cortical areas, such as Brodmann areas 4 and 3a, and entorhinal cortex (Augustinack et al. 2005; Fischl et al. 2004) has emphasized the very important point that structural images obtained using MRI are most useful for neuroscience purposes and most generalizable when they map the basic MRI parameters – proton density, T1 and T2*–rather than simply providing a traditional 'weighted' image,

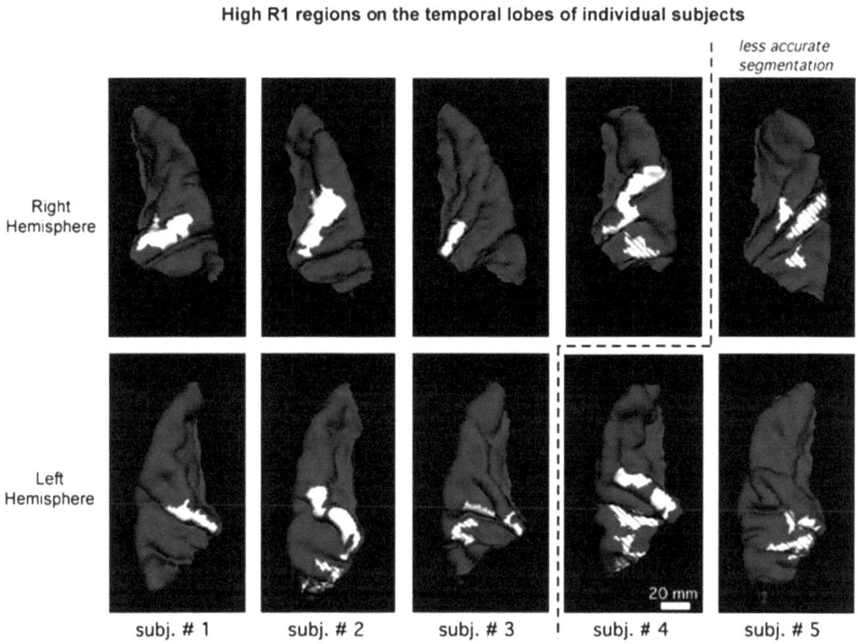

Fig. 7.6 (Reprinted from Sigalovsky et al. 2006, with permission). Regions of high R1 on the superior temporal lobe. Each panel shows either the *right* (*top*) or *left* (*bottom*) superior temporal lobe of a given subject. High R1 regions are *white* if they overlap first Heschl's gyrus and are shaded (with *diagonal lines*) if they do not

which depends strongly on sequence parameter details, field strength, scanner manufacturer, etc.

Further studies using the TSE sequence to detect the stria of Gennari have been pursued by Turner et al. (2008) and Trampel et al. (2011). The first of these papers investigated the scanning requirements at 3 T, with the necessary isotropic voxels, for observation of this feature in vivo throughout primary visual cortex. It was demonstrated that with the typical SNR obtained at 3 T, considerable signal averaging was needed, even when a carefully optimized version of this highly sensitive sequence was used.

An important development was contributed by Bock and colleagues (2009), who made a careful comparison between in vivo maps of T1 weighted image intensity in the cortex of marmoset brain, obtained using the MP-RAGE sequence, and myelin stained histological sections of cadaver marmoset brain. These showed a very precise correlation between areas of greater in T1-weighted signal and the locations of heavily myelinated cortical areas.

The later TSE-based study (Trampel et al. 2011) used the much higher sensitivity available at 7 T to address the question whether congenitally blind human subjects continue to possess a Stria of Gennari. Because primary visual cortex in these subjects has never been used for processing visual input, and because the Stria of Gennari is a reliable marker of V1 in sighted individuals, it was hypothesized that congenitally blind individuals would not show this feature. Somewhat surprisingly, in these subjects the Stria of Gennari was easily visible, and showed no statistical differences from that found in V1 of normally sighted subjects. A striking feature of this study was the consistently high quality of the very high resolution (0.5 mm isotropic voxels) data, obtained using a 2D Turbo-Spin-Echo sequence at 7 T, turbo factor $= 2$, with echo time TE $= 27$ ms. With this sequence, most of the grey-white matter contrast (see Fig. 7.1) arises from magnetization transfer effects.

Cortical maps corresponding to relative myelin content have been created recently by Geyer et al. (2011) and by Glasser and van Essen (2011). Geyer used quantitative maps of mean T1 averaged across the cortex, acquired using MP2RAGE at 7 T with 0.7 mm isotropic resolution, while Glasser employed a more ad hoc method, acquiring volumetric images of the brain with two sequences, the 3D MP-RAGE sequence and the TSE sequence, and taking the ratio to enhance the contrast due to myelin and to remove bias field effects. Each of these studies, discussed elsewhere in this volume, reveal a wealth of detail in the distribution of cortical myelin, even without layer-specific analysis. Areas corresponding to Brodmann areas 3a, 3b, 4, 41, and 17 can quite easily be identified, and the visual motion area V5/MT is also prominent. The Geyer paper also reports the use of T2* weighted images at 7 T with 0.4 mm isotropic voxels to allow the automated parcellation of primary visual cortex V1. A clustering algorithm was used to group cortical profiles that included the stria of Gennari.

7.5 High Field MRI: Difficulties and Benefits

From the above descriptions it is clear that the opportunity to acquire structural images of human brain at a magnetic field strength as high as 7 T has led to a dramatic gain in visibility of borders of some cortical areas in vivo. This opens the potential for creating native cortical maps for each individual volunteer subject or patient, which will supplant any less precise strategies for specifying cortical location, such as probabilistic atlases (see Eickhoff chapter). Such native cortical maps should in turn enable far more objective and testable hypotheses regarding the relationships between neural substrate, connectivity and function in human brain.

However, MRI at 7 T and higher fields has its own set of difficulties, the least of which is the significantly greater cost of the scanner. It is likely that mapping T1 and T2* at an isotropic spatial resolution of 0.5 mm and better (see Fig. 7.7) will provide extensive information regarding myelin density and myeloarchitectural layer structure in many cortical areas. However, much work is still needed to optimise MR acquisition techniques for these purposes.

To begin with, improved RF transmit and receive coils are needed, to reduce bias field effects and to maximise receive sensitivity. At the time of writing, it would be highly desirable to be able to use the combination of an 8-channel, two row transmit coil with a 96-channel receive coil (with surface-mounted preamplifiers).

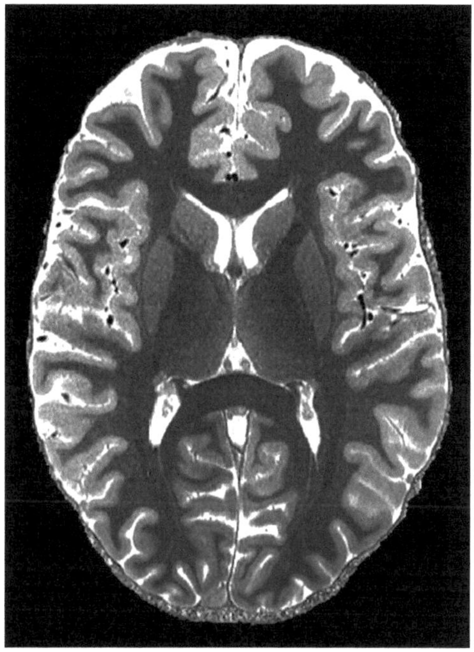

Fig. 7.7 T1 map of volunteer human brain, obtained using MP2RAGE sequence at 7 T, 0.5 mm isotropic resolution. 32-channel RF receiver coil (NOVA Medical, Massachusetts)

According to simulations and results from prototypes of the components of such a system, it could provide an almost uniform transmit RF field, together with a receiver sensitivity perhaps twice as high as with the 32 channel receiver coils currently available, at least in peripheral brain regions. Such an RF system has not yet been constructed.

Developments of MR sequence design are still at an early stage. While Bock et al. (2013) and others have taken already existing MR sequences and attempted to optimize the scan parameters for contrast-to-noise per unit time, very little research has gone into novel techniques that may provide much faster data acquisition rates, such as parallel and multiplexed acquisitions (Feinberg et al. 2010), and segmented echo-planar methods (Deichmann 2006). There are good reasons to believe that such methods can provide similar spatial resolution, with comparable CNR, in a fraction of the scanning time.

Going beyond T1 and T2* contrast, cortical areas defined by cyto and myeloarchitecture are known often to differ in other ways, such as vascular density (Zheng et al. 1991), cortical thickness (Meyer et al. 1996; MacDonald et al. 2000), and dendritic density (Jespersen et al. 2010; Dhital and Turner 2012). MRI methods for assessing each of these are under development, but have yet to deliver systematic results.

Further questions that should be addressed in the near future are as follows:

(a) Will myelin content based MRI contrast be enough to identify the majority of cortical areas?

(b) Can image analysis techniques be developed that can automatically parcellate cortex, taking account of the fact that the location of cortical layers depends strongly on the radii of curvature of the cortical sheet at any given point?

References

Alexander AL, Hurley SA, Samsonov AA, Adluru N, Hosseinbor AP, Mossahebi P, Tromp do PM, Zakszewski E, Field AS (2011) Characterization of cerebral white matter properties using quantitative magnetic resonance imaging stains. Brain Connect 1(6):423–446

Andrews-Shigaki BC, Armstrong BS, Zaitsev M, Ernst T (2011) Prospective motion correction for magnetic resonance spectroscopy using single camera Retro-Grate reflector optical tracking. J Magn Reson Imaging 33(2):498–504

Ashburner J, Friston KJ (2000) Voxel-based morphometry–the methods. Neuroimage 11 (6 Pt 1):805–821

Augustinack JC, van der Kouwe AJ, Blackwell ML, Salat DH, Wiggins CJ, Frosch MP, Wiggins GC, Potthast A, Wald LL, Fischl BR (2005) Detection of entorhinal layer II using 7 Tesla magnetic resonance imaging. Ann Neurol 57:489–494

Axer M, Amunts K, Grässel D, Palm C, Dammers J, Axer H, Pietrzyk U, Zilles K (2011) A novel approach to the human connectome: ultra-high resolution mapping of fiber tracts in the brain. Neuroimage 54(2):1091–1101

Barbier EL, Marrett S, Danek A, Vortmeyer A, van Gelderen P, Duyn J, Bandettini P, Grafman J, Koretsky AP (2002) Imaging cortical anatomy by high-resolution MR at 3.0T: detection of the stripe of Gennari in visual area 17. Magn Reson Med 48:735–738

Blackband SJ, Buckley DL, Bui JD, Phillips MI (1999) NMR microscopy–beginnings and new directions. MAGMA 9(3):112–116

Bock NA, Kocharyan A, Liu JV, Silva AC (2009) Visualizing the entire cortical myelination pattern in marmosets with magnetic resonance imaging. J Neurosci Methods 185(1):15–22

Bock NA, Hashim E, Janik R, Stanisz GJ, Weiss M, Turner R, Geyer S (2013) Optimizing T1-weighted imaging of cortical myelin content at 3.0 Tesla. Neuroimage 65:1–12

Bourgeois ME, Wajer FT, Roth M, Briguet A, Décorps M, van Ormondt D, Segebarth C, Graveron-Demilly D (2003) Retrospective intra-scan motion correction. J Magn Reson 163 (2):277–287

Bridge H, Clare S, Jenkinson M, Jezzard P, Parker AJ, Matthews PM (2005) Independent anatomical and functional measures of the V1/V2 boundary in human visual cortex. J Vis 5:93–102

Carmichael DW, Thomas DL, De Vita E, Fernández-Seara MA, Chhina N, Cooper M, Sunderland C, Randell C, Turner R, Ordidge RJ (2006) Improving whole brain structural MRI at 4.7 Tesla using 4 irregularly shaped receiver coils. Neuroimage 32(3):1176–1184

Clark VP, Courchesne E, Grafe M (1992) In vivo myeloarchitectonic analysis of human striate and extrastriate cortex using magnetic resonance imaging. Cereb Cortex 2:417–424

Cohen-Adad J, Polimeni JR, Helmer KG, Benner T, McNab JA, Wald LL, Rosen BR, Mainero C (2012) T_2* mapping and B_0 orientation-dependence at 7T reveal cyto- and myeloarchitecture organization of the human cortex. Neuroimage 60(2):1006–1014

Dasgupta S, Hogan EL (2001) Chromatographic resolution and quantitative assay of CNS tissue sphingoids and sphingolipids. J Lipid Res 42:301–308

De Vita E, Thomas DL, Roberts S, Parkes HG, Turner R, Kinchesh P, Shmueli K, Yousry TA, Ordidge RJ (2003) High resolution MRI of the brain at 4.7 Tesla using fast spin echo imaging. Br J Radiol 76(909):631–637

Deichmann R (2006) Fast structural brain imaging using an MDEFT sequence with a FLASH-EPI hybrid readout. Neuroimage 33(4):1066–1071

Denk W, Briggman KL, Helmstaedter M (2012) Structural neurobiology: missing link to a mechanistic understanding of neural computation. Nat Rev Neurosci 13(5):351–358
Deoni SC, Rutt BK, Arun T, Pierpaoli C, Jones DK (2008) Gleaning multicomponent T1 and T2 information from steady-state imaging data. Magn Reson Med 60(6):1372–1387
Dhital B, Turner R (2012) Comparison between kurtosis and biexponential models for diffusion-weighted brain imaging with high resolution and high b-factor. Abstract 3624, ISMRM 20th Annual Meeting, Wiley Publishing
Drayer B, Burger P, Darwin R, Riederer S, Herfkens R, Johnson GA (1986) MRI of brain iron. AJR Am J Roentgenol 147(1):103–110
Duflou H, Maenhaut W, De Reuck J (1989) Regional distribution of potassium, calcium, and six trace elements in normal human brain. Neurochem Res 14(11):1099–1112
Duyn JH (2011) High-field MRI of brain iron. Methods Mol Biol 711:239–249
Duyn JH, van Gelderen P, Li TQ, de Zwart JA, Koretsky AP, Fukunaga M (2007) High-field MRI of brain cortical substructure based on signal phase. Proc Natl Acad Sci USA 104 (28):11796–11801
Fatterpekar GM, Naidich TP, Delman BN, Aguinaldo JG, Gultekin SH, Sherwood CC, Hof PR, Drayer BP, Fayad ZA (2002) Cytoarchitecture of the human cerebral cortex: MR microscopy of excised specimens at 9.4 Tesla. AJNR Am J Neuroradiol 23(8):1313–1321
Feinberg DA, Moeller S, Smith SM, Auerbach E, Ramanna S, Gunther M, Glasser MF, Miller KL, Ugurbil K, Yacoub E (2010) Multiplexed echo planar imaging for sub-second whole brain FMRI and fast diffusion imaging. PLoS One 5(12):e15710
Fischl B, Salat DH, van der Kouwe AJ, Makris N, Ségonne F, Quinn BT, Dale AM (2004) Sequence-independent segmentation of magnetic resonance images. Neuroimage 23(Suppl 1): S69–S84
Frahm J, Haase A, Matthaei D (1986) Rapid three-dimensional MR imaging using the FLASH technique. J Comput Assist Tomogr 10(2):363–368
Fukunaga M, Li TQ, van Gelderen P, de Zwart JA, Shmueli K, Yao B, Lee J, Maric D, Aronova MA, Zhang G, Leapman RD, Schenck JF, Merkle H, Duyn JH (2010) Layer-specific variation of iron content in cerebral cortex as a source of MRI contrast. Proc Natl Acad Sci USA 107 (8):3834–3839
Geyer S, Weiss M, Reimann K, Lohmann G, Turner R (2011) Microstructural parcellation of the human cerebral cortex – from Brodmann's post-mortem map to in vivo mapping with high-field magnetic resonance imaging. Front Hum Neurosci 5:19
Glasser MF, Van Essen DC (2011) Mapping human cortical areas in vivo based on myelin content as revealed by T1- and T2-weighted MRI. J Neurosci 31(32):11597–11616
Gringel T, Schulz-Schaeffer W, Elolf E, Frölich A, Dechent P, Helms G (2009) Optimized high-resolution mapping of magnetization transfer (MT) at 3 Tesla for direct visualization of substructures of the human thalamus in clinically feasible measurement time. J Magn Reson Imaging 29:1285–1292
Helms G, Dathe H, Dechent P (2010) Modeling the influence of TR and excitation flip angle on the magnetization transfer ratio (MTR) in human brain obtained from 3D spoiled gradient echo MRI. Magn Reson Med 64(1):177–185
Hennig J, Nauerth A, Friedburg H (1986) RARE imaging: a fast imaging method for clinical MR. Magn Reson Med 3(6):823–833
Hinds OP, Rajendran N, Polimeni JR, Augustinack JC, Wiggins G, Wald LL, Diana Rosas H, Potthast A, Schwartz EL, Fischl B (2008) Accurate prediction of V1 location from cortical folds in a surface coordinate system. Neuroimage 39(4):1585–1599
Hopf A (1954) Die Myeloarchitektonik des Isocortex temporalis beim Menschen. J Hirnforsch 1:208–279
Hopf A (1955) Über die Verteilung myeloarchitektonischer Merkmale in der isokortikalen Schläfenlappenrinde beim Menschen. J Hirnforsch 2:36–54
Hopf A (1956) Über die Verteilung myeloarchitektonischer Merkmale in der Stirnhirnrinde beim Menschen. J Hirnforsch 2(4):311–333

Jespersen SN, Bjarkam CR, Nyengaard JR, Chakravarty MM, Hansen B, Vosegaard T, Østergaard L, Yablonskiy D, Nielsen NC, Vestergaard-Poulsen P (2010) Neurite density from magnetic resonance diffusion measurements at ultrahigh field: comparison with light microscopy and electron microscopy. Neuroimage 49(1):205–216

Kalantari S, Laule C, Bjarnason TA, Vavasour IM, MacKay AL (2011) Insight into in vivo magnetization exchange in human white matter regions. Magn Reson Med 66(4):1142–1151

Katscher U, Börnert P, Leussler C, van den Brink JS (2003) Transmit SENSE. Magn Reson Med 49(1):144–150

Koenig SH (1991) Cholesterol of myelin is the determinant of gray-white contrast in MRI of brain. Magn Reson Med 20(2):285–291

Kolind SH, Mädler B, Fischer S, Li DK, MacKay AL (2009) Myelin water imaging: implementation and development at 3.0T and comparison to 1.5T measurements. Magn Reson Med 62 (1):106–115

Kozlov M, Turner R (2011) Analysis of RF transmit performance for a 7T dual row multichannel MRI loop array. Conf Proc IEEE Eng Med Biol Soc 2011:547–553

Kucharczyk W, Macdonald PM, Stanisz GJ, Henkelman RM (1994) Relaxivity and magnetization transfer of white matter lipids at MR imaging: importance of cerebrosides and pH. Radiology 192(2):521–529

Lankford CL, Does MD (2012) On the inherent precision of mcDESPOT. Magn Reson Med. doi: 10.1002/mrm.24241, Mar 12 [Epub ahead of print]

Lee CC, Jack CR Jr, Grimm RC, Rossman PJ, Felmlee JP, Ehman RL, Riederer SJ (1996) Real-time adaptive motion correction in functional MRI. Magn Reson Med 36(3):436–444

Lee SC, Kim K, Kim J, Lee S, Han Yi J, Kim SW, Ha KS, Cheong C (2001) One micrometer resolution NMR microscopy. J Magn Reson 150(2):207–213

Lee J, van Gelderen P, Kuo LW, Merkle H, Silva AC, Duyn JH (2011) T2*-based fiber orientation mapping. Neuroimage 57(1):225–234

Li W, Wu B, Liu C (2011) Quantitative susceptibility mapping of human brain reflects spatial variation in tissue composition. Neuroimage 55(4):1645–1656

MacDonald D, Kabani N, Avis D, Evans AC (2000) Automated 3-D extraction of inner and outer surfaces of cerebral cortex from MRI. Neuroimage 12(3):340–356

MacKay A, Laule C, Vavasour I, Bjarnason T, Kolind S, Mädler B (2006) Insights into brain microstructure from the T2 distribution. Magn Reson Imaging 24(4):515–525

Maclaren J, Herbst M, Speck O, Zaitsev M (2013) Prospective motion correction in brain imaging: a review. Magn Reson Med 69(3):621–636

Mao W, Smith MB, Collins CM (2006) Exploring the limits of RF shimming for high-field MRI of the human head. Magn Reson Med 56(4):918–922

Marques JP, Maddage R, Mlynarik V, Gruetter R (2009) On the origin of the MR image phase contrast: an in vivo MR microscopy study of the rat brain at 14.1T. Neuroimage 46(2):345–352

Marques JP, Kober T, Krueger G, van der Zwaag W, Van de Moortele PF, Gruetter R (2010) MP2RAGE, a self bias-field corrected sequence for improved segmentation and T1-mapping at high field. Neuroimage 49(2):1271–1281

Meyer JR, Roychowdhury S, Russell EJ, Callahan C, Gitelman D, Mesulam MM (1996) Location of the central sulcus via cortical thickness of the precentral and postcentral gyri on MR. AJNR Am J Neuroradiol 17(9):1699–1706

Mugler JP 3rd, Brookeman JR (1990) Three-dimensional magnetization-prepared rapid gradient-echo imaging (3D MP RAGE). Magn Reson Med 15(1):152–157

Nishimura DG (2010) Principles of magnetic resonance imaging. Stanford University Press, Stanford

Ooi MB, Krueger S, Thomas WJ, Swaminathan SV, Brown TR (2009) Prospective real-time correction for arbitrary head motion using active markers. Magn Reson Med 62(4):943–954

Oros-Peusquens AM, Laurila M, Shah NJ (2008) Magnetic field dependence of the distribution of NMR relaxation times in the living human brain. MAGMA 21(1–2):131–147

Oros-Peusquens AM, Stoecker T, Amunts K, Zilles K, Shah NJ (2010) In vivo imaging of the human brain at 1.5T with 0.6-mm isotropic resolution. Magn Reson Imaging 28(3):329–340
Oshio K, Feinberg DA (1991) GRASE (Gradient- and spin-echo) imaging: a novel fast MRI technique. Magn Reson Med 20(2):344–349
Preibisch C, Deichmann R (2009) Influence of RF spoiling on the stability and accuracy of T1 mapping based on spoiled FLASH with varying flip angles. Magn Reson Med 61(1):125–135
Roemer PB, Edelstein WA, Hayes CE, Souza SP, Mueller OM (1990) The NMR phased array. Magn Reson Med 16(2):192–225
Sánchez-Panchuelo RM, Francis ST, Schluppeck D, Bowtell RW (2012) Correspondence of human visual areas identified using functional and anatomical MRI in vivo at 7T. J Magn Reson Imaging 35(2):287–299
Schleicher A, Amunts K, Geyer S, Morosan P, Zilles K (1999) Observer-independent method for microstructural parcellation of cerebral cortex: a quantitative approach to cytoarchitectonics. Neuroimage 9(1):165–177
Schulz J, Siegert T, Reimer E, Labadie C, Maclaren J, Herbst M, Zaitsev M, Turner R (2012) An embedded optical tracking system for motion-corrected magnetic resonance imaging at 7 T. MAGMA 25:443–453, Jun 13
Shah NJ, Ermer V, Oros-Peusquens AM (2011) Measuring the absolute water content of the brain using quantitative MRI. Methods Mol Biol 711:29–64
Siegel GJ, Agranoff BW, Albers RW, Fisher SK, Uhler MD (eds) (1999) Basic neurochemistry: molecular, cellular and medical aspects, 6th edn. Lippincott-Raven, Philadelphia
Sigalovsky IS, Fischl B, Melcher JR (2006) Mapping an intrinsic MR property of gray matter in auditory cortex of living humans: a possible marker for primary cortex and hemispheric differences. Neuroimage 32(4):1524–1537
Speck O, Hennig J, Zaitsev M (2006) Prospective real-time slice-by-slice motion correction for fMRI in freely moving subjects. MAGMA 19(2):55–61
Thomas DL, De Vita E, Roberts S, Turner R, Yousry TA, Ordidge RJ (2004) High-resolution fast spin echo imaging of the human brain at 4.7T: implementation and sequence characteristics. Magn Reson Med 51(6):1254–1264
Thomas DL, De Vita E, Deichmann R, Turner R, Ordidge RJ (2005) 3D MDEFT imaging of the human brain at 4.7T with reduced sensitivity to radiofrequency inhomogeneity. Magn Reson Med 53(6):1452–1458
Thudichum JLW (1884) A treatise on the chemical constitution of the brain. Bailliere, Tindall & Cox, London
Trampel R, Heidemann RM, Turner R (2010). Anatomical imaging at 7T using 2D GRASE – a comparison with 2D TSE. Poster presented at ISMRM 18th Annual Meeting, Stockholm
Trampel R, Ott DV, Turner R (2011) Do the congenitally blind have a stria of Gennari? First intracortical insights in vivo. Cereb Cortex 21(9):2075–2081
Turner R, Oros-Peusquens AM, Romanzetti S, Zilles K, Shah NJ (2008) Optimised in vivo visualisation of cortical structures in the human brain at 3T using IR-TSE. Magn Reson Imaging 26(7):935–942
Van de Moortele PF, Auerbach EJ, Olman C, Yacoub E, Uğurbil K, Moeller S (2009) T1 weighted brain images at 7 Tesla unbiased for proton density, T2* contrast and RF coil receive B1 sensitivity with simultaneous vessel visualization. Neuroimage 46(2):432–446
Vavasour IM, Whittall KP, MacKay AL, Li DKB, Vorobeychik G, Paty DW (1998) A comparison between magnetization transfer ratios and myelin water percentages in normals and multiple sclerosis patients. Magn Reson Med 40:763–768
Walters NB, Egan GF, Kril JJ, Kean M, Waley P, Jenkinson M, Watson JD (2003) In vivo identification of human cortical areas using high resolution MRI: an approach to cerebral structure-function correlation. Proc Natl Acad Sci USA 100:2981–2986
West J, Warntjes JB, Lundberg P (2012) Novel whole brain segmentation and volume estimation using quantitative MRI. Eur Radiol 22(5):998–1007

Whittall KP, MacKay AL, Graeb DA, Nugent RA, Li DK, Paty DW (1997) In vivo measurement of T2 distributions and water contents in normal human brain. Magn Reson Med 37(1):34–43
Wiggins GC, Polimeni JR, Potthast A, Schmitt M, Alagappan V, Wald LL (2009) 96-Channel receive-only head coil for 3 Tesla: design optimization and evaluation. Magn Reson Med 62 (3):754–762
Wolff SD, Balaban RS (1989) Magnetization transfer contrast (MTC) and tissue water proton relaxation in vivo. Magn Reson Med 10(1):135–144
Zheng D, LaMantia AS, Purves D (1991) Specialized vascularization of the primate visual cortex. J Neurosci 11(8):2622–2629

Chapter 8
Visualizing Myeloarchitecture In Vivo with Magnetic Resonance Imaging in Common Marmosets (*Callithrix jacchus*)

Nicholas A. Bock and Afonso C. Silva

Abstract This Chapter details the visualization of myeloarchitecture in vivo in small New World monkeys (common marmosets – *Callithrix jacchus*) using magnetic resonance imaging (MRI). The features of myeloarchitecture in marmosets are well described from traditional histology studies; here we use very high resolution MRI (160 μm isotropic) to visualize these features in living animals. Following processing, our images show the complete pattern of myelin content over the marmoset cortex, revealing the size, location, and spatial relationship between regions including the primary auditory, somatosensory, and visual regions, and visually-associated areas including the middle temporal and dorsomedial regions. For morphological studies, the surface areas of regions can be computed for individual animals, which in this Chapter reveals the large proportion of the marmoset cortex dedicated to vision. Finally, digital flattening of the surface further reveals fine details in the myeloarchitecture.

8.1 Introduction

The pattern of myelination across the cerebral cortex, termed myeloarchitecture, is an oft-used feature to investigate cortical organization with histology in a variety of primate species. The ability to image myeloarchitecture in vivo in primates is appealing because the location and extent of major cortical regions (including the primary somatosensory, auditory, visual, and motor cortices) can be visualized.

N.A. Bock (✉)
Medical Physics and Applied Radiation Sciences, McMaster University, 1280 Main Street West, L8S 4K1 Hamilton, ON, Canada
e-mail: bockn@mcmaster.ca

A.C. Silva
Cerebral Microcirculation Unit, Laboratory of Functional and Molecular Imaging, National Institute of Neurological Disorders and Stroke, National Institutes of Health, Bethesda, MD, USA

S. Geyer and R. Turner (eds.), *Microstructural Parcellation of the Human Cerebral Cortex*, DOI 10.1007/978-3-642-37824-9_8, © Springer-Verlag Berlin Heidelberg 2013

This is useful for studies of cortical morphology (for example, the influence of age or sex on the size of cortical fields) and cortical plasticity (the influence of disease and rehabilitation on cortical organization). Marmosets are lissencephalic animals that lack extensive cortical folding, which makes the species ideal for in vivo studies of myeloarchitecture. Since the marmoset cortex has been extensively mapped, there is a good known correspondence between cortical organization revealed by myeloarchitecture and cortical organization revealed by function. In this chapter, we describe how magnetic resonance imaging can be used to visualize myeloarchitecture in common marmoset monkeys.

8.2 Common Marmosets

Common marmosets (*Callithrix jacchus*) are New World monkeys native to Brazil that are easily bred in captivity, and hence, extensively used in primate research. They are small, with adult males measuring 188 mm high on average and weighing 256 g, and adult females measuring 185 mm high on average, and weighing 236 g (Rowe 1996). When bred in captivity, marmosets tend to produce twins or triplets (Tardif et al. 2003) making them ideal for studies where it is important to have age and sex-matched controls. The marmoset genera has a relatively primitive brain layout (Newman et al. 2009; Hikishima et al. 2011) which lacks the extensive cortical folding seen in Old World primate and human brains (Zilles et al. 1989). Marmosets, however, possess a sophisticated range of behaviours and are widely used as models of visual (Yu et al. 2010; Puller and Haverkamp 2011; Solomon et al. 2011) and auditory processing (Aitkin and Park 1993; Wang 2007; Reser et al. 2009), and vocal communication (King 2005; Eliades and Wang 2008) in the primate brain.

Beyond basic neuroscience studies, marmosets are often used as models of human brain disease to bridge the gap between preclinical studies in rodents and clinical studies in humans. The monkeys have well-developed white matter tracts, and are an important model of chronic multiple sclerosis following the induction of experimental autoimmune encephalitis (EAE) ('t Hart et al. 2004). Established marmoset models also exist for stroke (Bihel et al. 2010) and Parkinson's disease (Eslamboli 2005), and transgenic marmosets have recently been produced (Sasaki et al. 2009) opening the possibility of genetic studies of brain disease in primates. The rich suite of marmoset models for basic neuroscience and disease modeling warrants the adaptation of the many in vivo techniques for assessing brain structure and function in humans for use in small monkeys.

Owing to their small size, marmosets can be imaged using magnetic resonance imaging (MRI) on existing small animal imaging systems such as those used for rodent imaging. MRI has unparalleled soft tissue contrast in the brain and is the modality of choice for neuroimaging. Since it does not use ionizing radiation, it is suitable for longitudinal studies, allowing each monkey to serve as its own control, thereby reducing the number of animals needed to produce good statistics in an experiment. MRI can be tailored to study a variety of features of the marmoset brain, such as basic neuroanatomy (Sati et al. 2012), diffusion in white matter tracts (Yamada et al. 2008; Sati et al. 2012), pathology (Blezer et al. 2007; Diem et al. 2008; Bihel et al. 2011), and even function in awake, behaving animals using functional MRI (fMRI) (Meyer et al. 2006; Silva et al. 2011).

Because it is an inherently 3D imaging modality, MRI shows the shape and spatial relationships of features over the entire brain, free of distortion. Anatomical MRI is commonly performed in marmosets that have been anaesthetized to minimize motion during the relatively long times needed for making high resolution images. In this manner, it is possible to image the entire marmoset brain at a resolution between 100 and 150 μm isotropic in an hour or so at a field strength of 7 T (Liu et al. 2011). This is one order of magnitude better in linear dimensions (and three orders of magnitude in volume) than the 3D whole-brain isotropic resolution of 0.5–1 mm available for human clinical imaging at lower field strengths. Imaging marmosets thus provides a unique opportunity to investigate cortical microstructure with MRI. Since cortical thickness is largely preserved across primate species, but the available resolution for marmoset MRI is much higher than for human MRI, the marmoset cortex can be imaged over its layers in much greater detail than the human cortex can. In fact, the reported thickness of the primary visual cortex (V1) in marmosets of 1.5 mm (Missler et al. 1993) is quite close to the reported thickness in humans of 2.0 mm (Zilles and Amunts 2012); however, the available resolution for imaging is 10 times better in marmosets than in humans. This is also true in the primary somatosensory cortex (S1) with a reported thickness of 2.5 mm in marmosets (Gorrie et al. 2008) and 1.8 mm in humans (Zilles and Amunts 2012) and likely holds over all cortical regions. The cortical sheet in humans on the other hand is much larger in surface area at roughly 250 000 mm^2 (Peters and Jones 1984) than it is in marmosets at 2,000 mm^2 (Bock et al. 2011), and so the surface area of specific cortical regions in humans are larger than in marmosets and the lower resolution available for human imaging is not as detrimental to surface area measurements of cortical regions.

In vivo imaging of cortical microstructure in marmosets is further attractive in that the marmoset brain has been extensively mapped, both structurally using a wide array of histological techniques, and functionally, using electrophysiology. This means that there is extensive literature parcellating the marmoset cortex into many distinct and unambiguous regions for comparison with MRI results (Krubitzer and Kaas 1990; Rosa and Schmid 1995; Rosa and Elston 1998; de la Mothe et al. 2006; Burman et al. 2008). This is especially true for myeloarchitecture, whose pattern is more completely described in the marmoset than in humans.

8.3 Myeloarchitecture in Common Marmosets

While most myelinated axons in the brain are found in discrete white matter tracts, there is a smaller number present throughout the cortex that run either vertically (radially) or horizontally (tangentially) within the layers of the cortical gray matter.

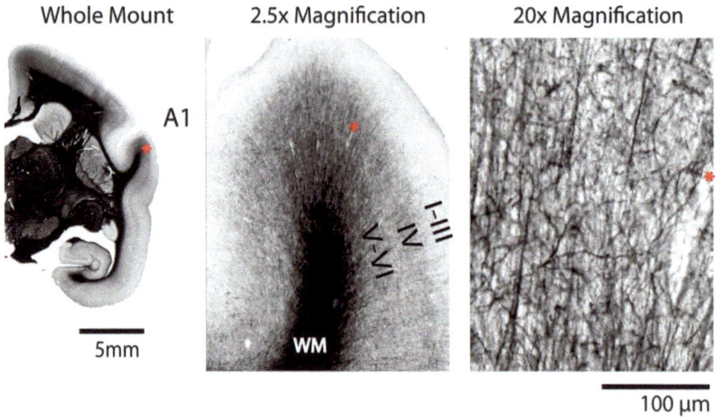

Fig. 8.1 Representative cortical myelination in the marmoset. A 40 µm–thick coronal histological section stained for myelin using a modified Gallyas silver staining method. At the whole mount level of magnification showing half of the brain, distinct dense areas of staining are seen that correspond to specific cortical areas (for example, the primary auditory cortex [A1]). At 2.5 times magnification, dark-stained fibers can be seen running vertically from the white matter through Layers VI and V. At 20 times magnification in Layer IV, the fibers are arborized, and vertical and horizontal stained branches can be seen. There is little myelin staining through Layers III-I (*WM*, white matter. The asterix denotes the same blood vessel in each magnification.) (From Bock et al. 2011)

The distribution and appearance of these fibres describes the myeloarchitecture of the cortex. These myelinated fibres speed the conduction of input and output signals through the cortex and their density is regionally dependent. Typically in animals, myeloarchitecture is visualized ex vivo in myelin-stained sections of the cortex at low power magnifications (Annese et al. 2004) or in whole mount images of flattened, stained cortex (Wong and Kaas 2010). The cortical patterns of myelination in the marmoset brain have been extensively characterized with these conventional histological approaches (Krubitzer and Kaas 1990; Pessoa et al. 1992; Rosa and Schmid 1995; Pistorio et al. 2006; Burman et al. 2008; Jeffs et al. 2009).

Cortical myelination can be detected on MRI, even though it has a much lower resolution than light microscopy, because the myelinated axons persist over several cortical layers. Figure 8.1 shows an example of cortical myelination seen with conventional histology in the primary auditory cortex (A1) of a common marmoset. The majority of the fibres run either tangentially from the white matter surface to Layer IV (which is the input layer of A1) or horizontally within Layer IV. The same pattern is seen in other myelinated regions of the marmoset cortex, save for the primary visual cortex (V1), where the density of myelination is highest in Layer IV (the Stripe of Gennari). As long as MRI can roughly resolve the cortical layers, it is possible to image patterns of myelination over the cortex. This makes myeloarchitecture a better candidate for in vivo imaging of cortical microstructure than cytoarchitecture, where the appearance of individual layers must be accurately resolved.

8.4 MRI of Myelination

MRI is sensitive to a number of different physical properties of tissue, including the longitudinal relaxation time of protons (T_1), the transverse relaxation time of protons (T_2), the magnitude and direction of water diffusion in the tissue, and the magnetic susceptibility of the tissue (detected as T_2* contrast). In the brain, the presence of myelin in white matter structures causes a large difference in the MR properties of the tissue in these structures relative to the properties in gray matter structures. These differences give rise to the exquisite contrast seen on anatomical MRIs that highlights gross brain anatomy. Imaging cortical myelin is based on the same MR contrast mechanism by which myelin in white matter tracts produces contrast between major gray and white matter structures of the central nervous system (CNS). In fact, it has been shown that the Stripe of Gennari, a heavily myelinated structure in V1, can be identified in humans on T_1-weighted (Barbier et al. 2002), T_2-weighted (Carmichael et al. 2006; Trampel et al. 2011), and T_2 *-weighted images (Fukunaga et al. 2010).

In marmosets, we have measured the difference in T_1 between regions of the cortex with a low myelin content and a high myelin content to be 270 ms at 7 T (Bock et al. 2009). Although this difference is smaller than the difference in T_1 between gray matter with a low myelin content and white matter of 610 ms, it is sufficient to generate intracortical contrast on an MRI. We image myeloarchitecture based on this T_1-weighted contrast using a three dimensional (3D) T_1-weighted MRI sequence commonly used to image gross neuroanatomy that we have optimized to increase the contrast for imaging differences in cortical myelination (Bock et al. 2009). We choose to image based on T_1 contrast because T_1-weighted sequences are highly amenable for 3D high resolution anatomical neuroimaging.

To visualize myeloarchitecture across the entire cortex, we image isoflurane-anaesthetized marmosets for roughly 4 h with our 3D sequence to produce an MRI of the whole brain at 150 μm isotropic resolution. In Fig. 8.2, we show 2D coronal slices from a representative 3D MRI of a marmoset and corresponding matched histological sections stained for myelin. There is a very good spatial correspondence between known highly myelinated regions of the marmoset cortex and signal enhancement in the MRI. This importantly suggests that MRI is sensitive to myelin in the cortex, although there is a remote possibility that we are visualizing another feature of the cortex that is highly co-localized with myelin (such as non-heme iron (Fukunaga et al. 2010)), which would also brighten a T_1-weighted MRI.

8.5 Visualization of MRI Data

While it is possible to explore myeloarchitecture in marmosets in 2D slices extracted from a 3D MRI, the patterns can be best appreciated if viewed across the cortical surface. This requires the MRI data to be surface rendered, which proceeds as follows (Fig. 8.3).

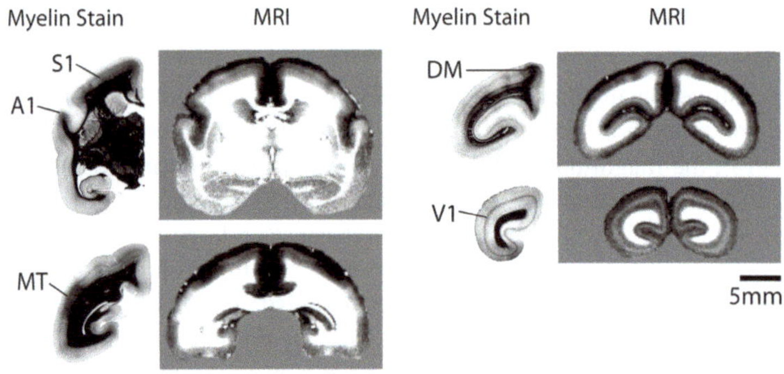

Fig. 8.2 40 μm thick coronal myelin-stained histology sections through the marmoset brain and corresponding 167 μm slices from an in vivo T_1-weighted MRI. The MRIs are masked to only show the brain and presented on a gray background to make the edges of the cortex visible. The number in the *upper right hand corner* of each MRI indicates the order of the slices from rostral to caudal. (*A1*, primary auditory cortex; *S1*, primary somatosensory cortex; *MT*, middle temporal area; *DM*, dorsomedial area; *V1*, primary visual cortex). Note that the contrast is reversed in the two types of images, with myelinated areas appearing dark in the histological sections and bright in the MRI (Adapted from Bock et al. 2009)

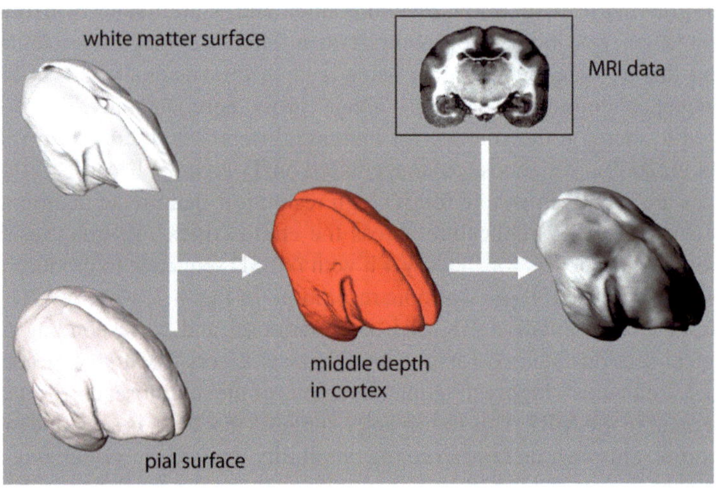

Fig. 8.3 Surface processing of 3D MRI data. The white matter surface and the pial surface are first obtained from the 3D MRI data using semi-automatic image segmentation. The middle distance between these surfaces is then computed to produce a new surface at a middle depth in the cortex. The MRI intensity data at this middle depth is then projected on the surface to form a map

First, the inner and outer surfaces of the cortex must be segmented (Tosun et al. 2004). In our marmoset images, we segment the boundary between the cortex and the underlying white matter based on intensity thresholding. This is robust because there is such high contrast between gray and white matter in our images. The pial

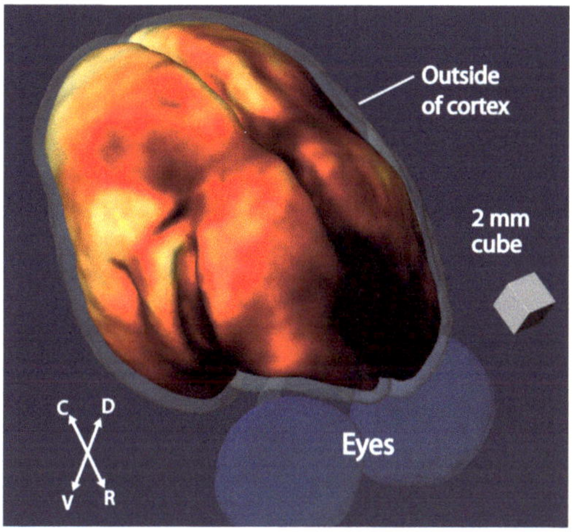

Fig. 8.4 3D map of myeloarchitecture in a representative 3-year-old male common marmoset. The figure shows a view of a 3D map of the cortex in a marmoset centered on the dorsal parietal cortex. Here, the MRI intensity data are displayed using an *orange* colormap to highlight contrast and areas of enhancement represent cortical areas with high myelin content. The map is placed at a middle depth in the cortex, and the surface corresponding to the outside of the cortex is shown in light transparent *blue*. (*C*, caudal; *R*, rostral; *V*, ventral; *D*, dorsal.)

boundary is more difficult to segment because there is less contrast in the images between cortical tissue and the cerebral spinal fluid (CSF) or dura matter. We begin with a rough atlas-based segmentation of this boundary, followed by a manual correction. Once the pial and white matter boundaries of the cortex are segmented we create a surface mesh of each, then calculate normals between these surfaces. Next, we compute a new surface at a distance between the pial and white matter surfaces along these normals which then lies at a middle depth in the myelinated layers of the cortex. For instance, this middle surface lies in the Stripe of Gennari in V1 and in Layer IV in S1. For display purposes, the MRI intensity data at the middle surface depth is displayed on the surface. To aid visualization, we display the grayscale MRI intensity data using an orange colourmap in our whole-cortex maps (Fig. 8.4). The brighter regions in the map represent regions of the marmoset cortex with a high myelin content. The surface-rendered map can be viewed from different angles (Fig. 8.5) to reveal myeloarchitecture over the entire cortex. This is perhaps the best way to appreciate the patterns of myeloarchitecture in the marmoset free of distortion, although a surface map can also be digitally flattened (Fig. 8.6) to show all heavily myelinated regions in a single view, which is similar to what one would see with flatmounting on histology.

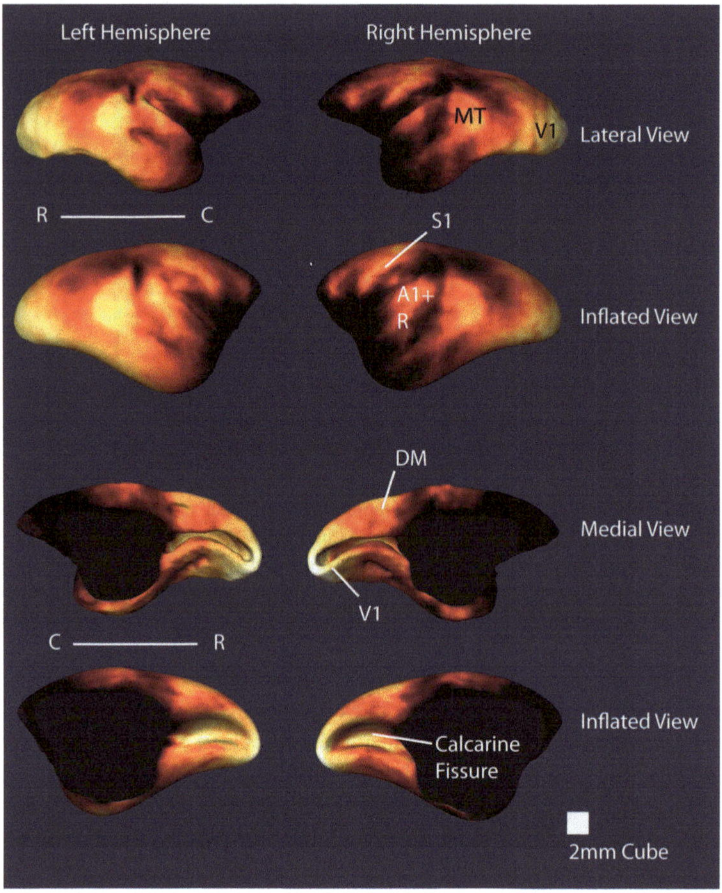

Fig. 8.5 Multiple views of the surface-rendered cortex from Fig. 8.4. Each view is paired with an inflated view of the cortex to show myeloarchitecture the cortical folds of the lateral sulcus and calcarine fissure. (*C*, caudal; *R*, rostral) (Several major myelinated cortical regions are labeled: *V1*, primary visual area; *MT*, middle temporal area; *A1*, primary auditory area; *R*, rostral auditory area; *S1*, primary somatosensory cortex; *DM*, dorsomedial area)

8.6 Overall Myeloarchitecture in the Marmoset

Overall, we observe in our maps of myeloarchitecture five highly-myelinated major cortical regions in the marmoset: the primary sensory regions A1, S1, and V1, and regions in the extrastriate visual pathway: the middle temporal region (MT) and the dorsomedial region (DM). It makes sense that the primary sensory regions would be highly myelinated so as to improve the speed of input signals. The axons traversing visual associated areas too are likely myelinated to also speed visual responses in the brain. The location and geometry of major myelinated regions agree well with

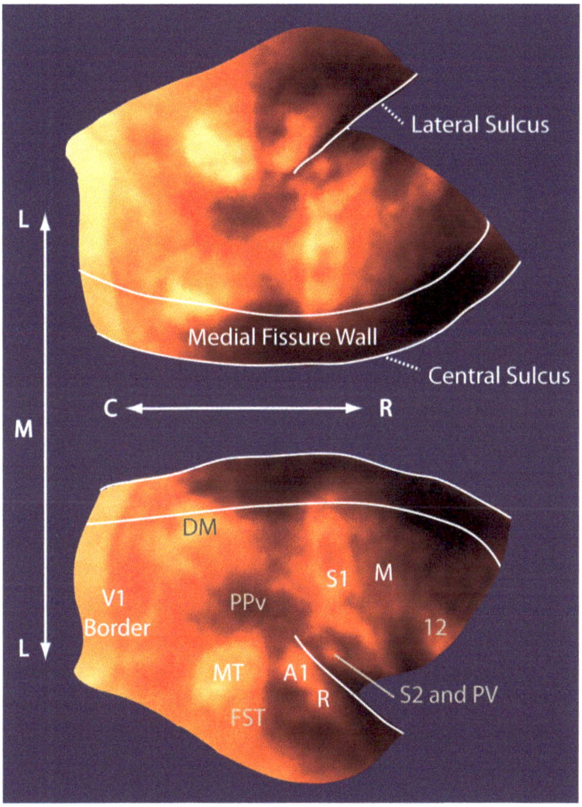

Fig. 8.6 Flattened map of myeloarchitecture in a representative 3-year old female common marmoset. The dorsal cortical surface from a 3D map is flattened with the major enhancing areas labeled. Note that the flattening produces spatial distortions in the surface; however, the flattening allows for easier visualization of cortical features and their spatial relationships. (*C*, caudal; *R*, rostral; *L*, lateral; *M*, medial) (Major myelinated cortical regions are labeled in *white*: *V1*, primary visual area; *MT*, middle temporal area; *A1*, primary auditory area; *R*, rostral auditory area; *S1*, primary somatosensory cortex; *M*, motor cortex including primary and premotor areas and the frontal eye fields) (Cortical features are labeled in *gray*: *DM*, dorsomedial area; *PPv*, ventral posterior parietal cortex; *FST*, fundus of the superior temporal area; *S2*, secondary somatosensory cortex; *PV*, parietal ventral area; *12*, area 12) (Adapted from Bock et al. 2011)

previous studies of myeloarchitecture in the marmoset brain. Interestingly, we do not see strong enhancement in the primary motor cortex (M1) in the marmoset, although this region is one of the most heavily myelinated in humans (Hopf 1956). This highlights potential differences in neuroanatomy in small, New World primates versus large Old World primates. Aside from strong enhancements in major cortical regions, there are also fainter enhancements in regions that can be identified because they are surrounded by cortex with a very low myelin content (e.g. area 12 in the frontal cortex) or are specifically identified by their low myelin content (e.g. the ventral posterior parietal cortex, PPv (Rosa et al. 2009)). Specific features of marmoset myeloarchitecture observable with MRI are discussed in further detail below.

8.7 Myeloarchitecture of Specific Regions

8.7.1 Auditory Regions

The primary auditory cortex, A1, in marmosets is a prominent myelinated feature of the temporal lobe (Pistorio et al. 2006), which is unsurprising as the marmoset has a well-developed repertoire of tonal vocalizations, a similar hearing range as humans (Aitkin and Park 1993), and can even discriminate the pitch of complex tones (Bendor and Wang 2005). We readily identify A1 in our images in the dorsal temporal cortex, extending into the lateral sulcus. Continuous with A1 is another myelinated auditory area, the rostral auditory area (R). Since there is no discernible border between A1 and R, they are treated as a single region in our maps. In humans, short T_1s have also been noted in A1 and related to a high myelin content (Sigalovsky et al. 2006). Interestingly, that study measured lower T_1s in Heschl's gyrus on the left side of the brain than on the right, suggesting that there are asymmetries in cortical myelination that may reflect specialized functional processing. It would be interesting in a future study to see if the same asymmetries are present in marmosets, either in the T_1 values, or in the size of A1.

8.7.2 Somatosensory Regions

Another highly myelinated primary sensory area in the marmoset is the primary somatosensory cortex S1 (also referred to as area 3b). This area features a characteristic somatotopic organization, with sub-areas corresponding to representations of different parts of the body within S1 being distinguished by patchy regions of high myelination surrounded by poorly myelinated regions (Krubitzer and Kaas 1990). S1 is strongly enhanced in our T_1-weighted images (Figs. 8.2 and 8.3), and the patchiness is also apparent. Within the medial wall of the lateral sulcus the much smaller secondary somatosensory area (Krubitzer and Kaas 1990), S2, can also be seen. This is continuous with another myelinated region, the parietal ventral area (PV) (Krubitzer and Kaas 1990) and since there is no discernible border again, the two regions are treated as one in our maps.

8.7.3 Visual Regions

V1 was one of the first cortical areas to be segregated based on its high myelin content (Gennari 1782). Since, the stripe of Gennari has been the major structural feature used to distinguish V1 from other cortical areas both in monkeys (Pessoa et al. 1992), as well in humans (Clarke and Miklossy 1990). V1 is the area that has been most extensively visualized with MRI in humans in vivo and ex vivo (Clark et al. 1992; Barbier et al. 2002; Walters et al. 2003; Eickhoff et al. 2005; Bridge et al. 2005; Bridge and Clare 2006; Hinds et al. 2008). From a technical standpoint, it is actually one of the most difficult myelinated areas to image because of the high spatial resolution needed to visualize the stripe of Gennari (Peters and Sethares 1996). All the other highly myelinated areas we identified contain myelinated fibers spanning multiple lower layers of the cortex which make them easier to image. In V1, the myelinated fibers are mostly confined to just layer IV, although there is light myelination of the deeper cortical layers too. The border between V1 and the secondary visual cortex (V2) is readily delineated since the density of myelin throughout the layers of V2 is far lower than in V1. The rostral border of V2 is not discernible in our map because there is not a great enough difference in myelination between V2 and adjoining rostral areas. The extent of V2 is traditionally visualized using cytochrome oxidase staining (Jeffs et al. 2009), rather than myelin staining, and our inability to delineate V2 in our map shows the limitation of describing cortical organization based on only one contrast mechanism.

Another visual area that has been extensively studied in monkeys based on myeloarchitecture is the middle temporal area, MT (Van Essen et al. 1981; Bourne et al. 2007), which has been implicated in processing visual motion (Ungerleider and Desimone 1986). In the marmoset, MT is a heavily myelinated region situated dorsal and posterior to the temporal sulcus. Since the myelinated fibers span from layer III to VI (Bourne et al. 2007), we were able to readily identify MT in both our T_1 maps and in our T_1-weighted images. The corresponding homologous area in humans, V5/MT, has been identified using both anatomical MRI with myelin-based contrast, and with functional MRI using a stimulation protocol selective for moving stimuli (Walters et al. 2003). Ventral and anterior to MT is an enhancing area corresponding to the fundus of the superior temporal area (FST) (Rosa and Elston 1998).

Another major enhancing region we identified in the visual pathway is the dorsomedial area (DM). The shape and extent of DM is not well defined in the literature based on its myelination; however, it has strong enhancement and we have labeled it in our map although its actual borders are indistinct. DM is similar in size to MT and receives projections from V1 (Rosa and Schmid 1995; Lyon and Kaas 2001).

8.7.4 *Motor Regions and Frontal Cortex*

In the parietal-frontal cortex of our map, the motor areas are visible. Even though these output areas contain less myelin than the primary sensory areas, their caudal border is readily identified because of contrast with S1 which is heavily myelinated and their rostral border is visible as there is very little myelin in the rest of the frontal cortex. Although the large motor area, M, we have identified in our map contains many subdivisions, including the primary motor cortex, the premotor areas, and the frontal eye fields (Burman et al. 2008; Burish et al. 2008), there is little contrast between these regions, so they must be grouped into a common area. Area 12 in the dorsolateral and orbital frontal cortex is relatively well-myelinated (Burman and Rosa 2009) and is easily identified in our map. Interestingly, this area is an important source of projections to the heavily myelinated MT area (Burman et al. 2006).

8.8 Implications

The implications of being able to visualize in vivo myeloarchitecture in marmosets revolve around our new ability to non-invasively and robustly observe cortical organization in these monkeys. While it has been possible to observe patterns of myeloarchitecture in marmosets previously using histology, it required sacrifice of the animal, which precluded longitudinal studies and increased the number of animals required for morphological studies.

We have performed preliminary morphological studies to show our technique can quantitatively summarize cortical organization in the marmoset. We measured the surface areas of the major enhancing areas in maps from four female monkeys: two three-year old young adults, and two eight-year old middle aged monkeys (Abbott et al. 2003) (see Fig. 8.7 and Table 8.1). The low standard deviations in our measurements suggest that the size of cortical areas may be largely preserved over different individuals, although a more statistically rigorous study with a large number of animals would be required to confirm this. This type of study is easily performed now, since no animals need to be sacrificed. Comparing measurements between hemispheres, we again see little variation, although more animals would also be needed to statistically confirm this.

What we do see in our surface area data is that a large proportion of the marmoset cortex is devoted to vision, since V1 is almost five times as large as all the motor areas combined, seven times as large as S1, and almost 22 times as large as A1+R. If we classify MT and DM as visual areas as well, then at least a quarter of the marmoset cortex is devoted to vision. Our data compares quite closely with data from studies performed with traditional histology: we found the average surface area of V1 to be 220 mm^2, whereas previous studies (Fritsches and Rosa 1996; Missler et al. 1993) have reported averages of 205 mm^2 and 200 mm^2 respectively. There is great utility; however, in our being able to quantitatively follow major cortical areas in vivo since we can perform longitudinal studies in disease or development and investigate possible cortical plasticity. Because of our ability to follow each animal longitudinally, the technique could be used to monitor post-natal maturation and development of the primary sensory areas (A1, S1, and V1), as well as MT, which have been shown to be the first to mature in the marmoset (Burman et al. 2007). As well, our technique can be used to study the reorganization of sensory regions that are known to occur following brain injuries associated with

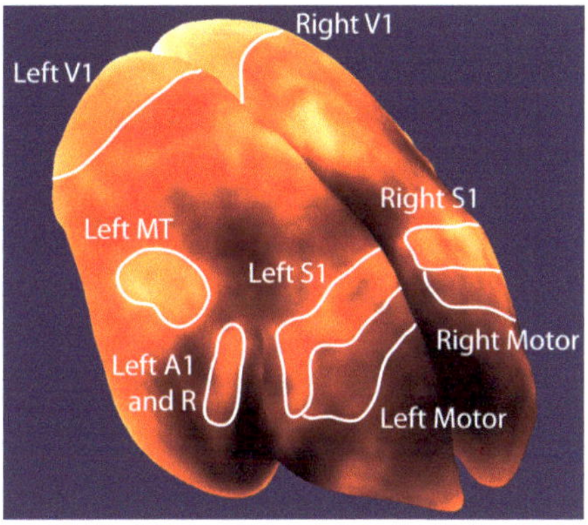

Fig. 8.7 Definition of cortical areas for surface area measurements (From Bock et al. 2011)

Table 8.1 Surface area measurements

Region	Surface area (mm^2)	Percent of total cortical surface area of hemisphere
Left cortex	1005 ± 21	–
Left V1	219 ± 12	22
Left motor	36 ± 2	4
Left S1	28 ± 4	3
Left MT	17 ± 3	2
Left A1 and R	11 ± 3	1
Left DM	8 ± 1	1
Right cortex	1007 ± 34	–
Right V1	222 ± 3	22
Right motor	37 ± 3	4
Right S1	30 ± 4	3
Right MT	19 ± 2	2
Right A1 and R	11 ± 3	1
Right DM	7 ± 1	1

The measurements of the left and right cortex include the entire cortical surface, ($n = 4$, surface areas reported as mean ± standard deviation)

diseases such as stroke (Jiang et al. 2010) and multiple sclerosis (Filippi et al. 2010) or in response to trauma (Zhang et al. 2010).

Our studies of myeloarchitecture in marmosets have important implications for in vivo studies in humans too. A major goal in human brain mapping is to establish a concordance between descriptions of cortical organization based on structure or function. There is thus an overall need to relate myeloarchitecture to better known cytoarchitecture and then to function. This is more easily realized in monkeys like the marmoset than in the human, since matched observations of function with electrophysiology, and myeloarchitecture and cytoarchitecture from histological staining have often been performed in the same animal so the concordance is well-known. Thus we are relatively confident in saying our visualization of a certain region based on myeloarchitecture, say V1, reflects the same region and this demonstration of correspondence bodes well for the human case.

A drawback of our technique is that is only allows visualization of major myelinated cortical areas and we lack the ability to delineate areas based on subtle variations in myelination patterns across the cortical areas. In reality, this level of resolution is probably beyond that of in vivo MRI, and highly detailed analyses of cortical organization are best left for traditional post-mortem histology. However, the usefulness of being able to visualize cortical organization in living monkeys using our technique can not be overstated.

References

Abbott DH, Barnett DK, Colman RJ, Yamamoto ME, Schultz-Darken NJ (2003) Aspects of common marmoset basic biology and life history important for biomedical research. Comp Med 53(4):339–350

Aitkin L, Park V (1993) Audition and the auditory pathway of a vocal New World primate, the common marmoset. Prog Neurobiol 41(3):345–367

Annese J, Pitiot A, Dinov ID, Toga AW (2004) A myelo-architectonic method for the structural classification of cortical areas. Neuroimage 21(1):15–26

Barbier E, Marrett S, Danek A, Vortmeyer A, Van GP, Duyn J, Bandettini P, Grafman J, Koretsky A (2002) Imaging cortical anatomy by high-resolution MR at 3.0 T: detection of the stripe of Gennari in visual area 17. Magn Reson Med 48(4):735–738

Bendor D, Wang X (2005) The neuronal representation of pitch in primate auditory cortex. Nature 436(7054):1161–1165

Bihel E, Pro-Sistiaga P, Letourneur A, Toutain J, Saulnier R, Insausti R, Bernaudin M, Roussel S, Touzani O (2010) Permanent or transient chronic ischemic stroke in the non-human primate: behavioral, neuroimaging, histological, and immunohistochemical investigations. J Cereb Blood Flow Metab 30(2):273–285

Bihel E, Roussel S, Toutain J, Bernaudin M, Touzani O (2011) Diffusion tensor MRI reveals chronic alterations in white matter despite the absence of a visible ischemic lesion on conventional MRI: a nonhuman primate study. Stroke 42(5):1412–1419

Blezer ELA, Bauer J, Brok HPM, Nicolay K, 't Hart BA (2007) Quantitative MRI-pathology correlations of brain white matter lesions developing in a non-human primate model of multiple sclerosis. NMR Biomed 20(2):90–103

Bock NA, Kocharyan A, Liu JV, Silva AC (2009) Visualizing the entire cortical myelination pattern in marmosets with magnetic resonance imaging. J Neurosci Methods 185(1):15–22

Bock NA, Hashim E, Kocharyan A, Silva AC (2011) Visualizing myeloarchitecture with magnetic resonance imaging in primates. Ann N Y Acad Sci 1225(Suppl 1):E171–E181

Bourne JA, Warner CE, Upton DJ, Rosa MGP (2007) Chemoarchitecture of the middle temporal visual area in the marmoset monkey (*Callithrix jacchus*): laminar distribution of calcium-binding proteins (calbindin, parvalbumin) and nonphosphorylated neurofilament. J Comp Neurol 500(5):832–849

Bridge H, Clare S (2006) High-resolution MRI: in vivo histology? Philos Trans R Soc Lond B Biol Sci 361(1465):137–146

Bridge H, Clare S, Jenkinson M, Jezzard P, Parker AJ, Matthews PM (2005) Independent anatomical and functional measures of the V1/V2 boundary in human visual cortex. J Vis 5 (2):93–102

Burish MJ, Stepniewska I, Kaas JH (2008) Microstimulation and architectonics of frontoparietal cortex in common marmosets (*Callithrix jacchus*). J Comp Neurol 507(2):1151–1168

Burman KJ, Rosa MGP (2009) Architectural subdivisions of medial and orbital frontal cortices in the marmoset monkey (*Callithrix jacchus*). J Comp Neurol 514(1):11–29
Burman KJ, Palmer SM, Gamberini M, Rosa MGP (2006) Cytoarchitectonic subdivisions of the dorsolateral frontal cortex of the marmoset monkey (*Callithrix jacchus*), and their projections to dorsal visual areas. J Comp Neurol 495(2):149–172
Burman KJ, Lui LL, Rosa MGP, Bourne JA (2007) Development of non-phosphorylated neurofilament protein expression in neurones of the New World monkey dorsolateral frontal cortex. Eur J Neurosci 25(6):1767–1779
Burman KJ, Palmer SM, Gamberini M, Spitzer MW, Rosa MGP (2008) Anatomical and physiological definition of the motor cortex of the marmoset monkey. J Comp Neurol 506(5):860–876
Carmichael DW, Thomas DL, De Vita E, Fernández-Seara MA, Chhina N, Cooper M, Sunderland C, Randell C, Turner R, Ordidge RJ (2006) Improving whole brain structural MRI at 4.7 Tesla using 4 irregularly shaped receiver coils. Neuroimage 32(3):1176–1184
Clark VP, Courchesne E, Grafe M (1992) In vivo myeloarchitectonic analysis of human striate and extrastriate cortex using magnetic resonance imaging. Cereb Cortex 2(5):417–424
Clarke S, Miklossy J (1990) Occipital cortex in man: organization of callosal connections, related myelo- and cytoarchitecture, and putative boundaries of functional visual areas. J Comp Neurol 298(2):188–214
de la Mothe LA, Blumell S, Kajikawa Y, Hackett TA (2006) Cortical connections of the auditory cortex in marmoset monkeys: core and medial belt regions. J Comp Neurol 496(1):27–71
Diem R, Demmer I, Boretius S, Merkler D, Schmelting B, Williams SK, Sättler MB, Bähr M, Michaelis T, Frahm J, Brück W, Fuchs E (2008) Autoimmune optic neuritis in the common marmoset monkey: comparison of visual evoked potentials with MRI and histopathology. Invest Ophthalmol Vis Sci 49(8):3707–3714
Eickhoff S, Walters NB, Schleicher A, Kril J, Egan GF, Zilles K, Watson JDG, Amunts K (2005) High-resolution MRI reflects myeloarchitecture and cytoarchitecture of human cerebral cortex. Hum Brain Mapp 24(3):206–215
Eliades SJ, Wang X (2008) Neural substrates of vocalization feedback monitoring in primate auditory cortex. Nature 453(7198):1102–1106
Eslamboli A (2005) Marmoset monkey models of Parkinson's disease: which model, when and why? Brain Res Bull 68(3):140–149
Filippi M, Rocca MA, Benedict RHB, Deluca J, Geurts JJG, Rombouts SARB, Ron M, Comi G (2010) The contribution of MRI in assessing cognitive impairment in multiple sclerosis. Neurology 75(23):2121–2128
Fritsches KA, Rosa MG (1996) Visuotopic organisation of striate cortex in the marmoset monkey (*Callithrix jacchus*). J Comp Neurol 372(2):264–282
Fukunaga M, Li T, van Gelderen P, de Zwart JA, Shmueli K, Yao B, Lee J, Maric D, Aronova MA, Zhang G, Leapman RD, Schenck JF, Merkle H, Duyn JH (2010) Layer-specific variation of iron content in cerebral cortex as a source of MRI contrast. Proc Natl Acad Sci USA 107 (8):3834–3839
Gennari F (1782) De Peculiari Structura Cerebri. Nonnullisque ejus morbis. Ex Regio Typographeo, Parma
Gorrie CA, Waite PME, Rogers LJ (2008) Correlations between hand preference and cortical thickness in the secondary somatosensory (SII) cortex of the common marmoset, *Callithrix jacchus*. Behav Neurosci 122(6):1343–1351
Hikishima K, Quallo MM, Komaki Y, Yamada M, Kawai K, Momoshima S, Okano HJ, Sasaki E, Tamaoki N, Lemon RN, Iriki A, Okano H (2011) Population-averaged standard template brain atlas for the common marmoset (*Callithrix jacchus*). Neuroimage 54(4):2741–2749
Hinds OP, Rajendran N, Polimeni JR, Augustinack JC, Wiggins G, Wald LL, Diana Rosas H, Potthast A, Schwartz EL, Fischl B (2008) Accurate prediction of V1 location from cortical folds in a surface coordinate system. Neuroimage 39(4):1585–1599
Hopf A (1956) Über die Verteilung myeloarchitektonischer Merkmale in der Stirnhirnrinde beim Menschen. J Hirnforsch 2:311–333

Jeffs J, Ichida JM, Federer F, Angelucci A (2009) Anatomical evidence for classical and extra-classical receptive field completion across the discontinuous horizontal meridian representation of primate area V2. Cereb Cortex 19(4):963–981

Jiang Q, Zhang ZG, Chopp M (2010) MRI evaluation of white matter recovery after brain injury. Stroke 41(Suppl 10):S112–S113

King AJ (2005) Auditory plasticity: vocal output shapes auditory cortex. Curr Biol 15(13): R503–R505

Krubitzer LA, Kaas JH (1990) The organization and connections of somatosensory cortex in marmosets. J Neurosci 10(3):952–974

Liu JV, Bock NA, Silva AC (2011) Rapid high-resolution three-dimensional mapping of T1 and age-dependent variations in the non-human primate brain using magnetization-prepared rapid gradient-echo (MPRAGE) sequence. Neuroimage 56(3):1154–1163

Lyon D, Kaas J (2001) Connectional and architectonic evidence for dorsal and ventral V3, and dorsomedial area in marmoset monkeys. J Neurosci 21(1):249–261

Meyer JS, Brevard ME, Piper BJ, Ali SF, Ferris CF (2006) Neural effects of MDMA as determined by functional magnetic resonance imaging and magnetic resonance spectroscopy in awake marmoset monkeys. Ann N Y Acad Sci 1074:365–376

Missler M, Wolff A, Merker HJ, Wolff JR (1993) Pre- and postnatal development of the primary visual cortex of the common marmoset. II. Formation, remodelling, and elimination of synapses as overlapping processes. J Comp Neurol 333(1):53–67

Newman JD, Kenkel WM, Aronoff EC, Bock NA, Zametkin MR, Silva AC (2009) A combined histological and MRI brain atlas of the common marmoset monkey, *Callithrix jacchus*. Brain Res Rev 62(1):1–18

Pessoa V, Abrahao J, Pacheco R, Pereira L, Magalhaes-Castro B, Saraiva P (1992) Relative sizes of cortical visual areas in marmosets: functional and phylogenetic implications. Exp Brain Res 88(2):459–462

Peters A, Jones E (1984) Cellular components of the cerebral cortex. Plenum, New York

Peters A, Sethares C (1996) Myelinated axons and the pyramidal cell modules in monkey primary visual cortex. J Comp Neurol 365(2):232–255

Pistorio AL, Hendry SH, Wang X (2006) A modified technique for high-resolution staining of myelin. J Neurosci Methods 153(1):135–146

Puller C, Haverkamp S (2011) Bipolar cell pathways for color vision in non-primate dichromats. Vis Neurosci 28(1):51–60

Reser DH, Burman KJ, Richardson KE, Spitzer MW, Rosa MGP (2009) Connections of the marmoset rostrotemporal auditory area: express pathways for analysis of affective content in hearing. Eur J Neurosci 30(4):578–592

Rosa MG, Elston GN (1998) Visuotopic organisation and neuronal response selectivity for direction of motion in visual areas of the caudal temporal lobe of the marmoset monkey (*Callithrix jacchus*): middle temporal area, middle temporal crescent, and surrounding cortex. J Comp Neurol 393(4):505–527

Rosa MG, Schmid LM (1995) Visual areas in the dorsal and medial extrastriate cortices of the marmoset. J Comp Neurol 359(2):272–299

Rosa MGP, Palmer SM, Gamberini M, Burman KJ, Yu H, Reser DH, Bourne JA, Tweedale R, Galletti C (2009) Connections of the dorsomedial visual area: pathways for early integration of dorsal and ventral streams in extrastriate cortex. J Neurosci 29(14):4548–4563

Rowe N (1996) The pictorial guide to the living primates. Pogonias Press, Charlestown

Sasaki E, Suemizu H, Shimada A, Hanazawa K, Oiwa R, Kamioka M, Tomioka I, Sotomaru Y, Hirakawa R, Eto T, Shiozawa S, Maeda T, Ito M, Ito R, Kito C, Yagihashi C, Kawai K, Miyoshi H, Tanioka Y, Tamaoki N, Habu S, Okano H, Nomura T (2009) Generation of transgenic non-human primates with germline transmission. Nature 459(7246):523–527

Sati P, Silva AC, van Gelderen P, Gaitan MI, Wohler JE, Jacobson S, Duyn JH, Reich DS (2012) In vivo quantification of T_2 anisotropy in white matter fibers in marmoset monkeys. Neuroimage 59(2):979–985

Sigalovsky IS, Fischl B, Melcher JR (2006) Mapping an intrinsic MR property of gray matter in auditory cortex of living humans: a possible marker for primary cortex and hemispheric differences. Neuroimage 32(4):1524–1537

Silva AC, Liu JV, Hirano Y, Leoni RF, Merkle H, Mackel JB, Zhang XF, Nascimento GC, Stefanovic B (2011) Longitudinal functional magnetic resonance imaging in animal models. Methods Mol Biol 711:281–302

Solomon SS, Tailby C, Gharaei S, Camp AJ, Bourne JA, Solomon SG (2011) Visual motion integration by neurons in the middle temporal area of a New World monkey, the marmoset. J Physiol 589(Pt 23):5741–5758

't Hart BA, Laman JD, Bauer J, Blezer E, van Kooyk Y, Hintzen RQ (2004) Modelling of multiple sclerosis: lessons learned in a non-human primate. Lancet Neurol 3(10):588–597

Tardif SD, Smucny DA, Abbott DH, Mansfield K, Schultz-Darken N, Yamamoto ME (2003) Reproduction in captive common marmosets (*Callithrix jacchus*). Comp Med 53(4):364–368

Tosun D, Rettmann ME, Han X, Tao X, Xu C, Resnick SM, Pham DL, Prince JL (2004) Cortical surface segmentation and mapping. Neuroimage 23(Suppl 1):S108–S118

Trampel R, Ott DVM, Turner R (2011) Do the congenitally blind have a stria of Gennari? First intracortical insights in vivo. Cereb Cortex 21(9):2075–2081

Ungerleider LG, Desimone R (1986) Cortical connections of visual area MT in the macaque. J Comp Neurol 248(2):190–222

Van Essen DC, Maunsell JH, Bixby JL (1981) The middle temporal visual area in the macaque: myeloarchitecture, connections, functional properties and topographic organization. J Comp Neurol 199(3):293–326

Walters NB, Egan GF, Kril JJ, Kean M, Waley P, Jenkinson M, Watson JDG (2003) In vivo identification of human cortical areas using high-resolution MRI: an approach to cerebral structure-function correlation. Proc Natl Acad Sci USA 100(5):2981–2986

Wang X (2007) Neural coding strategies in auditory cortex. Hear Res 229(1–2):81–93

Wong P, Kaas JH (2010) Architectonic subdivisions of neocortex in the Galago (Otolemur garnetti). Anat Rec (Hoboken) 293(6):1033–1069

Yamada M, Momoshima S, Masutani Y, Fujiyoshi K, Abe O, Nakamura M, Aoki S, Tamaoki N, Okano H (2008) Diffusion-tensor neuronal fiber tractography and manganese-enhanced MR imaging of primate visual pathway in the common marmoset: preliminary results. Radiology 249(3):855–864

Yu H, Verma R, Yang Y, Tibballs HA, Lui LL, Reser DH, Rosa MGP (2010) Spatial and temporal frequency tuning in striate cortex: functional uniformity and specializations related to receptive field eccentricity. Eur J Neurosci 31(6):1043–1062

Zhang Y, Xiong Y, Mahmood A, Meng Y, Liu Z, Qu C, Chopp M (2010) Sprouting of corticospinal tract axons from the contralateral hemisphere into the denervated side of the spinal cord is associated with functional recovery in adult rat after traumatic brain injury and erythropoietin treatment. Brain Res 1353:249–257

Zilles K, Amunts K (2012) Architecture of the cerebral cortex. In: Paxinos G, Mai JK (eds) The human nervous system. Academic, San Diego

Zilles K, Armstrong E, Moser KH, Schleicher A, Stephan H (1989) Gyrification in the cerebral cortex of primates. Brain Behav Evol 34(3):143–150

Chapter 9
High-Field Magnetic Resonance Mapping of the Border Between Primary Motor (Area 4) and Somatosensory (Area 3a) Cortex in Ex-Vivo and In-Vivo Human Brains

Stefan Geyer

Abstract Unraveling the functional properties of structural elements in the brain is one of the fundamental goals of neuroscientific research. In the cerebral cortex this is not so easy to accomplish, since cortical areas are defined microstructurally in post-mortem brains but functionally in living brains with electrophysiological or neuroimaging techniques – and cortical areas vary in their topographical properties across individual brains. To map both microstructure and function in the same brains noninvasively in vivo would represent a huge leap forward. In this chapter, we show our approach, based on a MP2RAGE sequence run on a 7 T Siemens MR scanner to produce in living human subjects quantitative T1 maps that reflect local microanatomy. On inflated surface maps of individual subjects, cortical areas known from post-mortem studies to be heavily myelinated are easily discernible from surrounding less myelinated regions. Classically, cortical areas (and their precise borders that are indispensable for a valid correlation with functional data) are defined by their myelo- and cytoarchitectonic pattern ex vivo, i.e., in post-mortem brains. Hence, with the same MP2RAGE sequence at 7 T we scan fixed tissue blocks of the human cortex, section them with a microtome, stain the sections for myelin sheaths or cell bodies, and correlate the MR architecture directly with myelo- and cytoarchitecture. This "triple jump" approach allows to (i) define a cortical area based on myelo- and cytoarchitecture, (ii) extract the "MR fingerprint" of this area ex vivo, and (iii) transfer this "fingerprint" to living brains and define this area in vivo. With this technique we mapped in living subjects the functionally important border between primary motor (Brodmann Area 4) and somatosensory (Brodmann Area 3a) cortex. This technology sets the stage for the development of an in vivo myeloarchitectonic brain map, with the enormous potential to make direct correlations between microstructure and function in living human brains.

S. Geyer (✉)
Department of Neurophysics, Max Planck Institute for Human Cognitive and Brain Sciences, Stephanstrasse 1a, 04103 Leipzig, Germany
e-mail: sgeyer@cbs.mpg.de

S. Geyer and R. Turner (eds.), *Microstructural Parcellation of the Human Cerebral Cortex*, DOI 10.1007/978-3-642-37824-9_9, © Springer-Verlag Berlin Heidelberg 2013

9.1 Introduction

Unraveling the functional properties of structural elements in the brain is one of the fundamental goals of neuroscientific research. Ideally, this is done noninvasively (to avoid problems related to invasive procedures, e.g., infections, bleedings, etc.) in individual (to avoid problems arising from group studies, e.g., smoothing, averaging, etc.) and living brains. However, this apparently trivial "gold standard" is not so easy to accomplish. Three points are worth considering in this context.

First, functional imaging studies correlate activations only with the brain's macroanatomy. At best, cortical activation foci from functional magnetic resonance imaging (fMRI) or positron emission tomography (PET) experiments are correlated with a single subject's individual gyral and sulcal anatomy (see, e.g., Moore et al. 2000). The underlying microanatomy, however, is not yet accessible to MR technology. A notable exception is the very prominent Stria of Gennari in primary visual cortex which was observed for the first time with MRI in vivo more than 20 years ago (Clark et al. 1992).

Second, many invasive electrophysiology studies in nonhuman primates and other laboratory animals have shown functional borders where microanatomy changes. In animals, microstructure can be directly correlated with function since when the electrophysiological experiments are completed the brains can be sectioned, stained for cell bodies or myelin sheaths, and stimulation or recording sites compared with the cyto- or myeloarchitectonic pattern. Many studies have found that sites with similar electrophysiological properties lie within a cortical region with similar microarchitectonic features. However, across an architectonic border, the electrophysiological properties change dramatically. This has been observed since the early days of electrophysiological experiments when the non-human primate cortex was stimulated with surface electrodes (e.g., Vogt and Vogt 1919) and replicated later with more refined techniques such as intracortical microstimulation and unit recording, e.g., in the somatosensory cortex (Iwamura et al. 1983a, b, 1985) or supplementary and pre-supplementary motor cortex (Luppino et al. 1991; Matelli et al. 1991).

Third, microanatomy – and thus the location and extent of a given cortical area – is topographically variable across brains (Amunts et al. 1999, 2000, 2005; Caspers et al. 2006; Eickhoff et al. 2006a, c; Geyer et al. 1996, 1999; Grefkes et al. 2001; Kurth et al. 2010; Malikovic et al. 2007; Rademacher et al. 1993, 2002; Scheperjans et al. 2008a, b). This means that structural-functional correlations based solely on macroanatomy are questionable and may account for at least some of the conflicting results functional imaging studies have provided in the past.

Talairach and Tournoux (1988, 1993) when establishing their widely used stereotaxic reference system for spatially normalizing imaging data, adopted Brodmann's (1909, 1914) nomenclature of cytoarchitectonic areas for parcellating the cortex of their standard brain in "Talairach Space". However, their atlas is also of very limited value. It is not based on microstructural data (the authors only transferred each area from Brodmann's schematic drawing to a corresponding position on the exposed cortical surface of their reference brain), the atlas indicates only the approximate position of each area (borders between areas are not marked), and it ignores the problem of interindividual variability (only one brain is depicted in the atlas).

One attempt to overcome this dilemma is to generate probabilistic cytoarchitectonic maps in standard anatomical (e.g., Montreal Neurological Institute, MNI) space (Roland and Zilles 1994, 1998; Roland et al. 1997; Toga et al. 2006). Cortical areas are defined cytoarchitectonically in post-mortem brains, reconstructed in 3-D, and spatially normalized. Superimposing the areas from, say ten brains gives a probabilistic description of each area's spatial variability. In this format, the areas can be matched with co-registered functional imaging data. However, due to interindividual variability, the population maps of adjacent areas overlap considerably and only after extensive thresholding (i.e., considering only voxels above a certain probability level) – and thus discarding structural information – is it possible to unequivocally assign a given voxel in standard space to the population map of a particular cortical area (Eickhoff et al. 2005, 2006b, 2007). Furthermore, the invasive nature of microanatomical studies precludes microstructure and function to be studied in the same brain – such correlations can only be probabilistic in nature.

It would be a huge leap forward if it were possible to remove this guesswork and generate an individual-specific map of cortical microstructure in vivo and correlate it with cortical function in the same brain. In recent years, two advances have brought us closer to this ambitious goal. The first is the dramatic improvement in the quality of in vivo MRI scans. With 7 T magnets and high sensitivity radiofrequency receive coils, the current state of the art allows structural images of entire brains to be obtained with 0.5 mm (Trampel et al. 2011) and functional blood oxygenation level dependent (BOLD) contrast changes with less than 1 mm isotropic resolution (Heidemann et al. 2012). The second advancement is based on the observation that maps of the longitudinal relaxation time T1 effectively indicate the presence of myelin and closely resemble myelin-stained histological sections (Dick et al. 2012; Geyer et al. 2011; Sereno et al. 2012), whereas differences in cytoarchitecture are detectable with MRI only in rare instances, e.g., in the case of the unique islands of large neurons in layer II of the entorhinal cortex (Augustinack et al. 2005). Research by Cécile and Oskar Vogt, two pioneers in the field of myeloarchitecture in the first half of the 20th century, has shown that there is a great degree of concordance between structural parcellations of the cortex based on differences in myeloarchitecture and differences in cytoarchitecture (Vogt and Vogt 1919). Myelo- and cytoarchitecture are not two parallel universes but two different views of the same universe.

"In vivo Brodmann mapping" (Geyer et al. 2011) exploits the longitudinal relaxation time T1 and reveals cortical microstructure by showing, similar to myelin-stained histological sections, differential grey matter myelination. Cortical areas known from post-mortem studies to be heavily myelinated, e.g., primary motor, somatosensory, auditory, visual cortex or area V5-MT (Clarke and Miklossy 1990; Hopf 1955, 1956; Hopf and Vitzthum 1957) are easily discernible from surrounding less myelinated regions. In addition, a "triple jump" approach allows to validate the myelin-based in vivo maps with "classical" histology data ex vivo: formalin-fixed post-mortem tissue blocks of the human cortex are scanned with a 7 T MR sequence that produces T1 maps, sectioned with a microtome, sections are stained for myelin sheaths or cell bodies, and the MR architecture correlated with myelo- and cytoarchitecture. This approach allows to (i) define a cortical area based on myelo- and cytoarchitecture, (ii) extract the "MR fingerprint" of this area ex vivo, and (iii) transfer this "fingerprint" to living brains and define this area in vivo.

In this chapter we describe the "triple jump"-based mapping and validation of the functionally important border between primary motor (area 4) and somatosensory (area 3a) cortex in the fundus of the central sulcus.

9.2 Materials and Methods

9.2.1 Ex Vivo Analysis

For ex vivo analysis we fixed post-mortem brains from subjects without neurological or psychiatric diseases in 4 % formalin for several weeks (males and females, 60–75 years, autopsy performed with consent of the patient's relatives, post-mortem interval before start of fixation 24–48 h, brains suspended by the basilar artery during fixation to avoid compression or distortions). We cut out tissue blocks including portions of the pre- and postcentral gyrus of the fixed brains and scanned the blocks with a 7 T whole-body MR scanner (MAGNETOM 7 T, Siemens, Erlangen, Germany) and a 24 channel phased array coil (Nova Medical Inc., Wilmington, MA, USA) or a custom-built single channel square-shaped single loop coil. We generated T1 maps using a MP2RAGE sequence (Marques et al. 2010) with the following parameters: TR = 3,000 ms, TE = 4.48 ms, TI_1 = 180 ms, TI_2 = 900 ms, α_1 = 8 deg, α_2 = 8 deg, bandwith = 230 Hz/pixel, voxel size = $(0.33\ \text{mm})^3$, 32 averages, acquisition time = 3 h 50 min, surrounding medium: Fomblin (Solvay Solexis, Bollate, Italy). For histological validation we cryo-protected the blocks by immersion in 30 % sucrose, sectioned them with a freezing microtome (SM2000 R, Leica Microsystems, Nussloch, Germany; section thickness: 30 μm) and stained the sections for cell bodies (Merker 1983) or myelin sheaths (rat monoclonal antibody against myelin basic protein (Abcam, Cambridge, UK), avidin-biotin-peroxidase complex (ABC) method, chromogen: 3,3'-diaminobenzidine (DAB) tetrahydrochloride and ammonium nickel(II) sulfate). We analyzed the cyto- and myeloarchitectonic pattern and correlated the histology-based laminar contrast with the ex vivo T1 contrast in the same tissue block.

9.2.2 *In Vivo Analysis*

For in vivo mapping we scanned young healthy volunteers (males and females, 20–30 years). All subjects had given written informed consent and the study protocol was consistent with guidelines of the Ethics Committee of the University of Leipzig, Germany. We scanned our subjects at 7 T with the identical hardware (Siemens MAGNETOM 7 T scanner, 24 channel Nova coil) and acquired T1 maps with the same MP2RAGE sequence (Marques et al. 2010) based on the following parameters: TR = 5,000 ms, TE = 3.84 ms, TI_1 = 900 ms, TI_2 = 2,750 ms, α_1 = 7 deg, α_2 = 5 deg, bandwith = 180 Hz/pixel, voxel size = $(0.6\ \text{mm})^3$, 3 averages, acquisition time = 60 min.

9.2.3 *"Triple Jump" Approach*

The idea behind the "triple jump" approach is to validate changes in the in vivo T1 contrast by correlating it with changes in the ex vivo T1 contrast and the myelo- and cytoarchitectonic pattern in corresponding regions of the tissue blocks. The three steps (summarized in Fig. 9.1) allow to (i) define a cortical area based on myelo- and cytoarchitecture, (ii) extract the "MR (T1) fingerprint" of this area ex vivo, and (iii) transfer this "T1 fingerprint" to living brains and define this area in vivo.

9.3 Results

Figure 9.2 depicts a 3D-rendered quantitative T1 map of a living subject in an anatomically "normal" view with gyri and sulci (Fig. 9.2a) and in an inflated view (Fig. 9.2b). The inflated surface map shows a very heterogeneous spatial T1 distribution on the lateral convexity with low T1 regions rostral and caudal to the central sulcus, in the planum temporale (lower bank of the Sylvian fissure), on the lateral occipital surface, and around the occipital pole extending further medio-rostrally along the calcarine sulcus (not shown). It is interesting to compare this T1 map with a map of myelin density in a post-mortem brain (Fig. 9.3). In the 1950s Adolf Hopf, working in the Institute for Brain Research and General Biology (Institut für Hirnforschung und allgemeine Biologie) in Neustadt in the Black Forest (directed at that time by Oskar Vogt) started to produce semi-quantitative maps of the density of myelinated fibers ("Gesamtfaserdichte") in the cerebral cortex. The map shown in Fig. 9.3 is a collage from three different publications by Hopf: a map of the frontal (Hopf 1956), parietal (Hopf and Vitzthum 1957), and temporal lobe (Hopf 1955). The maps are based on earlier qualitative myeloarchitectonic parcellations of the frontal (Strasburger 1937; Vogt 1910), parietal (Batsch 1956; Vogt 1911) and temporal cortex (Hopf 1954). The occipital cortex

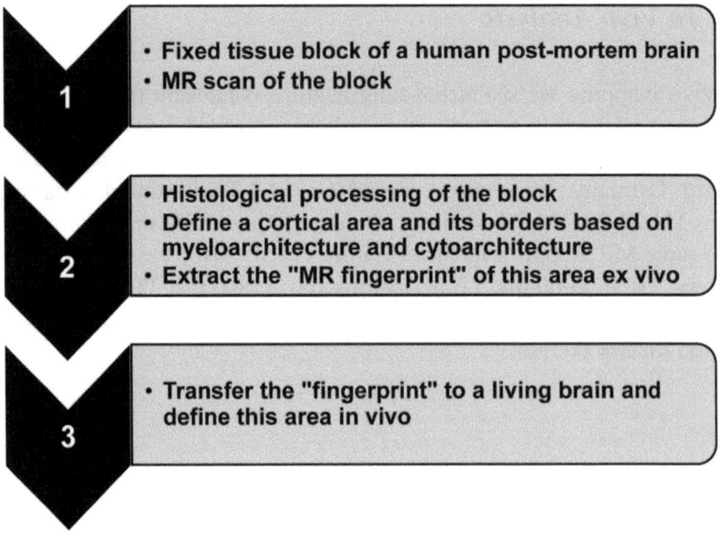

Fig. 9.1 Three steps of the "triple jump" approach

was never charted, so that part of the brain remains a "terra incognita". The dashed and dotted lines in Fig. 9.3 represent the boundaries of the myeloarchitectonic areas, the grey values are a semi-quantitative measure of the average density of myelinated fibers within each area. Dark grey represents a high, light grey a low mean fiber density. There is a striking correspondence between high fiber density (dark in Fig. 9.3) and low T1 values (dark in Fig. 9.2) rostral and caudal to the central sulcus and in the planum temporale. The most probable candidates for the "dark" regions rostral and caudal to the central sulcus are the primary motor (M1) and somatosensory (S1) cortex, respectively. The "dark" area in the planum temporale most probably belongs to the primary auditory cortex (A1). A plausible candidate for the "dark" region on the lateral occipital surface is area V5/MT, known from human post-mortem studies (Clarke and Miklossy 1990) to be heavily myelinated. The "dark" patch around the occipital pole extending on the mesial surface further rostrally along the calcarine sulcus most likely corresponds to the primary visual cortex (V1) that owes its high myelination to the very prominent myelin-rich Stria of Gennari.

This qualitative comparison between ex-vivo myelination and in-vivo T1 values shows the approximate location and extent of heavily myelinated, mainly primary areas on the cortical surface. However, what is ultimately needed for a valid correlation with functional imaging data, are *precise borders* between cortical areas – and here the "triple jump" approach comes into play since, classically, structural borders between areas are defined by their myelo- and cytoarchitectonic pattern ex vivo. Figure 9.4a, b shows a tissue block (pre- and postcentral gyrus, in situ in Fig. 9.4a and after dissection in Fig. 9.4b) from a post-mortem human brain prior to MR scanning and histological processing. M1 denotes the primary motor cortex in the posterior wall of the precentral gyrus, S1 the primary somatosensory

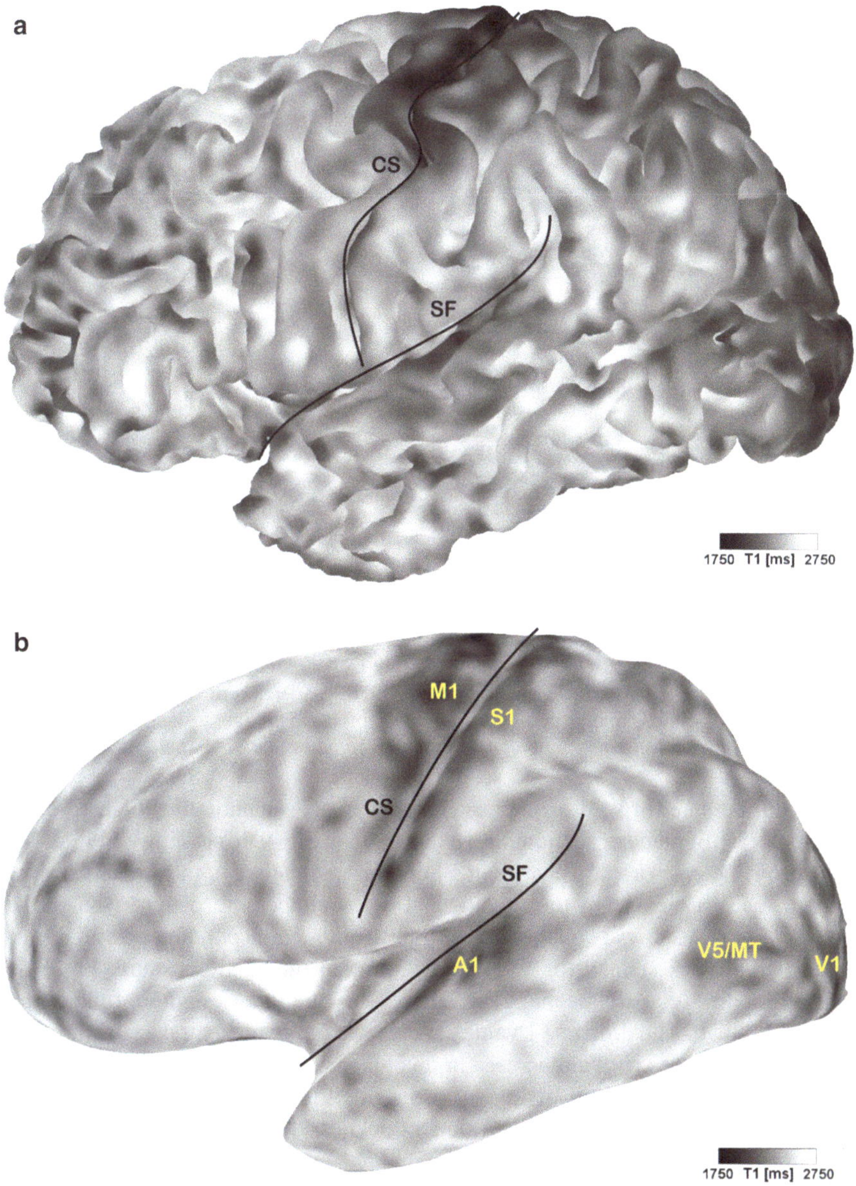

Fig. 9.2 In vivo quantitative T1 map in an anatomically "normal" (**a**) and an inflated (**b**) view. Dark regions with low T1 values reflect cortical areas with increased density of myelinated fibers (cf. Fig. 9.3). All major primary functional areas (*M1*, primary motor; *S1*, somatosensory; *A1*, auditory; and *V1*, visual area), known for their high myelination, can be identified. *CS*, central sulcus; *SF*, Sylvian fissure; *V5/MT*, motion selective visual area V5/MT

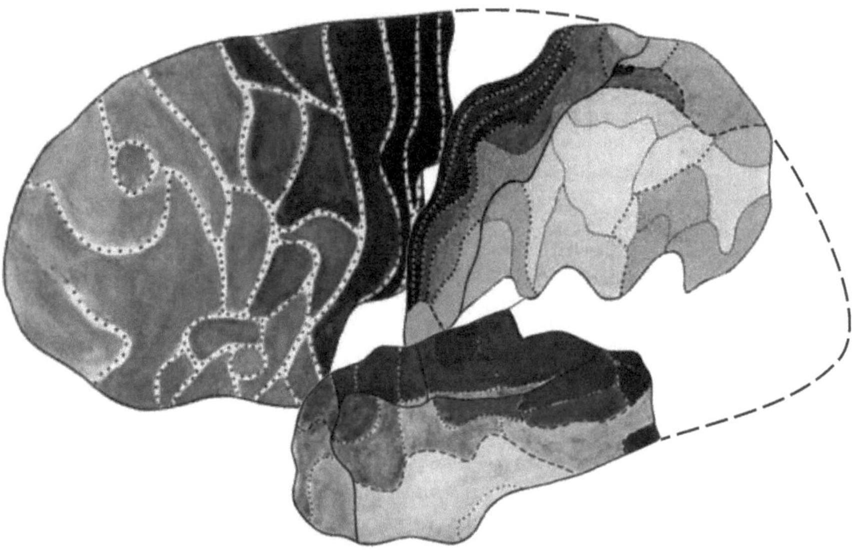

Fig. 9.3 Semi-quantitative map of the density of myelinated fibers based on microscopic analysis of myelin-stained sections from post-mortem brains. *Dashed* and *dotted* lines represent boundaries of myeloarchitectonic areas, *grey* values indicate average density of myelinated fibers within each area (*light gray*: low density, *dark gray*: high density). Collage from three publications by Adolf Hopf: frontal cortex map reproduced from Hopf (1956), parietal cortex map from Hopf and Vitzthum (1957), and temporal cortex map from Hopf (1955). Note striking correspondence between high fiber density (*dark* in Fig. 9.3) and low T1 values (*dark* in Fig. 9.2)

cortex in the anterior wall of the postcentral gyrus. A quantitative T1 map of the tissue block (for plane of sectioning see rectangle in Fig. 9.4b) is shown in Fig. 9.4c. The arrow indicates a sharp change in T1 contrast at the base of the precentral gyrus that matches a change in the myelo- and cytoarchitectonic pattern (see further below). Fig. 9.5 shows two adjacent sections from a corresponding position of the same tissue block stained for myelin basic protein (myeloarchitecture, Fig. 9.5a) and cell bodies (cytoarchitecture, Fig. 9.5b). Only the fundus region of the central sulcus is shown. The drop in T1 values at the base of the precentral gyrus (arrow in Fig. 9.4c) coincides with an increase in myelin basic protein immunostaining (line in Fig. 9.5a). In an accompanying section stained for cell bodies, this position is characterized by an increase in gray matter thickness, a disappearing inner granular layer (asterisks), and emerging giant pyramidal (Betz) cells (arrowheads). This transition (line in Fig. 9.5b) marks the border between area 3a (somatosensory cortex) and area 4 (primary motor cortex; Geyer et al. 1999). The final step of the "triple jump" approach is to transfer this microanatomically validated "T1 border" to living brains and define the same border in vivo. A T1 map of the central sulcus region in a living subject (coronal view of a brain section in Fig. 9.6a, latero-dorsal view of the 3-D rendered and inflated cortical surface in Fig. 9.6b) shows a sharp drop in T1 values and an increase in cortical thickness at the base of the precentral gyrus (arrows in Fig. 9.6a and dashed line in Fig. 9.6b). Position and MR parameters of this border in vivo match the corresponding border between area 3a and 4 ex vivo (cf. Figs. 9.4 and 9.5).

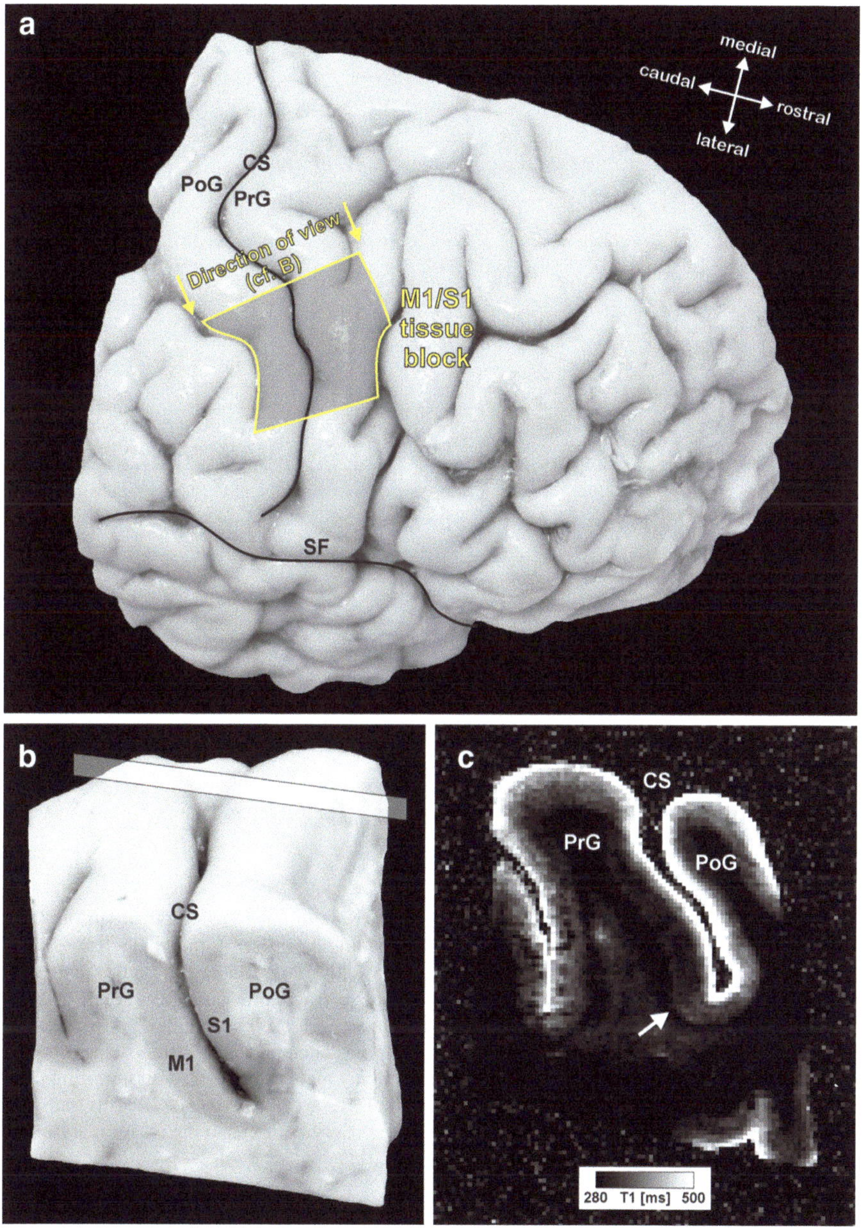

Fig. 9.4 Ex vivo mapping of the border between primary motor (area 4) and somatosensory (area 3a) cortex. (**a**, **b**) Tissue block of the precentral (PrG) and postcentral (PoG) gyrus in situ (**a**) and after dissection (**b**) from a post-mortem human brain. (**c**) Quantitative T1 map of the tissue block (for plane of sectioning see *rectangle* in **b**). *Arrow* indicates a sharp change in T1 contrast at the base of the precentral gyrus that matches a change in the myelo- and cytoarchitectonic pattern (see Fig. 9.5). *CS*, central sulcus; *M1*, primary motor cortex; *S1*, primary somatosensory cortex; *SF*, Sylvian fissure (Panel **b** and **c** reproduced from Geyer et al. 2011)

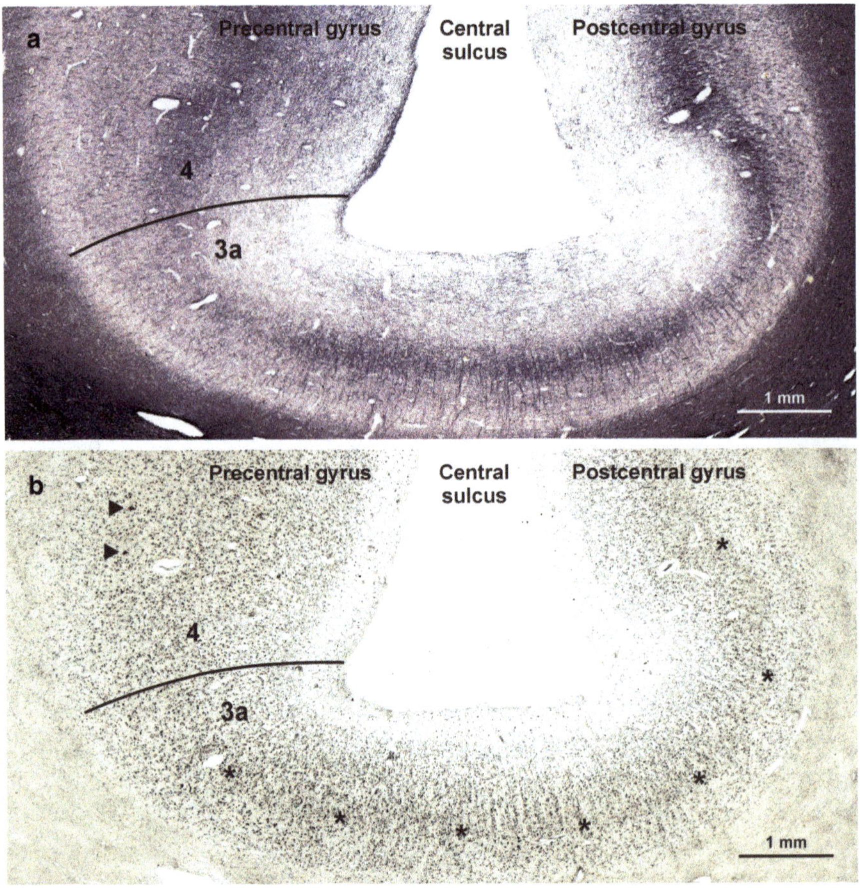

Fig. 9.5 Ex vivo mapping of the border between primary motor (area 4) and somatosensory (area 3a) cortex. Two adjacent sections from a corresponding position of the same tissue block (cf. Fig. 9.4c) stained for myelin basic protein (myeloarchitecture, **a**) and cell bodies (cytoarchitecture, **b**). Micrographs show fundus of the central sulcus (same orientation as in Figs. 9.4b, c). The drop in T1 values at the base of the precentral gyrus (cf. Fig. 9.4c) coincides with an increase in myelin basic protein immunostaining (*line* in **a**). In an accompanying section stained for cell bodies, this position is characterized by an increase in gray matter thickness, a disappearing inner granular layer (*asterisks*), and emerging giant pyramidal (Betz) cells (*arrowheads*). This transition (*line* in **b**) marks the border between area 3a (somatosensory cortex) and area 4 (primary motor cortex) (Panel **a** and **b** reproduced from Geyer et al. 2011)

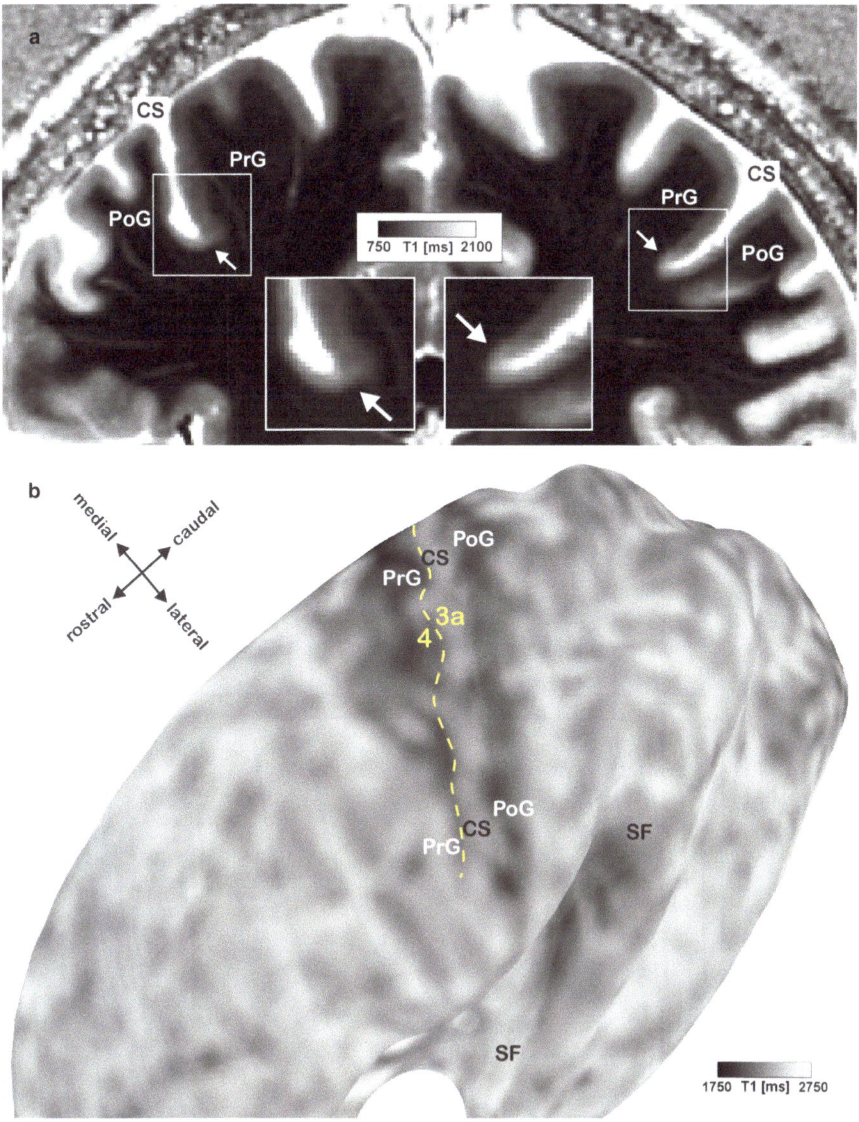

Fig. 9.6 In vivo mapping of the border between primary motor (area 4) and somatosensory (area 3a) cortex. Quantitative T1 map of the central sulcus region (coronal view in **a**, insets show fundus regions at higher magnification (x 1.5), latero-dorsal view of the inflated surface in **b**). *Arrows* in **a** and *dashed* line in **b** show a sharp drop in T1 values at the base of the precentral gyrus (PrG). Position and MR parameters of this border in vivo match the corresponding border between area 3a and 4 ex vivo (cf. Figs. 9.4 and 9.5). *CS*, central sulcus; *PoG*, postcentral gyrus; *SF*, Sylvian fissure (Panel **a** reproduced from Geyer et al. 2011)

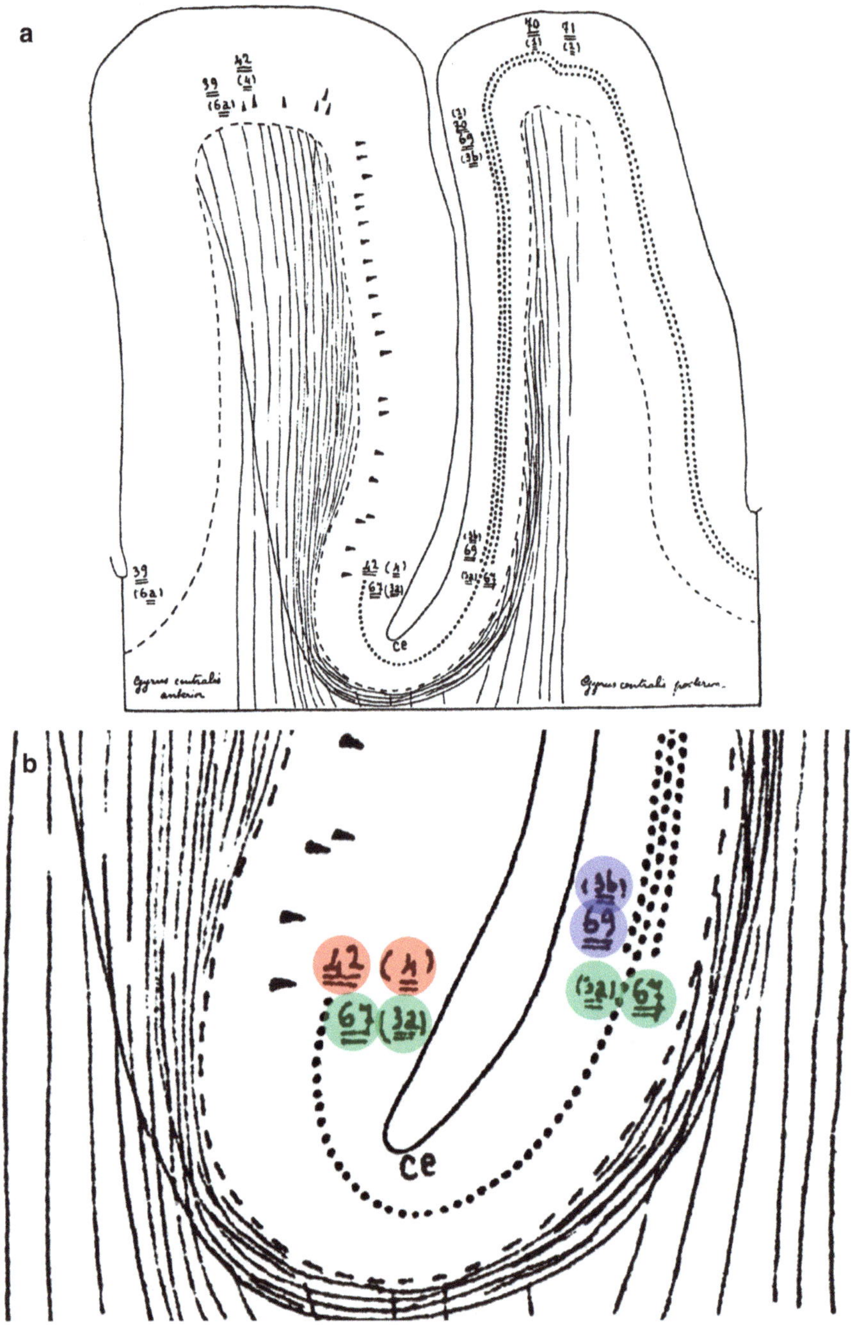

Fig. 9.7 Schematic drawing of the precentral (Gyrus centralis anterior, *left*) and postcentral (Gyrus centralis posterior, *right*) gyrus (**a**) and enlarged view of the fundus region of the central sulcus (**b**) with a synopsis of the topography of cortical areas defined by cytoarchitecture (areas 6a, 4, 3a, 3b, 1, 2; nomenclature according to Brodmann (1909) and Vogt and Vogt (1919)) and

9.4 Concluding Remarks

As already outlined in the introduction, differences in cytoarchitecture are detectable with MRI only in rare instances, but myelin is very easy to pick out. The main challenge is to find MR parameters that correlate well with the cortical myelin content. Among several candidates, e.g., myelin water fraction, magnetization transfer rate, transverse relaxation times T2 and T2*, the longitudinal relaxation time T1 seems to be a very promising candidate. Myelin and its biochemical components, e.g., cholesterol and galactocerebroside, substantially accelerate T1 relaxation by magnetization transfer from slowly relaxing free water protons to rapidly relaxing protons bound to large molecules in the cell membranes. Specifically designed MR sequences such as MP2RAGE (Marques et al. 2010) sample the MP RAGE (magnetization-prepared rapid gradient-echo; Mugler and Brookeman 1990) signal at two different inversion times and thus allow to map the longitudinal relaxation time T1 with reasonable precision. This is, in a nutshell, the MR basis for our myelin density-based mapping approach. Many more details on this topic can be found in the chapter "MRI Methods for In-Vivo Cortical Parcellation" by Robert Turner in this book.

The justification for terming this myelin-based approach "in vivo Brodmann mapping" (which clearly refers to cytoarchitecture, the technique used by Brodmann for mapping the cortex) comes from the writings of Cécile and Oskar Vogt who repeatedly stressed the great degree of topographical concordance between areal borders based on differences in myeloarchitecture and cytoarchitecture. Figure 9.7 shows a schematic drawing of the pre- and postcentral gyrus, published by the Vogts in 1919, in which they present a pictorial synopsis of the topography of cortical areas and their borders as defined by cyto- and myeloarchitecture. There is a precise spatial correspondence, e.g., in the fundus region of the central sulcus (enlarged in Fig. 9.7b) between cytoarchitectonic area 4 and myeloarchitectonic area 42 (red), cytoarchitectonic area 3a and myeloarchitectonic area 67 (green), and cytoarchitectonic area 3b and myeloarchitectonic area 69 (blue). Of course, this sketch represents only a tiny sample of the entire cerebral cortex and borders were defined solely by subjective visual inspection through the microscope – the only technique available at that time. Extending this concordance mapping between cyto- and myeloarchitecture to the entire cerebral cortex and verifying areal borders with objective (i.e., observer-independent) state-of-the-art analysis techniques will be a great challenge for the future.

Fig. 9.7 (continued) myeloarchitecture (areas 39, 42, 67, 69, 70, 71; nomenclature according to Vogt (1910, 1911)). Note precise spatial correspondence in borders between cytoarchitectonic and myeloarchitectonic areas, e.g., in the fundus region between cytoarchitectonic area 4 and myeloarchitectonic area 42 (*red* in **b**), cytoarchitectonic area 3a and myeloarchitectonic area 67 (*green*), and cytoarchitectonic area 3b and myeloarchitectonic area 69 (*blue*) (Figure reproduced from Vogt and Vogt 1919)

In conclusion, "in vivo Brodmann mapping" based on 7 T MRI is able to detect functionally important borders such as the one between primary motor (area 4) and somatosensory (area 3a) cortex ex vivo and – more importantly – also in vivo. This is the first step towards an individual-specific microanatomical brain map, with the great potential to eventually make direct correlations between microstructure and function in living human brains.

References

Amunts K, Schleicher A, Bürgel U, Mohlberg H, Uylings HBM, Zilles K (1999) Broca's region revisited: cytoarchitecture and intersubject variability. J Comp Neurol 412:319–341

Amunts K, Malikovic A, Mohlberg H, Schormann T, Zilles K (2000) Brodmann's areas 17 and 18 brought into stereotaxic space – where and how variable? Neuroimage 11:66–84

Amunts K, Kedo O, Kindler M, Pieperhoff P, Mohlberg H, Shah NJ, Habel U, Schneider F, Zilles K (2005) Cytoarchitectonic mapping of the human amygdala, hippocampal region and entorhinal cortex: Intersubject variability and probability maps. Anat Embryol (Berl) 210:343–352

Augustinack JC, van der Kouwe AJW, Blackwell ML, Salat DH, Wiggins CJ, Frosch MP, Wiggins GC, Potthast A, Wald LL, Fischl BR (2005) Detection of entorhinal layer II using 7 Tesla magnetic resonance imaging. Ann Neurol 57:489–494

Batsch EG (1956) Die myeloarchitektonische Untergliederung des Isocortex parietalis beim Menschen. J Hirnforsch 2:225–258

Brodmann K (1909) Vergleichende Lokalisationslehre der Großhirnrinde. Barth, Leipzig

Brodmann K (1914) Physiologie des Gehirns. In: Von Bruns P (ed) Neue Deutsche Chirurgie. Enke, Stuttgart

Caspers S, Geyer S, Schleicher A, Mohlberg H, Amunts K, Zilles K (2006) The human inferior parietal cortex: cytoarchitectonic parcellation and interindividual variability. Neuroimage 33:430–448

Clark VP, Courchesne E, Grafe M (1992) *In vivo* myeloarchitectonic analysis of human striate and extrastriate cortex using magnetic resonance imaging. Cereb Cortex 2:417–424

Clarke S, Miklossy J (1990) Occipital cortex in man: organization of callosal connections, related myelo- and cytoarchitecture, and putative boundaries of functional visual areas. J Comp Neurol 298:188–214

Dick F, Taylor Tierney A, Lutti A, Josephs O, Sereno MI, Weiskopf N (2012) *In vivo* functional and myeloarchitectonic mapping of human primary auditory areas. J Neurosci 32:16095–16105

Eickhoff SB, Stephan KE, Mohlberg H, Grefkes C, Fink GR, Amunts K, Zilles K (2005) A new SPM toolbox for combining probabilistic cytoarchitectonic maps and functional imaging data. Neuroimage 25:1325–1335

Eickhoff SB, Amunts K, Mohlberg H, Zilles K (2006a) The human parietal operculum. II. Stereotaxic maps and correlation with functional imaging results. Cereb Cortex 16:268–279

Eickhoff SB, Heim S, Zilles K, Amunts K (2006b) Testing anatomically specified hypotheses in functional imaging using cytoarchitectonic maps. Neuroimage 32:570–582

Eickhoff SB, Schleicher A, Zilles K, Amunts K (2006c) The human parietal operculum. I. Cytoarchitectonic mapping of subdivisions. Cereb Cortex 16:254–267

Eickhoff SB, Paus T, Caspers S, Grosbras MH, Evans AC, Zilles K, Amunts K (2007) Assignment of functional activations to probabilistic cytoarchitectonic areas revisited. Neuroimage 36:511–521

Geyer S, Ledberg A, Schleicher A, Kinomura S, Schormann T, Bürgel U, Klingberg T, Larsson J, Zilles K, Roland PE (1996) Two different areas within the primary motor cortex of man. Nature 382:805–807
Geyer S, Schleicher A, Zilles K (1999) Areas 3a, 3b, and 1 of human primary somatosensory cortex: 1. Microstructural organization and interindividual variability. Neuroimage 10:63–83
Geyer S, Weiss M, Reimann K, Lohmann G, Turner R (2011) Microstructural parcellation of the human cerebral cortex – from Brodmann's post-mortem map to in vivo mapping with high-field magnetic resonance imaging. Front Hum Neurosci 5:19
Grefkes C, Geyer S, Schormann T, Roland P, Zilles K (2001) Human somatosensory area 2: Observer-independent cytoarchitectonic mapping, interindividual variability, and population map. Neuroimage 14:617–631
Heidemann RM, Ivanov D, Trampel R, Fasano F, Meyer H, Pfeuffer J, Turner R (2012) Isotropic submillimeter fMRI in the human brain at 7 T: combining reduced field-of-view imaging and partially parallel acquisitions. Magn Reson Med 68:1506–1516
Hopf A (1954) Die Myeloarchitektonik des Isocortex temporalis beim Menschen. J Hirnforsch 1:208–279
Hopf A (1955) Über die Verteilung myeloarchitektonischer Merkmale in der isokortikalen Schläfenlappenrinde beim Menschen. J Hirnforsch 2:36–54
Hopf A (1956) Über die Verteilung myeloarchitektonischer Merkmale in der Stirnhirnrinde beim Menschen. J Hirnforsch 2:311–333
Hopf A, Vitzthum HG (1957) Über die Verteilung myeloarchitektonischer Merkmale in der Scheitellappenrinde beim Menschen. J Hirnforsch 3:79–104
Iwamura Y, Tanaka M, Sakamoto M, Hikosaka O (1983a) Functional subdivisions representing different finger regions in area 3 of the first somatosensory cortex of the conscious monkey. Exp Brain Res 51:315–326
Iwamura Y, Tanaka M, Sakamoto M, Hikosaka O (1983b) Converging patterns of finger representation and complex response properties of neurons in area 1 of the first somatosensory cortex of the conscious monkey. Exp Brain Res 51:327–337
Iwamura Y, Tanaka M, Sakamoto M, Hikosaka O (1985) Diversity in receptive field properties of vertical neuronal arrays in the crown of the postcentral gyrus of the conscious monkey. Exp Brain Res 58:400–411
Kurth F, Eickhoff SB, Schleicher A, Hoemke L, Zilles K, Amunts K (2010) Cytoarchitecture and probabilistic maps of the human posterior insular cortex. Cereb Cortex 20:1448–1461
Luppino G, Matelli M, Camarda RM, Gallese V, Rizzolatti G (1991) Multiple representations of body movements in mesial area 6 and the adjacent cingulate cortex: an intracortical microstimulation study in the macaque monkey. J Comp Neurol 311:463–482
Malikovic A, Amunts K, Schleicher A, Mohlberg H, Eickhoff SB, Wilms M, Palomero-Gallagher N, Armstrong E, Zilles K (2007) Cytoarchitectonic analysis of the human extrastriate cortex in the region of V5/MT+: A probabilistic, stereotaxic map of area hOc5. Cereb Cortex 17:562–574
Marques JP, Kober T, Krueger G, van der Zwaag W, van de Moortele PF, Gruetter R (2010) MP2RAGE, a self bias-field corrected sequence for improved segmentation and T_1-mapping at high field. Neuroimage 49:1271–1281
Matelli M, Luppino G, Rizzolatti G (1991) Architecture of superior and mesial area 6 and the adjacent cingulate cortex in the macaque monkey. J Comp Neurol 311:445–462
Merker B (1983) Silver staining of cell bodies by means of physical development. J Neurosci Methods 9:235–241
Moore CI, Stern CE, Corkin S, Fischl B, Gray AC, Rosen BR, Dale AM (2000) Segregation of somatosensory activation in the human rolandic cortex using fMRI. J Neurophysiol 84:558–569
Mugler JP, Brookeman JR (1990) Three-dimensional magnetization-prepared rapid gradient-echo imaging (3D MP RAGE). Magn Reson Med 15:152–157
Rademacher J, Caviness VS, Steinmetz H, Galaburda AM (1993) Topographical variation of the human primary cortices: Implications for neuroimaging, brain mapping, and neurobiology. Cereb Cortex 3:313–329

Rademacher J, Bürgel U, Zilles K (2002) Stereotaxic localization, intersubject variability, and interhemispheric differences of the human auditory thalamocortical system. Neuroimage 17:142–160

Roland PE, Zilles K (1994) Brain atlases – a new research tool. Trends Neurosci 17:458–467

Roland PE, Zilles K (1998) Structural divisions and functional fields in the human cerebral cortex. Brain Res Rev 26:87–105

Roland PE, Geyer S, Amunts K, Schormann T, Schleicher A, Malikovic A, Zilles K (1997) Cytoarchitectural maps of the human brain in standard anatomical space. Hum Brain Mapp 5:222–227

Scheperjans F, Eickhoff SB, Hömke L, Mohlberg H, Hermann K, Amunts K, Zilles K (2008a) Probabilistic maps, morphometry, and variability of cytoarchitectonic areas in the human superior parietal cortex. Cereb Cortex 18:2141–2157

Scheperjans F, Hermann K, Eickhoff SB, Amunts K, Schleicher A, Zilles K (2008b) Observer-independent cytoarchitectonic mapping of the human superior parietal cortex. Cereb Cortex 18:846–867

Sereno MI, Lutti A, Weiskopf N, Dick F (2012) Mapping the human cortical surface by combining quantitative T_1 with retinotopy. Cereb Cortex. doi:10.1093/cercor/bhs213

Strasburger E (1937) Die myeloarchitektonische Gliederung des Stirnhirns beim Menschen und Schimpansen. J Psychol Neurol 47(461–491):565–606

Talairach J, Tournoux P (1988) Co-planar stereotaxic atlas of the human brain. 3-Dimensional proportional system: an approach to cerebral imaging. Thieme, Stuttgart

Talairach J, Tournoux P (1993) Referentially oriented cerebral MRI anatomy. Atlas of stereotaxic anatomical correlations for gray and white matter. Thieme, Stuttgart

Toga AW, Thompson PM, Mori S, Amunts K, Zilles K (2006) Towards multimodal atlases of the human brain. Nat Rev Neurosci 7:952–966

Trampel R, Ott DVM, Turner R (2011) Do the congenitally blind have a Stria of Gennari? First intracortical insights in vivo. Cereb Cortex 21:2075–2081

Vogt O (1910) Die myeloarchitektonische Felderung des menschlichen Stirnhirns. J Psychol Neurol 15:221–232

Vogt O (1911) Die Myeloarchitektonik des Isocortex parietalis. J Psychol Neurol 18:379–390

Vogt C, Vogt O (1919) Allgemeinere Ergebnisse unserer Hirnforschung. J Psychol Neurol 25:279–461

Index

S. Geyer and R. Turner (eds.), *Microstructural Parcellation of the Human Cerebral Cortex*, DOI 10.1007/978-3-642-37824-9,

MIX
Papier aus verantwortungsvollen Quellen
Paper from responsible sources
FSC® C105338

If you have any concerns about our products,
you can contact us on
ProductSafety@springernature.com

In case Publisher is established outside the EU,
the EU authorized representative is:
Springer Nature Customer Service Center GmbH
Europaplatz 3, 69115 Heidelberg, Germany

Printed by Libri Plureos GmbH
in Hamburg, Germany